Lagrange Memorial Library
19 MUR c.1
Murray, Bruc Journey into space

36379000116968

D1047441

LG1 6/90

919
MUR
     Murray, Bruce C.
       Journey into
      space : the first
   three decades of
     space exploration.

TROUP - HARRIS - COWETA
REGIONAL LIBRARY
LaGrange, Georgia   30240

✓ LaGrange Memorial Library
__ Bookmobile
__ Ethel W. Kight Library
__ Harris County Library
__ Hogansville Library
__ Newnan-Coweta County Library
__ Senoia Area Public Library

BOOKS BY BRUCE MURRAY

JOURNEY INTO SPACE

EARTHLIKE PLANETS, *with Ronald Greeley
and Michael C. Malin*

FLIGHT TO MERCURY, *with Eric Burgess*

NAVIGATING THE FUTURE

MARS AND THE MIND OF MAN, *with
Ray Bradbury, Arthur C. Clarke,
Carl Sagan, and Walter Sullivan*

THE VIEW FROM SPACE, *with
Merton E. Davies*

Journey

INTO

Space

Bruce Murray

Journey
INTO
Space

The First Three Decades
of Space Exploration

W · W · NORTON & COMPANY

NEW YORK · LONDON

Copyright © 1989 by Bruce C. Murray. "If Only We Had Taller Been" reprinted by permission of Don Congdon Associates, Inc. Copyright © 1973 by Ray Bradbury. *All rights reserved.* Published simultaneously in Canada by Penguin Books Canada Ltd, 2801 John Street, Markham, Ontario L3R 1B4. Printed in the United States of America.

FIRST EDITION

The text of this book is composed in Times Roman, with display type set in Typositor Deepdene. Composition and manufacturing by the Maple-Vail Book Manufacturing Group. Book design by Marjorie J. Flock.

Library of Congress Cataloging in Publication Data
Murray, Bruce C.
  Journey into space: the first three decades of space exploration
  by Bruce Murray.—1st ed.
    p.   cm.
  Bibliography: p.
  Includes index.
  1. Outer space—Exploration.   I. Title.
TL793.M84   1989
919.9'04—dc19   88–26579

ISBN 0-393-02675-2

W. W. Norton & Company, Inc., 500 Fifth Avenue, New York, N. Y. 10110
W. W. Norton & Company Ltd., 37 Great Russell Street, London WC1B 3NU

1 2 3 4 5 6 7 8 9 0

TO SUZANNE

Contents

Author's Preface

THE EXPLORATION OF SPACE has been one of this century's most lustrous accomplishments—and a vivid American drama as well. Heroic achievements against great odds led to a brief golden age.

Like the wicked queen's mirror in the story of Snow White, space plays no favorites. It coldly displays national achievements and styles independently of language, history, territorial boundaries, or domestic fantasy. Space for a time reflected our past greatness, as it now reflects confusion and lack of purpose.

I have been involved from the beginning, witnessing the events from within. That view has often been exhilarating, sometimes devastating, but always fascinating. One is immersed in a swirling interplay of historic discovery, technological excellence, daring engineering, the dreams and rivalries of nations and of institutions and of individuals, the Byzantine world of domestic politics, and the sound and fury of the media.

For many years I have wanted to share this multifaceted story not only with those enthralled by space exploration but also with those in the larger community who struggle to grasp who we are in this country today—and wonder what we will be tomorrow. Notes, outlines, drafts, news items, articles, letters, speeches, publications, and memos have accumulated in my files, waiting for the time when both I and the readership would be ready. Finally, on January 28, 1986, *Challenger*'s fiery blast cracked the cocoon of domestic fantasy about space in which America had sealed itself for more than a decade. NASA—and America—emerged from the glare of subsequent investigations as adrift. The time had come for me to put into literary form that view from within.

For two and a half years now, I have worked to re-create an authentic picture of the highs and lows in the American journey into space and to identify the factors responsible for both. The story builds upon my personal experiences and is, necessarily, a memoir. I focus on selected incidents and on a few dozen key individuals, leaving to academic historians the more complete reporting of the important activities of thousands of others—scientists, engineers, NASA managers, elected officials, and committed citizens.

To encompass the many concurrent, but disparate, events that make up the story, major exploratory thrusts are traced separately in six relatively independent sections. These sections necessarily overlap in time. To aid the reader in keeping track, chronological displays immediately precede each section.

I have been greatly aided by critical reviews of early versions of various parts of this book by more than forty specialists. They are individually identified in the acknowledgments. I am deeply grateful for such expert help, generously carved out of busy lives. Of course, an ample supply of the author's personal biases and misconceptions remains, and errors of fact are difficult to eradicate entirely.

After three decades of exploring space, I remain as passionate an advocate of this pursuit as ever. But the United States must soon come to terms with its past in space, or the first golden age of American space exploration will surely be its last.

We can once again illuminate the heavens with our accomplishments, if we have the will.

It is my hope that this story will help reawaken that will.

Pasadena, California
August 1988

Journey
INTO
Space

Prologue

From Belligerence to Triumph

SPACE EXPLORATION burst forth amid open belligerence and armed confrontation between the United States and the Soviet Union. The flights of *Sputnik 1* and of *Explorer 1*, of Yuri Gagarin and of John Glenn, punctuated the crisis atmosphere surrounding the 1956 uprisings in Eastern Europe, Khrushchev's 1958 threat to rain missiles on Western Europe, the 1960 U-2 incident, the 1961 Berlin crisis, and, finally, the 1962 Cuban missile crisis. President Dwight D. Eisenhower, wary of the growing "military-industrial complex," tried to downplay space rivalry as "pie in the sky." His successor, John F. Kennedy, instead saw peaceful space exploration as a race we could—and must—win. Kennedy's Apollo legacy helped carry President Richard Nixon to the 1972 summit with the Soviet leader Leonid Brezhnev. Their SALT 1 and détente signaled a new era of increased restraint on the part of both superpowers. By providing the lunar *destination* for Americans in space, Kennedy transformed technological muscle-flexing into historic achievement and a major American foreign policy success.

But no such inspired presidential leadership materialized in the 1970s or 1980s to once again focus America's sights on meaningful accomplishments in space. Without leadership at the top, space fantasy replaced space policy, and overblown promises replaced meaningful achievements.

Mediocrity followed.

And finally catastrophe.

1 BY THE ROCKET'S RED GLARE • The launch of *Sputnik 1* on October 4, 1957, startled the world and alarmed the West. The rocket responsible for this historic achievement, termed the "A-1" in the West, is shown still enclosed by its gantry tower. A series of improvements quickly increased the payload of successive launches. By early 1961 the A-1 powered Yuri Gagarin into orbit, another historic first. All Soviet cosmonauts have been launched by versions of the A-1, without, so far as is known in the West, a single casualty. A diagram of the A-1 is included in figure 2.2, on p. 53. *NASA/Johnson Space Center photo*

2

A Belligerent Beginning

U.S.	USSR

———1964———

Valentina Tereshkova
First Woman in Space

———1963———

Mariner 2
First Planetary Flight   **Cuban Missile Crisis**

Vostok 3 'and 4
Dual Flight

John Glenn
First American in Orbit

———1962———

Second Manned Flight

Kennedy's
Apollo Speech

Yuri Gagarin
Berlin Blockade   First Human in Orbit

———1961———

Summit Collapses
Tiros   **U-2 Spy Plane Downed over Russia**
First Weather Satellite

———1960———

Luna 3 Photographs
Backside of Moon

———1959———

Khrushchev Threatens Nuclear Missile War

Sputnik 3

Explorer Success

———1958———

Vanguard Failure

Sputnik 2

Sputnik 1
First Satellite

———1957———

Uprising in Poland, East Germany, and Hungary

———1956———

3 ᴀᴍᴇʀɪᴄᴀ ᴏɴ ᴛʜᴇ ᴍᴏᴏɴ • The astronaut Charles M. Duke, Jr., salutes on the lunar surface beside the *Apollo 16* Lunar Excursion Module on April 21, 1972. *NASA/Johnson Space Center photo*

4
The Apollo Triumph

U.S.	USSR

Apollo 17

Apollo-Soyuz Test Project Started
SALT I Treaty

Apollo 16

Nixon Starts Space Shuttle

—— *1972* ——

Apollo 15

Salyut
First Space Station

Apollo 14

—— *1971* ——

Luna 17 Rover

Luna 16
Automated Sample Return

—— *1970* ——

Apollo 12

Apollo 11- First Manned
Lunar Landing
Armstrong-Aldrin-Collins

First Luna 15 Failure

—— *1969* ——

Apollo 8- First Manned
Lunar Flight

Venera 4 - 'First Entry
into Venus Atmosphere

Zond 6 - First Lunar
Flyby and Return

—— *1968* ——

Apollo Fire
Grissom-White-Chaffee
Killed

Soyuz1
Komarov Killed

—— *1967* ——

Surveyor 1
First U.S. Moon Landing

First Luna 9
First Moon Landing

—— *1966* ——

Mariner 4
First Mars Probe

Voskhod 2
First Space Walk

—— *1965* ——

From Fantasy to Catastrophe

January 26, 1986. The small imaging-team room at Pasadena's Jet Propulsion Laboratory was darkened so that every detail on the TV monitors stood out. I crowded in there that Saturday with a dozen of my associates and former students who had put years of work into the Voyager project. Across our excited faces flickered light from a sight never before seen. Television monitors displayed intricately detailed images of the strange surface of Miranda, a small moon circling the planet Uranus two billion miles away. Miranda's existence had been revealed only in 1948, as a barely detectable smudge on a telescope's photographic plate. Now Voyager, JPL's greatest engineering accomplishment, magnified that image fifty thousand times.

JPL had been my research base for more than two decades, and I had been its director for six years, from 1976 to 1982. It felt good to be back, as a visitor this time, for another planetary encounter. JPL had put America into space on January 31, 1958, with *Explorer 1*. And JPL was, for me, still the best space laboratory in the world. It had created Voyager to explore Jupiter and Saturn and their many moons. But engineers and scientists at JPL (and some farsighted people at NASA and in Congress way back in 1972) did more. They expanded Voyager's exploratory capability to encompass distant Uranus as well.

Now that farsighted enhancement of Voyager's journey seemed especially timely. No U.S. spacecraft had departed from Earth toward the planets since 1978. As a consequence, America had explored no new worlds since this same Voyager spacecraft flew past Saturn in 1981. By contrast, in the 1960s and 1970s a long succession of American robot explorers had treated the world to new visions of our planetary neighbors every year or two.

Finally, however, the painful and confining hiatus was due to end. In just five months, the Space Shuttle would fly at full throttle for the first time to lift a modified Centaur rocket into low orbit. There, the Centaur would be released. Then its liquid hydrogen–liquid oxygen rocket would be fired to launch a major European Space Agency probe deep into space. Ulysses had been delayed for an embarrassing three years by the Shuttle's development difficulties. It would be a relief to finally get it on its way. A month later, the same Shuttle / Centaur process would propel the even-longer-delayed Galileo mission to Jupiter. Then, in September, the full power of the Shuttle would be used again for science to place into low Earth orbit NASA's premier scientific endeavor of the 1980s, the twelve-ton Hubble Space Telescope.

American space exploration, once the pride of its citizens and the envy of the world, could bloom again after a decade of neglect. NASA touted 1986 as "the Year of Space Science."

Development of the Space Shuttle had ruled the roost in NASA since 1972 and the ending of the fabulous Apollo journey of men to the moon. Apollo's national focus on space exploration sparked robotic as well as human exploration. Not so with NASA's long Shuttle development, which concentrated solely on human utilization of Earth's orbit. Worst of all, NASA insisted that new space efforts plan to use astronauts, whether they were needed or not. As JPL's director, I watched with great sorrow and frustration as NASA sacrificed the great American enterprise of robotic planetary exploration in its desperate attempt to make good on overblown Shuttle expectations. It had been a personal relief for me to resign from JPL in 1982.

But the truism is true: time heals. Just before this Voyager encounter with Uranus in January 1986, I wound up a professional paper about the American space program with a constructive proposal for the future. After describing and analyzing NASA's unbalanced Shuttle-only program, I advocated an imaginative, balanced, long-range space goal for NASA— Americans to Mars with Russians. The article had been well received in draft form by many colleagues. Even my friend James Beggs, President Reagan's head of NASA, had penciled in his agreement with some of its contentious conclusions.

Unfortunately, Jim Beggs had little influence at NASA just then. He was, in fact, on leave, having been indicted with three others on federal charges of manipulating defense contracts stemming from his previous career at General Dynamics. Those who had worked with Beggs at NASA were incredulous at the charges (which were dropped by a red-faced Department of Justice two years later). But the harm to Jim and his role at NASA was done. Worse still for NASA, Dr. William Graham, Beggs's deputy, had arrived at NASA only two months earlier. Now he was suddenly the acting administrator, still virtually unknown to NASA personnel, to congressional committees, to university scientists, and to NASA contractors. Graham's career had centered on defense policy analysis, not on the civilian aspects of space.

Graham was not present to share our excitement at JPL over the Voyager encounter. Instead, he made a flying trip to the Kennedy Space Center, in Florida, to oversee NASA publicity efforts trumpeting the delayed twenty-fifth flight of the Space Shuttle. The novelty this time was Christa McAuliffe, an especially appealing and amiable teacher chosen in an intense nationwide teachers' contest. In an imaginative use of human space flight, she would speak to America's schoolchildren live from space. And she would probably also speak to the President from space, assuring broad public recognition of Reagan's association with this adventure.

Never had NASA been more vulnerable. Never had America's expectations in space been more at risk.

Forty-five hours later, nine miles above the Kennedy Space Center, *Challenger* exploded in a fiery ball. The scene was repeated with stunning clarity, again and again, on every American television set. It was seeing it over and over that I will never forget. The casualties included Christa McAuliffe and her six companions—and American innocence about space.

The American people, including their President, were shocked to learn so dramatically that space is not safe. Unprecedented national grief over the loss of the *Challenger* seven turned to dismay and horror as a sense of reality emerged from televised coverage of the investigation and of wreckage brought up from the bottom of the Atlantic Ocean. The once glistening Shuttle emerged as an inadequately maintained and operationally marginal aerospace plane of uncertain value. After basking too long in the golden afterglow of the Apollo Moon flights, NASA suddenly appeared aged and mediocre. The awful possibility loomed that the Shuttle's promise of cheap, easy access to space for the fulfillment of American frontier aspirations reflected more fantasy than solid experience and sound judgment.

Presidential theater at the *Challenger* memorial services in Houston, Texas, brought tears to the nation's eyes. "Ron, we'll build that Space Station for you," President Reagan promised, after mentioning the *Challenger* astronaut Ron McNair's desire to play the saxophone in the Space Station. But the White House turned tentative when serious questioning began. For what purpose should more American lives be risked in space? To create an enormously expensive inhabited edifice in the hostile environment of space? There must surely be more compelling national reasons for a space station than playing a saxophone in zero gravity. But NASA's Space Station concept had been spun uncritically out of the Shuttle fantasy. It was simply another expensive, newly proposed technology searching unsuccessfully for justification and purpose. The military services declared themselves uninterested and unwilling to contribute. Most space scientists saw the Space Station as a potential white elephant in space. And men of commerce did not care to share the financial risk with the taxpayers.

The immediate question of whether to replace *Challenger* immobilized the White House decision-making process. Obviously, a new Shuttle orbiter would be built—so went the initial murmurings. But then internal dissension arose, and properly so, over how to pay for it. The myth of the Shuttle as a sensible way to launch unmanned satellites to distant orbits had also perished in the *Challenger* explosion. What does the United States do now that the fundamental tenet of its space program has been exploded? Every alternative—Shuttle or not—will require billions of extra dollars. And still we will not be able to catch up with our military and civil space commitments for many years.

Finally, Ronald Reagan's visceral commitment to human adventure (which he shared with John Kennedy) overruled the vain search for economic justification by his White House staff. Billions for another orbiter to replace *Challenger*. Stay the course, regardless of the cost. The dream must be fulfilled. But the nimble John Kennedy had repudiated his predecessor's failed space policies and stamped out his own bold direction—to the Moon with Apollo. Reagan stuck stubbornly to the flawed theatrical visions he embraced from the past.

There were more Shuttle delays, and NASA stayed grounded for thirty-two months. Its Space Station future grew more expensive, less interesting, and ever more weakly supported. Once again NASA mortgaged America's future in space in a desperate bid to retain the Shuttle as its means and the ill-conceived Space Station as its end.

In truth, since 1972 the myth of the Shuttle's promise had been substituted for thoughtful goals in space. This country and its Presidents had not faced the question ''Where are we going in space, and what's in it for the country?''

But the Russians had.

In 1986 the USSR once again became the unquestioned world leader of peaceful space exploration, a distinction it had lost to the United States in the 1960s after President Kennedy put America on the path to the Moon. On February 20 the USSR successfully launched Mir, the central element in an ambitious, continuously manned space station. Russia now possessed the capability to do almost anything in space that NASA might want to do with its station ten years hence. How could *the* American goal in space be an expensive facility of dubious utility to be deployed ten years after Russia had deployed a comparable one? What had happened to the Apollo spirit?

Mir, whose name derives from the Russian word for ''peace,'' came as no surprise. Relying on routine use of proven expendable rockets, the USSR pursued a step-by-step, systematic approach to manned space flight. NASA, in contrast, gambled that Shuttle technology would provide a great leap forward in cheap, safe access to space—and America lost.

Nor was the Soviet Union's recapture of space leadership limited to manned space flight. On March 6, and again on March 12, 1986, two large Soviet VEGA spacecraft led an international flotilla of Earth's first comet explorers in spectacular encounters with Halley's comet. Scientists and journalists from all over the world jammed the Institute for Space Research in Moscow to hear the personable and articulate Dr. Roald Sagdeev carefully describe the initial results. Then the excitement moved to Darmstadt, Germany, and the findings from Giotto, the European Space Agency's comet probe. Giotto flew even closer than the VEGAs to the unexpectedly dark and irregularly shaped nucleus of Halley's comet. America had been first to Venus, Mars, Mercury, Jupiter, Saturn, and Uranus. But America did not even go to Halley, much less lead the way.

In July 1988, two and one-half years after *Challenger*'s explosion and with America still grounded, the next Soviet planetary mission was launched. The target this time was the Martian moonlet Phobos, where the Soviets plan to have their spacecraft touch down in April of 1989.* Scientists from fourteen nations (up from VEGA's eight) will converge on the Institute for Space Research in Moscow to carry out their joint investigations with the Phobos 1988 spacecraft. Russia, not America, leads in the peaceful international exploration of space.

Meanwhile, lonely old *Voyager 2*, now a ten-year veteran of space travel, doggedly continues to probe the unknown (mightily assisted by its JPL ground team). In August 1989 it will skim over distant Neptune, extending humankind's vision to the very edge of the Solar System. Otherwise, no new American spacecraft will scout other planets until the 1990s. The Hubble Space Telescope remains grounded. American space exploration, put on hold for a decade because of the Shuttle's painful development, still suffers on, a helpless hostage to NASA's failed Shuttle fantasy.

How did we achieve space exploration leadership and the Apollo triumph after the shock of *Sputnik 1* in 1957?

How did we drift from the Apollo reality to the Shuttle fantasy and, finally, to catastrophe?

How can we once again rise to greatness in space?

These are the questions pursued in the pages that follow.

*In fact, they failed to reach Phobos, but did acquire important new data about Mars.

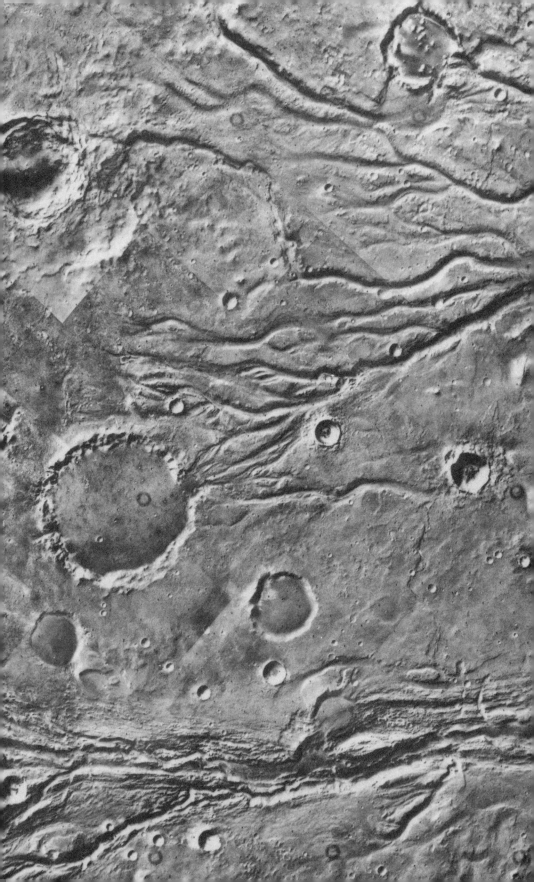

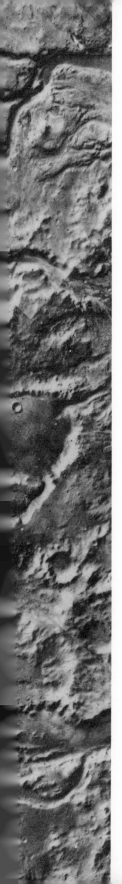

PART ONE

The Search for Life on Mars

1.1 THE SURPRISING FACE OF MARS • This mosaic of Viking orbiter TV pictures acquired in August 1976 covers an equatorial region of Mars centered at 17° N, 55° W, and is about 95 miles (150 kilometers) in width. Billions of years ago catastrophic flooding, probably released from subsurface water reservoirs to the west (left), scoured Mars's ancient cratered surface and produced giant channels. *Viking Lander 1* sampled the soil on the Chryse Plain immediately to the east (right) of this photo. Sunlight is coming from the upper left, causing the craters to be bright on the lower right rim and shadowed on the upper left rim. *NASA/JPL photo*

1.2

The Exploration of Mars

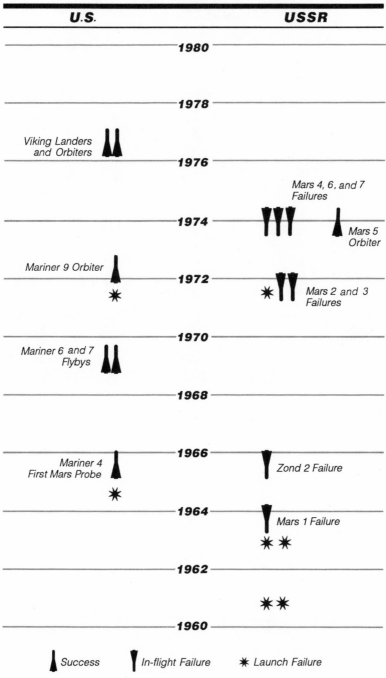

U.S.	USSR

Viking Landers and Orbiters

Mariner 9 Orbiter

Mariner 6 and 7 Flybys

Mariner 4 First Mars Probe

Mars 4, 6, and 7 Failures

Mars 5 Orbiter

Mars 2 and 3 Failures

Zond 2 Failure

Mars 1 Failure

1980

1978

1976

1974

1972

1970

1968

1966

1964

1962

1960

Success   In-flight Failure   ✳ Launch Failure

1.3 FIRST VIEW OF MARS • *Mariner 4*'s historic first look at Mars in 1965 displayed a heavily cratered, lunar-like surface with no hint of the water-eroded, volcanic, and novel landforms that later missions revealed. The scene is about 155 miles (250 kilometers) across. Sunlight is from the left. *NASA/JPL photo*

1. The Ghost of Percival Lowell

I FIRST SAW MARS through a telescope on a cool October night in 1960. Alone in darkness I perched on a tiny platform high up beside the sixty-inch telescope at Mt. Wilson Observatory. Mars's red, yellow, and white image glowed like a Japanese lantern through the eyepiece. In profound silence and solitude, completely absorbed in the shimmering image of something unimaginably far away, I felt transported to Mars itself, temporarily disconnected from my native world. That night, at the age of twenty-nine, I fell in love with observational astronomy—and with planet Mars.

My presence there that night was itself remarkable. I had never visited a professional observatory before or taken a course in astronomy. But Caltech is a special place for following one's scientific curiosity. Only two months earlier I still wore the lieutenant's bars of an Air Force officer while completing a two-year tour of active duty. (That stint, normally due right after the completing of my Ph.D. in geology at MIT in October 1955, was delayed, so I had spent three years looking for oil off the shore of Louisiana.)

Within a month of my arrival at Caltech as a postdoctoral research fellow, my sponsor, the geochemist Harrison Brown, recommended me to the crusty director of Mt. Wilson Observatory, Ike Bowen. Bowen, who helped pioneer the 200-inch telescope, had learned his optics under the Nobel Prize winner A. A. Michelson, whose key measurements of the speed of light paved the way for Einstein and the theory of relativity. Bowen brooked no amateurism with his telescopes. But after testing my newly won knowledge of the relevant astronomical literature and becoming satisfied with the seriousness of my planetary research, he helped tremendously. Hardly anywhere else in the world could such productive cross-disciplinary associations develop so rapidly and freely. Furthermore, the staccato of Sputnik, *Explorer 1,* and Yuri Gagarin was propelling America to unleash unprecedented resources for new rockets and spacecraft, including some that were intended to make a close-up exploration of the shimmering image of Mars that lay before me.

The Mars I pondered in 1960 through the sixty-inch telescope was Earth-like. Of course, the picture of an inhabited Mars proposed by Percival Lowell at the turn of the century had long since been dismissed as fanciful by serious scientists. Lowell had described Martian "canals" that, he speculated, supported a dying civilization whose heroic inhabitants struggled vainly against global drought. His romantic "scientific" vision fired the imagination of writers ranging from Edgar Rice Burroughs to Ray Bradbury, Robert Heinlein, and Arthur Clarke. And in 1938 Orson Welles catapulted himself to fame (and created a new awareness of radio as the emerging medium) with a terrifyingly authentic radio dramatization of H. G. Wells's *War of the Worlds.* As "news" spread of "Martian landings in New Jersey," radio listeners across the nation bolted their doors and prepared for the worst.

By 1960 scientists knew that Mars was far, far drier than Earth and devoid of oceans as well. However, it certainly seemed Earth-like in many other ways. For example, unlike our other planetary neighbors, Mars spins about an axis tilted 24° off its orbital plane, almost exactly like Earth. This geometric orientation (called obliquity) controls the alternations of seasons on a planet. Mars's winter and summer seasons alternate hemispheres, just as they do on Earth. Furthermore, in the winter hemisphere on Mars, broad white caps form and disappear, just as snow and ice seasonally develop and disappear in polar latitudes on Earth.

Observing such regular behavior through telescopes, pre-space-age planetary astronomers reasoned that the frost caps on Mars were probably thin deposits of water frost. What else could they be? The astronomers also concluded that Mars's atmospheric pressure is only one-tenth that of Earth's—terribly rarefied by Earth standards. Even so, modest amounts of surface liquid water and moisture in the soil might still exist. Most exciting, Mars exhibited an Earth-like rhythm of seasonal changes in regional color and contrast. These cyclic variations observed through telescopes relentlessly marched with the seasons on Mars. What could this mean? Probably simple plant regeneration—colored regenerative plant growth in the spring, followed by colorless dormancy from late summer through winter.

It makes the special similarity of Mars to Earth seem even more compelling that Mars rotates with an Earth-like daily period, once every twenty-four hours and thirty-seven minutes. No other planet has a rate of rotation close to this one. Obviously, Earth and Mars must be "sister" planets, somehow sharing a close geological history.

Then an even more tantalizing hint of Martian plant life emerged. The astronomer William Sinton, working first at Lowell's own observatory in Flagstaff and then, in 1960, with the 200-inch telescope at Mt. Palomar, reported faint spectral absorptions in the sunlight reflected from Mars in the vicinity of 3.5 microns wavelength (five times longer than that of red light). He proposed that these spectral features arose from organic chemicals on Mars exhibiting life-related carbon–hydrogen bonding and that they probably signified plant life currently on the planet. But in 1965 George Pimental, a renowned infrared spectroscopist at the University of California at Berkeley, recognized that Sinton's spectra matched those of the gaseous form of HDO (hydrogen deuterium oxide)—the heavy water in the atom bomb. Pimental wondered why Mars's atmosphere, compared with Earth's, was so much more enriched in heavy water over normal water. He concluded that the selective accumulation of the heavy deuterium isotope of hydrogen in the water vapor of Mars's atmosphere resulted from eons of evaporation of a once giant body of water—an ancient ocean. Mars's Earth-like past seemed confirmed by hard analytical data.

In fact, we know now that *all* of these Earth-like indications are extraordinary coincidences or misinterpretations. The Earth-Mars similarity is entirely fortuitous. There is no evidence of an Earth-like Mars, much less one with life.

Earth's rotation has been slowing over billions of years through Earth's gravitational (tidal) interaction with the Moon. In the Devonian period, Earth's day was twenty-two hours long. Mars has no large moon, and therefore its day has been of constant duration. Only because we happen to be making this comparison with Mars now—not a billion years earlier or later—are the durations of day so close. Similarly, the tilt of Mars's axis of rotation, now 24°, varies over time from as little as 15° to as much as 35°. This "nodding" motion of Mars occurs in cycles of hundreds of thousands to millions of years arising from faint, synchronized, gravitational tugs of giant Jupiter and Saturn.

It is sheer coincidence that we should study Mars at this point in its axial orientation. If *Homo sapiens*'s sudden intellectual development had occurred fifty thousand years earlier or later than it did, Mars's and Earth's obliquities would have differed dramatically when man first studied Mars through telescopes.

What about the seasonal white frost caps? What else could mimic water ice? Very dry solid carbon dioxide—the "dry ice" used to keep ice cream cold on a hot summer day. Mars's atmosphere is so rarefied that water in the liquid state cannot exist there. Instead, carbon dioxide freezes out of the atmosphere seasonally. And, as a final insult to Earthlings' intuition, the seasonal variations in surface markings record the annual cycle of dust storms and associated wind-caused changes on the surface of Mars. Inanimate dust, not plant life, causes Mars's seasonal changes in the brightness of different regions on its surface.

What, then, are we to make of the "Sinton bands," the faint spectral features in the infrared? Careful reexamination of Sinton's work (by Sinton and by others) confirmed that the heavy water is indeed atmospheric. But it is in *Earth's* atmosphere, not in Mars's. Sinton had been led astray by very subtle observational difficulties in measuring such an ephemeral property of that same small, shimmering image that mesmerized me in 1960. And Pimentel, while correctly recognizing the spectral signature of heavy water, was perhaps also lured into an acceptance of oceanic evaporation by the ghost of Percival Lowell.*

We know now that Mars's surface is an incredibly hostile environment, devoid of moisture and colder than anywhere on Earth. Life as we know it cannot develop or survive there. But in the 1960s Lowell's legacy still shaped scientific expectations for Mars exploration. The Space Science Board of the National Academy of Sciences strongly advocated Mars as the central focus of American planetary exploration because it felt there was some possibility of microbial life there. NASA accepted this scientific direction with enthusiasm, sensing the mythic power of the search for alien life forms.

I got caught up in the early Mars exploration. Six months after my first, absorbing look at Mars through the sixty-inch telescope on Mt. Wilson, the chance to help create the world's first Mars television system arose. The key figure was the Caltech physicist Robert Leighton, an experimentalist already famous for discoveries about cosmic rays and the "granular" motions of the Sun's atmosphere. In addition, he had built the world's first image-stabilized camera for use on a major telescope. In 1956, a great year to observe Mars, he took superb time-lapse pictures, which he made into a breathtaking "movie" of that glowing Japanese lantern rotating slowly and majestically.

Fortunately for me, in 1961 Gerry Neugebauer, a graduating Ph.D. student from Caltech, started his tour of duty in the U.S. Army—at JPL—

---

*Ironically, the HDO issue on Mars has surfaced again, as a result of indirect observations and interpretation.

a happy legacy of JPL's pre-*Explorer 1* status as an Army lab. (Gerry is now a leading infrared astronomer and the director of the Mt. Palomar Observatory.) Lieutenant Neugebauer's orders were to organize the scientific equipment for America's first probe of Mars. Gerry naturally thought of Leighton, his former professor and a Mars picture buff, and urged him to propose to NASA a spacecraft TV camera. Leighton, in turn, invited Dr. Robert P. Sharp, chairman of Caltech's Division of Geological Sciences and a world authority on landforms, to help him explore Mars close up. That was where I came in, because Bob Sharp, in turn, invited me to join them as a junior colleague. Quickly, we proposed a joint effort with JPL to develop, build, and then interpret the pictures to be taken from a miniature television camera to ride the first U.S. Mars probe. JPL had the engineers, the laboratories, the test facilities, and the experience and project discipline to tackle such a task.

President Kennedy's Apollo man-to-the-Moon initiative in May 1961 implicitly provided NASA carte blanche to race to the planets with robots as well. It was then national policy to be first in space exploration. Initially, NASA projected very large robotic payloads lofted to Mars within just a few years by an early version of the Saturn Moon rocket. But reality quickly intruded. By 1962 the proposed Mars mission had to be reduced to fit a new midsize rocket under development, the Atlas / Centaur. (See figure 2.2, on p. 52.)

The Atlas / Centaur development soon fell far behind schedule. The Centaur (upper) stage would burn liquid hydrogen and liquid oxygen fuels together to achieve the highest efficiency of any rocket in the world. But liquid hydrogen—tricky enough to control in terrestrial laboratories—proved an especially difficult substance to handle in a space rocket. So by February 1963 NASA had to scale down the first Mars spacecraft to fit onto an existing and much smaller launch vehicle heretofore used by the Air Force to loft early spy satellites, the Atlas / Agena. (See figure 2.2.) Suddenly, only twenty-one months before the planned launch date, our Mars spacecraft shrank to 575 pounds from the 2,200 pounds expected for Atlas / Centaur (the weight class available to our Soviet competitors). In order to relay back television pictures, a tape recorder, camera, and scan platform would nonetheless have to be included. Furthermore, this diminutive new spacecraft would have to operate for almost a year and communicate over a far greater distance in space than any craft had ever done before.

Some good news arrived in December 1962. The United States's *Mariner 2* spacecraft became the world's first successful planetary spacecraft, reaching Venus four months after launch on an Atlas / Agena. The bad news was that, even without complex television equipment, it barely survived that four-month trip—not a good omen for its younger Martian sibling.

Also in 1962 the Russians launched their first Mars probe, *Mars 1*. But that ambitious endeavor failed halfway to Mars.

So our first try for Mars posed a staggering engineering challenge, even if it was launched on the more capable Atlas / Centaur as originally planned. No earthly thing had survived to where we intended to travel. Forced to shrink the spacecraft to fit the Atlas / Agena, we faced overwhelming odds, the direct consequence of America's visible, even painful deficit in launch capability at the beginning of the space age. Yet early adversity had two happy consequences in the long run.

First, the United States designed its spacecraft to operate in the vacuum environment of space. Pressure vessels and other heavy internal systems necessary to maintain Earth's atmospheric temperature and pressure aboard a spacecraft simply could not be included in the tiny American payloads. America was forced to adapt to the space environment, while Russia opted to bring Earth laboratory conditions along inside its heavier spacecraft. Ultimately, adaptation proved one key to long life and reliability in the space environment. Second, because so few launch attempts could be expected, American spacecraft *had* to be reliable. There was no alternative to heavy investment in preflight ground testing of every component and subsystem of each spacecraft.

Soviet spacecraft development, in contrast, drew upon traditional aircraft flight-testing practice. In the testing of new airplanes, individual technical problems are gradually solved through repeated flights—from taxiing to short hops off the runway and a quick up-and-down flight. But there is no small first step for a spacecraft. Everything has to work right the first time. Because of its emphasis on reliability, the United States rang up a conspicuous string of successes from 1962 on, despite its late start in the development of launch vehicles. The Soviet Union, in contrast, launched a much larger number of heavier, more capable, and more expensive spacecraft to the planets but had no success until 1967. (See table 2.2, on p. 58, and the chronology on p. 28, and figure 2.3 on p. 60.)

By November 1964 it was time to launch for Mars—ready or not. As we awaited the news from Cape Canaveral, many of us remembered how *Mariner 1* had plunged into the Atlantic Ocean two years earlier, leaving *Mariner 2* to go on alone to Venus. Now, in unhappy imitation, the first launch to Mars ended up out of control and left *Mariner 3* in a useless orbit. Quick analysis of the radio signals from this devastating setback showed that a special lightweight covering of the spacecraft—a "shroud" to protect it from aerodynamic friction during its rapid ascent through Earth's atmosphere—had collapsed shortly after launch. That shroud, intended to save a few extra pounds, instead caused a catastrophic failure.

Only three weeks remained to launch *Mariner 3*'s twin, *Mariner 4*, before Earth and Mars moved out of suitable positions. During those crisis-filled weeks, JPL engineers (with their counterparts at the Lockheed Aircraft Corporation, supplier of the Agena rocket) diagnosed why the *Mariner 3* shroud had failed, frantically designed a new one, built it, tested it, and installed it on the Atlas / Agena rocket waiting in Florida. On November 28, 1964, *Mariner 4* rocketed smoothly out of Earth's gravity and sped toward Mars. (See figure 1.4.)

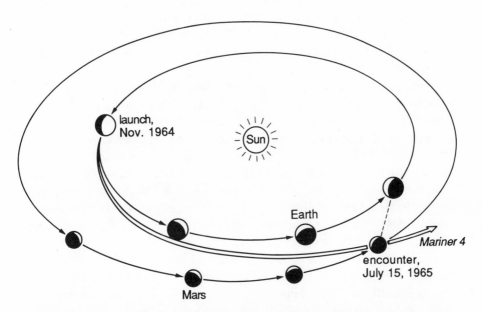

1.4 THE FAST TRACK TO MARS • The path of *Mariner 4* from Earth to Mars is shown, along with successive positions of Earth and Mars. At encounter on July 15, 1965, Earth and Mars were about 60 million miles (100 million kilometers) apart. *Mariner 4* remains in a permanent orbit about the Sun.

Mariner 4 had company in that fast lane. Two days after *Mariner 4* left Florida, the Soviet Union also launched a Martian probe, after two earlier failures. The Soviet Union, perhaps wary of failure, did not acknowledge this Mars mission, simply calling the spacecraft *Zond 2*. (*Zond* means "probe" in Russian.) However, it was obvious to us from the direction of its flight that *Zond 2* was targeted right at Mars, and probably intended to parachute instruments through the atmosphere. (*Venera 4* did just that at Venus three years later.) Russia badly wanted to be first to the surface of Mars—and to every cosmic body.

But *Zond 2*'s communications system failed about the same distance from the Earth as had *Mars 1* in 1962. Because the Russians had no way to send commands that would fine-tune its path, the silent *Zond 2* whizzed by Mars, probably close to its surface.[1] There is a slight chance that it actually hit Mars. Perhaps future explorers will puzzle over a fresh crater there, with metal fragments splattered about.

Noting the tendency of Soviet spacecraft to fail halfway to Mars, the thirty-year-old John Casani, then the JPL deputy spacecraft manager for *Mariner 4* and an irrepressible spirit and natural leader, began to joke about the "Great Galactic Ghoul," a mythical deity who lurked halfway to Mars, waiting to consume exploratory spacecraft. John's humor took an unexpected twist when *Mariner 4* likewise experienced an anomaly at about the same distance from Mars. Fortunately, *Mariner 4* rode it out and continued safely on its way.

JPL engineers, part of the team that had put America in space with *Explorer 1* in 1958, were not satisfied merely to be first to fly by Mars and take close-up pictures. They pushed a daring new kind of radio experiment to measure the true atmospheric pressure on Mars. This endeavor required the spacecraft to pass *behind* Mars as seen from Earth and from the Sun. (See figure 1.5.) The radio signal to Earth would thus be blocked (occulted) and silenced by the edge of the Martian disk. The detailed way in which the radio signal faded could provide a sensitive measurement of the pressure profile of Mars's atmosphere right down to the surface. Passing behind Mars, however, meant silencing radio communications with Earth and also interrupting *Mariner 4*'s Sun-directed orientation and the illumination on its solar cells. Our tiny pioneering spacecraft would have to rely solely on its batteries and gyroscopes for more than an hour—right after taping our close-up pictures—while it flew above the silent blackness of the Martian night. What if something went wrong while *Mariner 4* was out of touch with Earth, and we couldn't reestablish communications? All the pictures on the tape recorder would be lost, and our mission would be a failure. But the experiment was urgent. The value of *Mariner 4* as a novel atmospheric probe had gained great importance in 1964 after three Caltech and JPL astronomers reported spectroscopic indications with the hundred-inch telescope that Mars's atmosphere was much thinner than anyone thought.

Mariner 4's challenges mounted. Midway through the flight, in early 1965, a radio astronomer doing a routine sky survey detected an unexpected natural radio emission from the location of Mars. Enhanced radio signals from Mars (still believed to be Earth-like) probably meant the existence of strong radiation belts like those of Jupiter and Earth (where they are termed Van Allen belts, in honor of their discoverer). Indeed, *Mariner 4* carried instruments specifically designed to search for Earth-like magnetism and electrical fields—even for small radiation belts. But radiation belts strong enough to emit such radio signals would pose a critical threat to *Mariner 4*'s electronic components. (A decade later such radiation belts severely tested the specially designed Pioneer and Voyager spacecraft passing Jupiter.) Fortunately, as the months passed, the radio

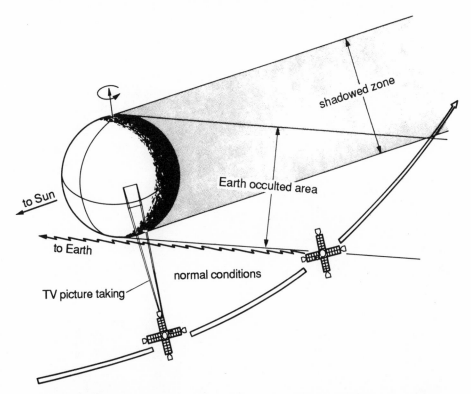

1.5 A DARING FIRST PASSAGE BY MARS • While passing high above the sunlit equatorial regions of Mars, *Mariner 4* snapped twenty-one TV framelets and automatically recorded them on its tape recorder. Then its motion carried it behind Mars, as seen from Earth (Earth occulted zone), severing its radio link with Earth for over one hour. The detailed variation of the fading radio signal provided a crucial measurement of Mars's unexpectedly rarefied atmosphere. Next, the spacecraft passed into the shadowed zone, where Mars's disk temporarily blocked out the sunlight that powered *Mariner 4*'s solar panels and provided its primary orientation. *Mariner 4* had to rely totally on automatic gyro control and batteries through this period, then automatically reestablish communication with Earth before those precious first TV pictures could be broadcast back to Earth.

"bright spot" detected in the survey proved to be part of the celestial background. It gradually separated in celestial position from Mars. Realizing that it was only a coincidence, we breathed easier. Mars, in its orbital track across the celestial background, had by chance passed in front of a minor peak in natural radio emission, probably from some distant galaxy.

Mariner 4 pushed farther and farther into space, setting communications records each day. Our anticipation grew. Finally, on July 15, 1965, it swooped past the red planet, into its shadow, and then back into the sunlight and communication with JPL's antenna at Goldstone, in the Mojave Desert. Our cheering reflected relief as much as pride. Two hours later the long-awaited first close-up pictures of Mars started trickling into JPL at a rate of slightly more than one tiny picture element each second. (Each frame consisted of 40,000 picture elements.)

With excruciating anticipation, Leighton, Sharp, and I, and the JPL

GLEEFUL SUCCESS • The JPL engineers John Casani (*left*) and John Hunter pose with *Mariner 4* data printouts while celebrating its successful first encounter with Mars on July 15, 1965. *NASA / JPL photo*

TV engineers slowly pieced together the first image—and when we had it, we couldn't recognize a thing! We couldn't even be sure we were looking at Mars. Evidently, some dust had collected on the optics. This created instrumental glare. Combined with the hazy Martian atmosphere and high-Sun viewing, this glare yielded Martian images of extremely low contrast. Nothing could be identified. Soberly, we began to invent computer image processing with which to remove the overriding glare and haze and, we hoped, to extract the underlying images of Mars.

Meanwhile, outside JPL's gates reporters from the three major television networks shouted at Frank Colella, JPL's press liaison. Percival Lowell, Orson Welles, Ray Bradbury, and a century of Martian speculation and imagery had aroused extraordinary public curiosity about the first close-up views of Earth's sister planet. Taxpayers' money had bought these pictures, and the media men knew we sat now behind locked doors examining them. Who did we think we were? "They're threatening to get the *President* to order those pictures released," groaned Frank. Meanwhile, we struggled around the clock with rudimentary image processing, even as additional images arrived, one every eight hours. (In those days, JPL had no food available at night. Our only source of nourishment was an ice cream machine, which led to a weight gain of about ten pounds per Mars encounter.)

I had been laboring over enlargements of the first three computer-processed pictures. These images were timed to overlap the edge of Mars's sunlit disk, *Mariner 4*'s first "landfall" after leaving Earth. Using a sheet of tracing paper, I tried to see how they might fit together. Gradually, I realized that a dark smudge in the corner of one image corresponded exactly to a dark smudge in the side of another overlapping image. Suddenly, it was clear—we were seeing the surface of Mars. At last we could start to construct a representation of Mars's surface.

Later pictures flowed in from their fifty-million-mile journey with much more favorable illumination of surface features, and our computer manipulations improved. Finally, we saw the *real* Mars—a big surprise. Huge craters dominated the scene. Mars looked more like the Moon than anything else. (See figure 1.3.)

Craters that size, hundreds of miles in diameter, could only be scars from a period of giant impacts very early in the history of the Solar System. Such ancient cratered scars still survive on the Moon, where they constitute the bright lunar uplands. (We learned later from the Apollo astronauts' rock samples that this violent period ended on the Moon about four billion years ago.) Similar scars once covered Earth but have been erased many times over by the combined action of atmospheric erosion and collisions

ANALYZING THE FIRST MARS PICTURES • Robert Leighton of Caltech, principal investigator of the *Mariner 4* TV experiment, makes a point to the author in late July 1965. The three overlapping frames tacked to the wall behind the seated author were the first received and are mentioned on p. 41. The large photo behind Leighton was used in the planning of the Surveyor automated lunar-landing missions then also underway at JPL. *NASA / JPL photo*

between crustal plates. Staring at the first *Mariner 4* pictures, we realized that for such old craters to have survived on Mars in recognizable form meant that our sister planet, like the Moon, *never* had an Earth-like atmosphere or oceans.

Other instrument results reinforced these unsettling and unanticipated implications. No Mars magnetism or radiation belts excited *Mariner 4*'s instruments, proving that Mars lacked an Earth-like dynamic metal core. Most important, the daring radio-occultation experiment paid off in a big way. When *Mariner 4* sailed behind Mars under gyro control and reappeared intact, its radio signals had faded and returned abruptly, little dispersed by Mars's atmosphere. Calculations revealed that the atmosphere was even thinner than observations with Mt. Wilson's hundred-inch telescope had suggested in 1964. Mars was enveloped by an atmosphere of only about 0.5 percent of the surface pressure of Earth. Its atmosphere was so thin that any liquid water there would immediately vaporize. Liquid water, even soil moisture, cannot exist on Mars's surface.

So much for Lowell's canals and seasonal plant life. Furthermore, that same radio occultation revealed only a very thin ionosphere high above the surface, much thinner than Earth's protective ozone layer. Hence, on Mars, harmful ionizing ultraviolet radiation from the Sun impinges directly on the surface—another hazard for would-be Martian microbes.

The Mars we had found was just a big Moon with a thin atmosphere and no life. There were no Martians, no canals, no water, no plants, no surface characteristics that even faintly resembled Earth's. In jarring contrast with our expectations, we had just discovered a Mars incredibly inhospitable to life.[2]

The Russians were watching too. The Russian radio astronomer Arkady Kuzmin visited Caltech as part of an early Soviet-U.S. scientific exchange. Of course, in the mid-1960s any Soviet scientist permitted to live alone in the United States was exceptionally well trusted and was naturally collecting new scientific information for his government. (Even in 1987, ordinary Soviet scientists often could not travel overnight alone to scientific meetings within the United States.) With his crewcut hair topping a thin, intense face, Arkady displayed special interest at our first seminar on the Caltech campus when the *Mariner 4* results were unofficially released. I had never before seen anyone photograph projected slides in a darkened lecture hall. But Kuzmin's 35-mm camera clicked away when the key graphs appeared showing unmistakable evidence of Mars's unexpectedly thin atmosphere.

Those results simply could not wait for normal scientific publication to be circulated in Russia. They made the Russian probe design of *Zond 2* obsolete. The Martian atmosphere was too thin for the Soviet entry payload to be slowed enough to reach the surface safely. Even if it survived all that distance between Earth and Mars, *Zond 2* would have whacked into Mars much faster than a speeding jet plane and disintegrated.

Russia did not try to go to Mars again for six more years, and then only with the much larger Proton rocket propelling a larger and much more sophisticated spacecraft, one designed to land on Mars safely despite such a thin atmosphere.

Lyndon Johnson kept us waiting a week in Washington, while he struggled with how to tell the American people that, election promises notwithstanding, America would win the war in Vietnam by sending in more American troops. That was the bad news. The good news was the *Mariner 4* Mars television success. In preparation for this nationally televised press event, we were ably coached by James Webb, head of NASA and a true professional both in government management and in public relations. Webb was a close personal business associate and protégé of Senator Robert Kerr, chairman of the Senate Space Committee (and the most powerful man in the Senate as far as space was concerned). Webb's public image in Washington was that of a cornball southern politician, as in Tom Wolfe's parody in *The Right Stuff*. Like Wolfe's stereotyped characterization of Johnson, that is an unfair portrayal. Webb was in fact shrewd and skilled in negotiating Washington's corridors of power. I was amused at how effortlessly his manner of speech changed abruptly from a public "Oklahoma" vernacular to a crisp "Washington, D.C." in the privacy of our NASA meeting room. He left no question as to who was in charge and what was expected of us.

In the East Room of the White House, in late July 1965, I saw a president up close for the first time. Johnson looked awful—red faced and haggard. Only later did I realize how he must have been struggling then over his Vietnam decision. He squinted to read the words on the telepromp-

LYNDON JOHNSON ENCOUNTERS MARS • William Pickering, director of JPL, Oran Nicks, a NASA official, and Jack James, the *Mariner 4* manager, hold a framed version of picture number 11 (figure 1.3, p. 30) for Johnson, who is flanked by a buoyant NASA administrator, James Webb, shortly before the televised presentation described above. *NASA / JPL photo*

ter extolling America's success with the first pictures from Mars. Then Bob Leighton smoothly presented the pictures, now completely cleaned up, to the man who more than any other politician had from the beginning insisted that America be first in space.

We went on to Congress that afternoon, to mesmerize the House Space Committee. The next morning it fell to me to go through the ritual with the Senate Space Committee. Senator Stennis from Mississippi was presiding while I discussed the significance of these few frames of historic photography. As I finished, Stennis rose and said in a grand manner, "Son, Ah didn't unduhstand a wud you sayed—but it was great!"

And indeed it was. Great for science, and great for America in its peaceful contest with the Soviet Union.

It had been a wonderful five years since my first view of Mars through the sixty-inch telescope on a magical October night in 1960.

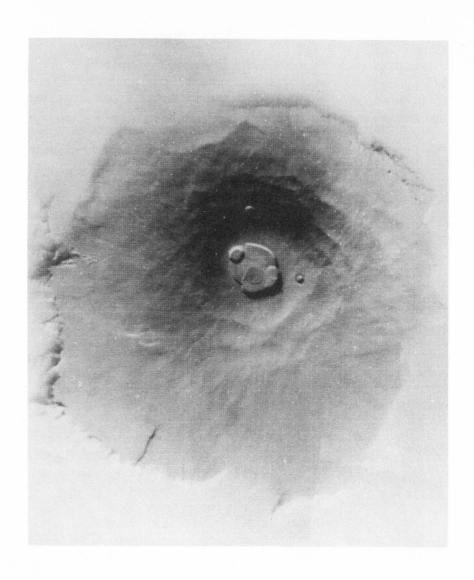

2.1 THE LARGEST VOLCANO KNOWN IN THE SOLAR SYSTEM • This computer-processed *Mariner 9* picture captures the full extent of Olympus Mons, which is about 310 to 370 miles (500 to 600 kilometers) across and rises 14 to 15 miles (22 to 26 kilometers) above the surrounding plain. The illumination is from the bottom right. Hawaii's Mauna Loa, the largest volcano and also the largest single mountain on Earth, is only one-fifth the areal size and one-twentieth the volume. The summit of Olympus Mons is an enormous caldera 50 miles (80 kilometers) across, bigger than the entire island of Hawaii.

2. The Changing Face of Mars

BY THE TIME LITTLE *Mariner 4* captured Mars's first portrait, in 1965, the long-delayed Atlas / Centaur rocket had successfully sent the American Surveyor spacecraft to the Moon's surface. The payoff was 11,500 close-up images of the lunar surface and many other kinds of scientific data. The Moon's surface was becoming familiar to engineer and citizen alike—as necessarily it had to before a manned landing there could be risked in 1969. Obviously, the Atlas / Centaur now stood ready to propel heavier and more capable Mars spacecraft, as NASA had originally planned. Furthermore, in 1969 and 1971 Mars and Earth would pass especially close to each other, permitting the Atlas / Centaur to launch a long-lived spacecraft that could *orbit* Mars and systematically map the entire planet. That was the view Caltech scientists offered NASA.

The dialogue developed because JPL belongs to both NASA and Caltech. It is not a free-standing entity. All JPL employees are full Caltech employees, as are the members of the research staff of the Caltech campus. But unlike the campus, JPL receives from NASA not only the money for salaries, buildings, spacecraft, and paperclips but also much direction on the character and pace of its projects. Organizationally, JPL is simultaneously a division of Caltech and a full-fledged NASA center.* Such a three-sided institutional relationship has inevitably generated a good deal

*It is the only NASA center not staffed by government employees. Other government-university partnerships include MIT's Lincoln Laboratory, which works for the Air Force under a relationship analogous to that of Caltech and NASA; the University of California's Livermore and Los Alamos Laboratories, which are under Department of Energy sponsorship; and the Applied Physics Laboratory of Johns Hopkins University, which works for the Navy. All these institutional ties are relics of World War II, when external threats drove both government and universities to find new ways to work together for the common good.

of creative tension over the years (which has stimulated Caltech and NASA as well as JPL).

In the early 1960s a brash, burgeoning, youthful NASA did not know what to make of the Caltech connection it had inherited in 1958 from the U.S. Army, JPL's original federal sponsor. NASA Administrator James Webb openly questioned what specific benefits NASA received for the fee it paid to aloof Caltech to manage JPL. Caltech, a fiercely independent private university, remained wary of the federal government's indirect presence through JPL. To the faculty and administration of Caltech, Webb was the caricature of a pushy government bureaucrat, insensitive to and uncaring about intrinsic university functions in his zeal to expand NASA's influence.

Into this breach stepped Dr. John Naugle, a key official for NASA's unmanned space efforts. Naugle was uncommonly sensitive to the intrinsic differences between the university and the government cultures and worked constructively to bridge the gap. He sensed the importance of getting Caltech visibly to perform some special services to NASA that exploited the scientific expertise of Caltech.

So, in late 1965 Naugle made a proposal. Could Caltech help plan a planetary program? Lee DuBridge, Caltech's highly regarded president, said of course. DuBridge then asked me to lead a JPL-faculty study for NASA. Our task was to lay out a scientifically sound planetary exploration program. We came up with a step-by-step plan exploiting the new Atlas / Centaur for successive flights to Mars as well as for new ones to Venus. This recommendation was to be a specific instance of how NASA benefited from its Caltech connection.

But NASA was not motivated to explore planets systematically. This was because the Saturn 5 production line, then running full blast producing Apollo Moon rockets, would soon need new orders. What NASA really wanted, we at Caltech were dismayed to realize, was Mars missions that would be such huge endeavors that they would require giant Saturn rockets. Never mind that *Mariner 4* had just revealed a Moon-like Mars with a distressingly thin atmosphere that greatly complicated any landing there. Never mind that much basic knowledge of Mars's atmosphere and surface was needed before really ambitious new efforts to explore Mars should proceed. NASA instead promoted Saturn 5 to Mars as the next giant step. In one wild leap, *Mariner 4*'s 575-pound spacecraft would be succeeded in NASA's plan by 50,000-pound Mars spaceships launched with the Saturn 5. (In fact, no robotic spacecraft this big has *ever* been launched by the United States or the USSR for any purpose.)

And what about the much less costly, evolutionary approach we advocated, using the reliable Atlas / Centaur? NASA decided to declare that rocket unavailable for expeditions to the planets, even though it would be used regularly for other space missions. This was my first encounter with NASA's obsession with the development of massive new space systems, usually at the expense of focusing resources on actual new achievements in space.

How could NASA secure official scientific approval for this brazen attempt to foist off on Mars exploration the vastly oversize and overexpensive Saturn 5 rocket? As bait for scientists, NASA proposed to build a huge, multibillion-dollar automated biological laboratory to be set down gently on the Martian surface in the early 1970s. Back on Earth, so the fantasy ran, Nobel Prize–winning biologists—and future prize winners— would sit comfortably in university laboratories, conducting a leisurely remote-control search for microbes on Mars.

NASA's clinching appeal to the Space Science Board (SSB) of the National Academy of Sciences was delivered by NASA's associate administrator Dr. Homer Newell. Newell was likeable, modest, self-effacing, believable, and earnest—perfectly suited to extract from the self-assured committee members an endorsement reflective more of their scientific self-interest than of their sound critical judgment. To help them rationalize their endorsement, he pleaded for their support in his continuing competition for NASA resources with the powerful manned-space-flight faction headed by Dr. George Mueller. "Fellows," he said, "if you don't help me, George will get all of those Saturn 5's." So the prestigious Space Science Board endorsed the idea of Saturn 5's to Mars with automated biological laboratories, and this extravagant fantasy became the *only* American planetary future.

During the *Mariner 4* experience JPL's John Casani and I had become close friends. Afterward we bemoaned the gigantic scale and ill-defined concept of NASA's "great leap forward," sometimes over a beer in a quiet little bar just outside the laboratory grounds. One afternoon in late 1965, after exchanging happy anecdotes from *Mariner 4*'s historic encounter with Mars just a few months before, John paused and said pensively, "That was the last good mission." He captured our feelings perfectly.

But help came from an unexpected source—the Budget Bureau (BoB, now the Office of Management and Budget, OMB). Contrary to popular opinion, the Budget Bureau (until David Stockman) was not strictly a budget-cutting force. Sometimes the BoB supplied enlightened support as well as necessary restraint to NASA's tempestuous planetary planning. In mid-December 1965 it killed the Saturn 5 Mars project outright.

NASA's Mars ambitions fell back to square one. Finally, NASA acknowledged that Atlas / Centaur vehicles could be available for probing the planets after all. New plans quickly followed. Two Atlas / Centaurs were scheduled for new Mariner flybys of Mars in 1969, two more to launch long-lived Mariner orbiters of Mars in 1971, and one for a single Mariner look at Venus and Mercury in 1973. In addition, in 1972 and 1973 two more Atlas / Centaurs would project spinning Pioneer spacecraft to make the first close-up observations of Jupiter.

Furthermore, in 1972 a German-American spacecraft, Helios, designed to probe solar emanations, would be propelled sunward on the first flight of a new, more powerful launch vehicle, the Titan / Centaur (see figure 2.2). Then, in 1973, two of those new Titan / Centaurs would launch the ambitious Viking orbiters and landers to Mars. With Viking the direct search for life on Mars would finally take place—a huge challenge, but not the absurdly huge Saturn 5 version promoted in 1965. (Viking would, even so, become the most sophisticated and expensive automated scientific spacecraft ever flown, down to the present.)

In 1966, less than six months after Casani's lament and NASA's rejection of Caltech's evolutionary approach, a golden age of American planetary exploration seemed likely once again. American engineers and scientists could explore, limited only by their own intelligence and capacities, not by bureaucratic constraints.

In the midst of these ups and downs in late 1965, there were also exciting new Mars facts to think about. The TV team leader Bob Leighton pondered a striking natural coincidence. It had to do with the amount of carbon dioxide gas on Mars, which was evident in early spectral studies through Earth-based telescopes. There is about thirty times more carbon dioxide in a column of Mars's atmosphere than in a terrestrial one (where carbon dioxide constitutes only 0.033 percent abundance). But the low surface pressure on Mars discovered by *Mariner 4* meant that carbon dioxide could not be just a minor constituent there. Instead, Mars's atmosphere must be nearly pure carbon dioxide to account for the amount of it measured from Earth. But why is Mars's atmosphere nearly pure carbon dioxide? And how might a pure carbon dioxide atmosphere differ from Earth's largely nitrogen one?

Leighton began to calculate the average temperature of the white polar caps on Mars, taking into account that they are warmed by sunlight during the daytime and through the long polar summer, and then cooled by radiation into space at night and through the dark polar winter. During the winter the excess heat stored from the summer sunlight is radiated back out into space. Leighton's calculations indicated that the polar caps on Mars should be very cold on the average, about 125 degrees centigrade below the freezing point of water. "Is there any significance to such a low temperature?" he asked himself. So he checked the published laboratory data concerning the natural pressure of carbon dioxide gas when in contact with solid carbon dioxide (dry ice) at such temperatures. Here he found a surprising "coincidence." The pressure of carbon dioxide gas that will come to equilibrium with carbon dioxide frost at minus 125 degrees centigrade in a laboratory bell jar is just about the actual carbon dioxide pressure on Mars.

Suddenly the truth dawned. The frost caps are not water ice, as everyone had thought. Rather, they are carbon dioxide frost. The carbon dioxide of Mars's atmosphere is in equilibrium with, and its pressure is controlled by, solid carbon dioxide on the surface. The five millibars pressure of Mars's carbon dioxide atmosphere is not a coincidence but is governed ultimately by the average amount of sunlight reaching the poles of Mars and controlling the average temperature of the frost there.

I joined Leighton early in this study, and in 1966 we published a paper predicting that Mars's seasonal frost caps, long believed to be water ice in some form, were in reality solid carbon dioxide, that is, dry ice. Furthermore, we speculated that an excess deposit of solid dry ice exists year-round on Mars and controls the pressure of its atmosphere. These predictions, which made Mars seem even more hostile to life, were received with great interest (and some skepticism) by our professional colleagues. How to test such predictions? Fortunately, the new Atlas / Centaur Mariner flybys for 1969 would carry a pair of instruments that together could settle the matter.

Much work lay ahead, however, before any new Mars television results would appear. In anticipation of the enlarged challenge, we expanded the original *Mariner 4* television team for the 1969 flybys by adding, among others, our Caltech biochemist colleague Norman Horowitz (who a decade later held the key to Viking's search for life on Mars); Bradford Smith, a New Mexico–based specialist in the photography of the planets through telescopes (who later headed Voyager's photographic exploration of the outer planets); and my friend Merton Davies, a veteran space-imaging specialist from the Rand Corporation.

Mert is a tall stringbean of a man blessed with an unusually sunny disposition. We first met because, in those days, my Caltech salary as a beginning professor was barely adequate to support a wife and three young children. Highly paid consulting for an aerospace company would have helped, but I sensed a conflict of interest. To consult for companies that were competing for NASA contracts in space exploration seemed in conflict with advising NASA internally on planetary missions and policy.

U.S. AND EUROPEAN SPACE EXPLORATION ROCKETS

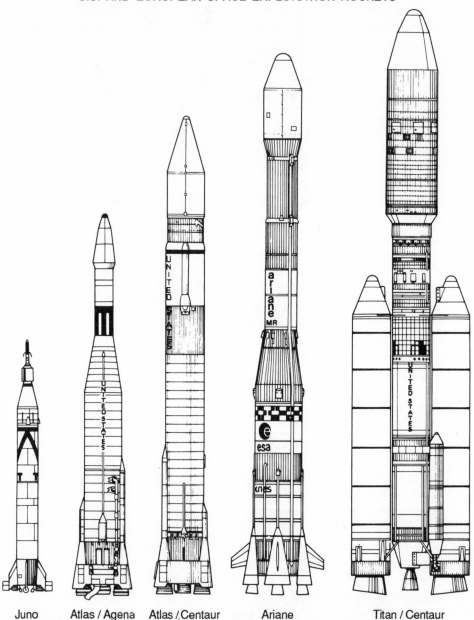

Juno   Atlas / Agena   Atlas / Centaur   Ariane   Titan / Centaur

2.2 DEEP-SPACE LAUNCH VEHICLES • On the left-hand page are the four U.S. rocket systems that have projected robotic payloads to the Moon and the planets, as well as ESA's Ariane rocket, which launched Giotto to Halley's comet. On the right-hand page are the four rocket systems the USSR used to send probes to the Moon, Mars, and Venus. The U.S. Titan 3E Centaur is 160 feet (49 meters) long, and the USSR Proton is 195 feet (60 meters) long. Not

USSR SPACE EXPLORATION ROCKETS

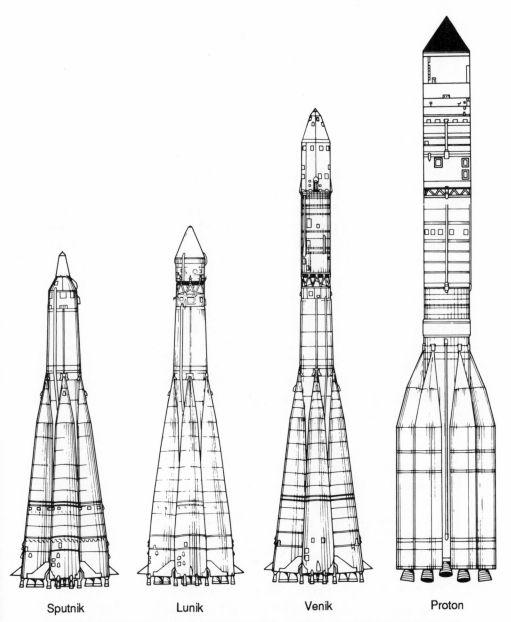

Sputnik   Lunik   Venik   Proton

shown is the Saturn 5 Apollo launcher, 363 feet (111 meters) long, and the Japanese Mu-3S, which launched two small cometary probes and which is 92 feet (28 meters) long. The Space Shuttle, equipped with an upper stage, is scheduled to launch its first planetary probe in 1989. Photographs of the Titan/Centaur, Ariane, Sputnik, and Proton can be found on pp. 134, 264, 16, and 252, respectively.

Hence, it was with relief and anticipation that in 1963 I started consulting instead on *military* space activities for the nonprofit Rand Corporation, in Santa Monica. Equally important to me in the long run, Rand (and service, later, on government advisory committees) provided a window on the dark side of space. Earth orbit was from the beginning an arena for reconnaissance and other military activities, a domain of deadly serious strategic rivalry between the United States and the Soviet Union. Rand began my education on that subject.

Thus, at Rand in 1963 Mert Davies and I began a professional collaboration that has continued informally to this day. Highly regarded for his contributions to early reconnaissance satellites,[3] he became excited about how I was reconnoitering Mars from Mariner probes. And he had some interesting new ideas about using TV pictures of the planetary surfaces to create a grid of small topographic features, natural "benchmarks," on the surfaces of each newly explored planetary body. Mert's geographical control points have since become the bases for all space-age maps of Mars, Mercury, and the moons of the outer planets.

The book jacket of *The View from Space,*[4] which Mert and I were writing in late 1969, noted that we were not only analyzing close-up Mars pictures from *Mariners* 6 and 7 but at the same time also helping design NASA photographic probes to *all* the other planets in the Solar System. What an extraordinary and future-oriented time that was! Nothing seemed beyond reach. In subsequent years, I chose to focus intently on exploring Earth's closest neighbors Mars, Venus, and Mercury. But Mert never wavered from our original course. He is one of only two people* in the world who have taken close-up spacecraft pictures of all the planets except Neptune and Pluto. And Mert pushes on still as part of the Voyager imaging team, preparing for Neptune's portrait in August 1989.

Our enlarged team designed a bigger and much more powerful TV camera for the 1969 Mars flybys. The trip to Mars the second time proved unexpectedly difficult. During preparations for launch at Cape Kennedy, the Atlas rocket propping up its Centaur upper stage partially collapsed just before the new Mariner flyby spacecraft was installed. Hasty reworking restored the Atlas/Centaur to flight readiness but added to the normal tension before a launch. Then, on February 25, and again on March 27, 1969, while Mars and Earth remained within favorable orbital positions, *Mariners* 6 and 7 were powered successfully on their way to Mars.

In 1969 we could take advantage of Atlas / Centaur's greater mass

---

*The other is my Caltech colleague Ed Danielson, whose exploits are touched on in chapters 5, 6, 19, and 22.

capacity. We not only incorporated a much better camera than was possible on our tiny *Mariner 4* in 1965 but also increased the amount of data returned to Earth. In addition, a new instrument, an infrared photometer for sensing the heat emitted from Mars's surface, was aboard the spacecraft. This new instrument was the brainchild of Gerry Neugebauer, now a beginning professor at Caltech, having completed his military tour of duty at JPL. On board, *Mariners 6* and *7* also was a second infrared instrument that sensed the spectral distribution of the heat energy emitted from Mars. This infrared spectrometer was developed by George Pimentel, who four years earlier had recognized that the "Sinton bands" in the Mars spectrum actually sprang from "heavy" water (DHO) in vapor form and not from organic compounds on Mars's surface. Together, this pair of infrared sensors could test our theory of dry ice frost caps on Mars.

On July 30, 1969, ten days after humans first walked on the Moon, *Mariner 6* encountered Mars and swept its camera view across cratered equatorial regions, just as we had planned. But, suddenly, catastrophe struck *Mariner 7*, which was following seven days behind and was programmed to view the Martian polar regions. Without warning, *Mariner 7*'s radio signals died. Frantic calls went from tracking station to tracking station on JPL's worldwide Deep Space Net. Finally, faint communication with *Mariner 7* was reestablished, and gradually it became evident that a rechargeable battery on board had exploded, destroying some of the engineering instrumentation and other critical parts of the spacecraft. *Mariner 7* had literally "taken a hit," from exploding shrapnel, like a bomber in wartime. Once again the Great Galactic Ghoul proved no joke to my friend Casani, who was now second in command of *Mariners 6* and *7*. The path to Mars seemed a mysterious and menacing route.

Within that critical week before *Mariner 7*'s swift passage by Mars, the engineers at JPL diagnosed the problem, gradually regained control of the spacecraft, and reprogrammed the on-board computer to bypass the damaged portions of the spacecraft. A nearly normal flyby provided good TV views of the southern seasonal frost cap of Mars—and the long-awaited opportunity to test with infrared data our hypothesis of dry ice frost caps. The data from Neugebauer's infrared radiometer were quickly processed, and they suggested that the frost caps were indeed as cold as we had predicted. Some ambiguity remained, however.* Hence, we waited for the spectral signature measurements to be presented by Pimentel. His instrument should directly detect any dry ice on Mars by recording its

---

*Neugebauer's instrument sensed all the heat from a large target region on the planet, which included some of the warmer bare ground surrounding the frost cap. The heat from those warm regions contaminated the cold polar cap signal.

unique infrared signature. However, spectral measurements of a complex and unfamiliar natural surface acquired from a moving spacecraft are much more difficult to interpret than controlled laboratory studies. Pimentel's team struggled through the night to understand the puzzling data *Mariner 7* had radioed back.

Finally, the next morning, at the crowded postencounter press conference at JPL, Pimentel dropped a scientific bombshell: the key life-related gases ammonia and methane had shown up in his infrared spectra from the edge of the south polar cap of Mars. They were just where Lowell might have prophesied, at the margin where moisture from melting water ice caps might be released seasonally. This meant that microbial life, perhaps even macroscopic plant life, was now a very real possibility.

But Lowell's reborn legacy flickered only briefly. More refined analysis by other investigators, and by Pimentel himself, revealed that the promising spectra corresponded, in fact, to sunlight reflected from exceptionally "dry" dry ice on Mars; there was no methane and ammonia. Indeed, the Pimentel spectra became the final nail in the coffin of an Earth-like Mars. Once again, Mars's surface was found to be colder and more hostile to life than imagined.

Television photographs revealed more lunar-like craters, and in much better detail than those of *Mariner 4*. An ultraviolet sensing instrument aboard both Mariners showed that an abundance of damaging radiation from the sun in that invisible region of the spectrum beyond the blue actually reached the surface of Mars. This was like baseball's third strike. Any living thing on Mars would not only have to survive in a waterless environment, under killingly low temperatures, but also have to shield itself from the lethal portion of Mars's sunlight.

How could anyone take life on Mars seriously any longer? But Mars had many more surprises in store for robotic explorers. And, the next time, it was my turn to be fooled by my preconceptions.

What about Soviet efforts? Despite complete failures in 1960, 1962, and 1964 (see tables 2.1 and 2.2), the Russians approached rocketry as they fought World War II—with brute force and perseverance. So they started over again. By 1971 they had readied a powerful new armada of large rockets and spaceships designed to reach the surface of Mars.

After a decade in development a large new booster called Proton was operational. Seven Protons were allocated to launch thirty-five tons of spacecraft toward Mars during the years 1971–73. Included in that tonnage were four large Mars orbiters, each carrying a full complement of remote sensing instruments and, most dramatically, five robotic packages intended to touch down on the surface of Mars before America's Vikings arrived. Viking's first try was now slated for July 4, 1976. The Russian landers carried cameras as well as instruments to reveal the chemical composition of the soil and atmosphere.

Only in the almost impossible biological task of searching directly for life on Mars were the Russian landers outpaced by the American Vikings' greater scientific ambition. This boldly conceived second round of Soviet exploration of Mars exploited the especially favorable orbital positions of Earth and Mars in 1971, with repeated attempts scheduled for 1973 if necessary. Soviet preoccupation with being first to reach each new planetary surface showed through clearly.

Six of the seven Protons projected their payloads toward Mars successfully. Nonetheless, the Russian reach exceeded the Russian grasp. (See table 2.2.) The *Mars 2* lander failed in the atmosphere before reaching Mars's surface. Then the *Mars 3* lander actually touched down on the surface intact on December 2, 1971. And radio transmissions from the surface began. But twenty seconds later—silence. Perhaps the parachute designed to slow the final descent failed to detach properly, causing the

TABLE *2.1*

U.S. EXPLORATION OF MARS

Launch	Spacecraft name	Spacecraft weight (kg)	Mission objective	Results
1964				
Nov. 5	*Mariner 3*	261	Flyby	Launch failure
Nov. 28	*Mariner 4*	261	Flyby	SUCCESS—returned 22 pictures and other data
1969				
Feb. 25	*Mariner 6*	380	Flyby	SUCCESS—returned 24 pictures and other data
Mar. 27	*Mariner 7*	380	Flyby	SUCCESS—returned 31 pictures and key polar data
1971				
May 8	*Mariner 8*	1,029	Orbit	Launch failure
May 30	*Mariner 9*	1,030	Orbit	SUCCESS—returned 6,786 pictures of Mars
1975				
Aug. 20	*Viking 1*	3,400	Orbit & soft-lander	SUCCESS—life detection attempted
Sept. 9	*Viking 2*	3,400	Orbit & soft-lander	SUCCESS—life detection attempted

TABLE **2.2**

SOVIET EXPLORATION OF MARS

Launch	Spacecraft name	Spacecraft weight (kg)	Mission objective	Results
			1960	
Oct. 10		640?	Flyby	Launch failure
Oct. 14		640?	Flyby	Launch failure
			1962	
Oct. 24		890?	Flyby	Launch failure
Nov. 1	Mars 1	894	Flyby	Flight failure—communications failed; passed Mars at 191,000 km
Nov. 4		890?	Flyby	Launch failure
			1964	
Nov. 30	Zond 2	890?	Flyby and probe(?)	Flight failure—communications failed; passed Mars at 1,500 km
			1965	
July 18	Zond 3	890?	Mars test to lunar distance	SUCCESS—returned 25 pictures
			1971	
May 10	Kosmos 419	4,650?	Orbit and soft-lander	Launch failure
May 19	Mars 2	4,650	Orbit and soft-lander	Partial—returned data from orbiter, but lander destroyed
May 28	Mars 3	4,650	Orbit and soft-lander	Partial—returned orbital data and survived landing for 20 seconds
			1973	
July 21	Mars 4	4,150?	Orbit	Flight failure—did not enter orbit
July 25	Mars 5	4,150?	Orbit	SUCCESS—returned data & pictures
Aug. 5	Mars 6	4,150?	Flyby and soft-lander	Partial—returned data from flyby, but lander signals ceased
Aug. 9	Mars 7	4,150?	Flyby and soft-lander	Partial—returned data from flyby, but lander missed by 1,300 km
			1988	
July 7	Phobos 1	~6,100	Orbit Mars and graze Phobos	Flight failure—command error
July 12	Phobos 2	~6,100	Orbit Mars and graze Phobos	Flight failure—just before grazing Mars

lander to tumble over in the high Martian winds that prevailed at the time. No one really knows, not even the Russians. Next, on March 12, 1973, the signals from the *Mars 6* lander continued until just before touchdown. Then silence once again. The *Mars 7* landing attempt missed the planet altogether. The giant armada revealed nothing from the surface and added only limited knowledge from orbit.

Ironically, in the same time period, the Soviets achieved steady success in exploring Venus—the result of another large space effort. Few in the West realize even now how hard the Russians tried to be first on Mars as well. The magnitude of the Soviet planetary effort is illustrated in figure 2.3.

NASA also targeted the favorable period in 1971 for a new Mars mission. Twin-orbiting spacecraft were commissioned to carry a modestly expanded complement of instruments derived from the preceding Mariner flybys. One orbiter would swing into circular polar orbit—like an Earth reconnaissance satellite—to create, we hoped, a high-resolution map of the whole planet. Photographic exploration of the entire surface is comparable to exploring the entire land area of Earth. So the preceding Mariner television team needed expansion once again. Fortunately, the U.S. Geological Survey (USGS) had gained unique expertise in mapping the Moon in support of the Apollo program. Now NASA pulled key members of that group over to help map Mars.

My primary role in the 1971 mission was shepherding the evolution of the TV cameras from the *Mariner 6* and *7* flyby version into the new Mars orbiter configuration. I was also in charge of polar imaging.* At the same time, I led the development of the camera for the proposed new Mariner flight to Mercury. (See chapters 5 and 6.) Harold Masursky of the USGS was designated overall leader of the 1971 television team.

NASA's second orbiting spacecraft was to be maneuvered into a more equatorial orbit, from which the still mysterious seasonal variations in light and dark surface markings on Mars could be systematically monitored. Previous Mariner flights had done little more than raise the level of speculation about those enigmatic features. The principal advocate for studying those markings was my friend Carl Sagan. He reasoned—correctly, as it turned out—that the seasonal markings were not caused by plant life but were rather the consequence of high winds blowing dust seasonally on the Martian surface. But before *Mariner 9* no one really knew.

Sagan had become the chief spokesman for the possibility of extraterrestrial life everywhere, and especially on Mars. Carl's devotion to the

---

*The term "imaging" includes pictures acquired with television systems or even with radar. "Photography" is technically restricted to the use of film.

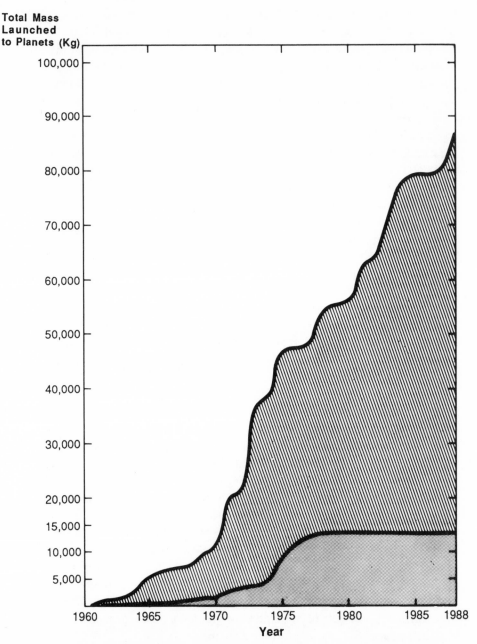

Total Mass Launched to Planets (Kg)

Year

2.3 U.S. AND USSR EFFORTS TO EXPLORE THE PLANETS • The much larger total mass of spacecraft launched toward the planets by the USSR (top curve) compared with that of the United States is evident. The rapid rise of the Soviet total beginning in 1970 reflects the introduction of the large Proton launcher. The plateau in U.S. efforts after 1978 is the consequence of the "Shuttle only" policy for planetary missions. This graph brings out the magnitude of the Soviet effort (rather than the relative achievement of the two efforts, in which U.S. efforts overall have overshadowed those of the USSR).

search for life sprang from his belief that that is the fundamental underlying question in solar system exploration, not because it was necessarily likely. In public and in private, he elaborated tirelessly on speculative possibilities not *absolutely* excluded by the exceedingly cold, dry, irradiated surface revealed by *Mariners 4, 6,* and *7.* Inevitably, Carl and I were drawn into scientific conflict. To me, the extraordinarily hostile environment revealed by the Mariner flybys made life there so unlikely that public expectations should not be raised. In one private moment of sharp debate, Carl charged, "Bruce, you at Caltech live on the side of pessimism." I thought to myself, "And, Carl, you at Cornell live on the side of optimism."

Ultimately, America's Viking mission vindicated my pessimism. But before then *Mariner 9* gave Carl much new ammunition for his hopes.

First, though, there was another setback. *Mariner 8,* like *Mariner 3,* never made it beyond Earth's atmosphere. The Atlas / Centaur rocket in 1971 was still not wholly reliable. Our beautiful spacecraft ended up on the bottom of the Atlantic Ocean. *Mariner 9,* however, roared off the pad on May 30, 1971, and kept going on its way to orbit Mars. But *what* orbit, now that *Mariner 8* was gone? Quickly, the entire sequence of orbital operations about the planet had to be reprogrammed so that one spacecraft could do the work of two. To make matters worse, an all-enveloping dust storm greeted our craft's arrival at Mars. The first pictures were completely featureless. The next big step in Mars exploration was not going well.

Even so, there was still time for fun and philosophy. I organized a soirée at Caltech's Beckman Auditorium on the night of November 12, 1971, just prior to *Mariner 9*'s arrival at Mars. It featured Arthur Clarke, Ray Bradbury, Carl Sagan, and myself as far-ranging and speculative panelists and the *New York Times* journalist Walter Sullivan as moderator. I didn't know what "Mars and the Mind of Man" would turn out to be. It was billed as a free-form, spontaneous public event to compare the imagery and literary significance of Mars with the emerging scientific reality.* In an emotional finale, Ray Bradbury read his newest poem, which perfectly expressed how I felt—and still feel.

If Only We Had Taller Been

> The fence we walked between the years
> Did balance us serene;
> It was a place half in the sky where
> In the green of leaf and promising of peach
> We'd reach our hands to touch and almost touch
>   that lie,

---

*It succeeded beyond my expectations. I converted it into a book later.[5]

That blue that was not really blue.
If we could reach and touch, we said,
'Twould teach us, somehow, never to be dead.

We ached, we almost touched that stuff;
Our reach was never quite enough.
So, Thomas, we are doomed to die.
O, Tom, as I have often said,
How sad we're both so short in bed.
If only we had taller been,
And touched God's cuff, His hem,
We would not have to sleep away and go with them
Who've gone before,
A billion give or take a million boys or more
Who, short as we, stood tall as they could stand
And hoped by stretching thus to keep their land,
Their home, their hearth, their flesh and soul.
But they, like us, were standing in a hole.

O, Thomas, will a Race one day stand really tall
Across the Void, across the Universe and all?
And, measured out with rocket fire,
At last put Adam's finger forth

As on the Sistine Ceiling,
And God's great hand come down the other way
To measure Man and find him Good,
And Gift him with Forever's Day?
I work for that.
Short man. Large dream. I send my rockets forth
     between my ears,
Hoping an inch of Will is worth a pound of years.
Aching to hear a voice cry back along the universal Mall:
We've reached Alpha Centauri!
We're tall, O God, we're *tall!*

"If Only We Had Taller Been" captured that half-conscious search for immortality that drives us so hard to view, even to touch, places never before seen. Long after our burned-out bodies have turned to dust, when our individual names are remembered only by genealogically minded descendants, our first look at our planetary neighbors will be collectively remembered. Faintly, we too will survive as part of humanity's future—and outwit, a little, our own mortality.

For weeks, the dust storm almost completely obscured the surface of Mars as we peered down via television. Only the outline of the residual

MARS AND THE MIND OF MAN • The science and symbolism of Mars formed the subject of a memorable soiree held in Caltech's Beckman Auditorium on November 12, 1971. *Mariner 9* went into orbit the next day, revealing a Mars surprisingly different from the one that had been expected. *From left,* Philip Morrison of MIT, Carl Sagan, Walter Sullivan of the *New York Times,* the author, and Ray Bradbury. *NASA / JPL photo*

south polar cap showed through, and four mysterious dark spots in the equatorial regions. As the planet cleared with frustrating slowness, the four dark spots gradually became huge crater-like forms 40 to 50 miles in diameter. Strangely, each of the four craters was located on top of a high mountain, the largest eventually measuring more than 350 miles across. Why should the ancient impacts necessary to produce such giant craters have occurred on top of mountains? What a puzzle! This was not the Mars we had seen with Mariner flybys in 1965 and 1969.

Within hours after computer-processed images of these dark crater-like spots finally became clear, the TV chief Hal Masursky addressed a news-hungry press corps, three hundred strong, jammed into JPL's von Karman Auditorium. Several other scientists and I sat near him on the stage to provide follow-up commentary. To my astonishment, Masursky called these crater-like forms volcanoes. Our three previous flyby missions had turned up no volcanic features at all. Now, suddenly, four such features were the first ones observed by *Mariner 9?* This seemed implausible to me.

But so they were.

And giant volcanoes were just the beginning. The next thing to be revealed was an enormous canyon, thousands of miles long, up to a hundred miles wide, and nearly three miles deep. Valles Marineris is a gigantic scar in the Martian crust with great branching valleys suggestive of erosion caused by the seepage of underground water. It is far, far larger than any comparable feature on Earth.

Then came the channels. As the dust storm cleared completely, the images revealed sinuously curving valleys, often with tributary valleys leading into them. These large valleys, now dry, must be the result of ancient flood erosion on a grand scale. But how could there ever have been enough water flowing on Mars to carve these fantastic channels and yet leave the ancient cratered terrain observed by *Mariners 4, 6,* and *7* relatively unaffected?

This started a series of debates. Ultimately, most scientific thinking converged on the view that Mars had a relatively brief but extraordinary aqueous phase, billions of years ago. Catastrophic flooding from *subterranean* sources scoured the older terrains of Mars. (Similar catastrophic flooding happened at the end of the last Ice Age in what is now eastern Washington and western Idaho, where the headwaters of the Columbia River lie.) Then the environment of Mars changed profoundly, and the aqueous phase ended (although a few smaller subsequent floods from subterranean sources probably occurred). For billions of years since, Mars has been bone dry, gripped in a permanent drought. When the Viking mission of 1976 sent back clear pictures of small craters on lava plains perhaps three billion years old, their sharp, uneroded topographic features stood as mute evidence against any global rainfall over those billions of years.

In the polar areas, my particular scientific specialty, a magnificent prize awaited us. Surrounding and underlying the residual frost caps were enormous thicknesses of delicately layered, smooth-surfaced sedimentary deposits. We quickly realized that these were global wind deposits, gently spread onto the Martian polar regions in response to periodic climatic fluctuations over tens—even hundreds—of millions of years. What could have caused this?

The answer flowed to us through a serendipitous personal association of the kind that characterizes science in general and Caltech in particular. My Caltech colleague Peter Goldreich had overseen a Ph.D. thesis in celestial mechanics by one William Ward. For a few months before moving on to another academic job, Ward was available in a postdoctoral research capacity. Did I have a need for him, and a way to pay? Goldreich asked. Fortunately, I had both.

I had been using a Caltech undergraduate to compute some changing properties of Mars's orbit. We needed greater expertise to know how to interpret these changes. Ward and I soon identified a pattern in the periodic and eccentric fluctuations in the shape of Mars's orbit. The cyclical variations were clearly the work of gravitational forces from Jupiter and Saturn, both much larger planets. As a consequence, each Martian hemisphere experiences alternating epochs of extremely hot summers and cold winters, followed by periods of mild summers and cool winters. These climatic alternations may control a corresponding cycle of giant dust storms, in turn influencing the accumulation of polar dust strata.

Ward, after helping me, moved on to Harvard, where he soon discovered a more important effect. The *tilt* of Mars's axis is also affected by Jupiter and Saturn and, he found, oscillates wildly—between 15° and 35°. It is 24° at present, close to Earth's 23.5°. Thus, the astronomical basis for global climatic fluctuations on Mars recorded in the polar-layered deposits emerged from an analysis of Mars's orbital history.* When sedimentary evidence of such fluctuations stood out unmistakably from the *Mariner 9* images of Mars's polar region, orbital and climatic theory could catch up. Planetary science is fundamentally an empirical science.

By the end of the *Mariner 9* mission, in 1972, our view of Mars had completely changed—once again. Lowell's Earth-like Mars was forever gone, but so was the Moon-like Mars portrayed by our first three flyby missions, *Mariners 4, 6,* and *7*. The Mars revealed by *Mariner 9* was not one-dimensional; it was an intriguingly varied planet with a mysterious history. The possibility of early life once more emerged.

Perhaps during Mars's brief, early aqueous phase, the chemical means for life to develop did appear. Perhaps microbial life crept forth at that point and then progressively adapted to the much more hostile conditions of the present. Perhaps microbes still lived on Mars. At least, this is what Sagan speculated as NASA prepared for the Viking mission of 1976.

We were peeling Mars like an onion, and I felt exhilarated. However, in the finale I would participate not as a workaday scientist but as an administrator.

---

*A much less extreme form of these astronomically controlled fluctuations had been postulated for Earth to explain glacial retreats and advances.

3.1 *Viking* SCOOPS UP MARTIAN SOIL • In the left-hand image the robotic arm of the *Viking Lander 2* scoops up a sample of the soil adjacent to a boulder and leaves behind the furrow seen in the right-hand image. The soil sample was transported back to the spacecraft and dropped into the entrance of an automated chemical and biological laboratory. *NASA/JPL photo*

3. Martian Microbes?

APRIL 1975. For fifteen years I had continuously pursued planetary discovery following that cool October night in 1960 when I first gazed at Mars. The Mariner Mars probes were history. *Mariner 10* had gone to Venus, then to Mercury. Needing a break, I opted to forgo future spacecraft roles and, aided by a Guggenheim Fellowship, contemplated eighteen months of professional freedom in La Jolla, California. There I aimed to resume personal science, away from big science, and write a textbook describing the surfaces of Mars, Venus, Earth, and the Moon.[6]

No longer would I need to worry about Congress, the OMB, or White House politics. In short, I could moult professionally in a low-pressure environment. Or so I thought.

On April 5, 1975, Dr. Harold Brown, then president of Caltech, asked me to become the new director of the Jet Propulsion Laboratory. Dr. William Pickering had reached Caltech's mandatory retirement age. The search for his successor had converged—entirely without my awareness—on me.

This unexpected development generated a personal conflict. Psychologically, I was already on the beach at La Jolla, shedding the occupational anxieties of space flight and making long lists for general reading. Yet I was suddenly being offered a unique opportunity to affect the character and direction of American planetary exploration.

So, I agreed to become the director of JPL—after a nine-month mini-sabbatical in La Jolla. It was really just a chance for me to prepare for the new role, but very necessary.

On April 1, 1976, I took the reins of the world's premier space exploration laboratory from Bill Pickering. Bill had been one of the pioneers in space flight and JPL's director for twenty-three years, from the days of early rocket accomplishments, through JPL's development of the first American spacecraft *Explorer 1,* and on to the world's first looks at Mars, Venus, and Mercury. Most JPL employees had never known another boss.

The Viking mission to land on Mars and search for microbial life was practically under way. Viking was (and still is) the most expensive and complex scientific space mission ever undertaken, and JPL was the site for all the Viking mission operations and scientific analyses. But JPL had not played the central role in the conception and development of Viking; NASA's Langley Research Center, in Virginia, had done that.

Demonstrating the kind of managerial leadership that enabled NASA in the 1960s to succeed with Apollo, a few dozen key Langley personnel supervised the design, development, and flight of the complex Mars soft-landers, equipped with sophisticated biological and organic chemical apparatus. Furthermore, the entire landing system had to be heat sterilized before launch to prevent any terrestrial microbes from hitchhiking along.

A multihundred-million-dollar contract with the Martin Marietta Corporation provided the Langley Research Center with the technical manpower and system management to pursue its giant task. JPL was to design and build the orbiting spacecraft and its instruments, drawing upon its remote-sensing experience from the earlier Martian missions to Mars. In addition, JPL tackled the toughest lander instruments.

The first lander, *Viking 1,* was scheduled to touch down on Mars three months hence, on July 4, 1976—the American bicentennial. So Viking at JPL was certain to be of major national significance. But for me there were personal complexities in this first challenge as JPL's director.

Viking looked modest by comparison with NASA's failed 1965 scheme for a Saturn 5 biological laboratory to Mars. Even so, Viking still represented a giant and abrupt escalation in Mars exploration, a complete departure from the step-by-step approach I advocated. I felt that Viking focused too much on a scientifically unjustified and hugely expensive search for Martian microbes. And, compounding that exaggeration, it was not instrumented to reveal even elementary mineralogical facts about Martian soil. Biological questions, I thought, should be pursued *after* critical environmental knowledge had been gained. Even before *Mariner 4* reached Mars, I strongly advocated that the first objective of Mars surface experimentation should be not the search for life but the search for clues to its surface geochemical cycles.[7] Where are the surface repositories for water vapor, for carbon dioxide, and for organic compounds? How do they exchange with the atmosphere? What is the soil chemistry and "ecology" in which microbial life could flourish?

But Viking was committed to skip over such geological soil questions and use biochemical tests to search directly for life. As a result, long before my JPL appointment I had avoided any involvement with Viking. Postulating the existence of Martian life publicly as a means of developing support for Viking made me uncomfortable. I was a critic, not a supporter, of Viking. The Viking project leadership, and that of the Langley Research Center, certainly must have fretted when I took over at JPL just before Viking's big moment.

But Viking at JPL worked out just fine. Indeed, within six months I was trying to use Viking's visibility at JPL to sell new planetary missions. As a working scientist, I had learned to fend off NASA's overly ambitious plans for planetary exploration. As JPL's director I found myself on the other side, dealing instead with a NASA that now gave low priority to the nourishing of new planetary endeavors, even while basking in the worldwide acclaim the Viking (and, later, Voyager) generated.

A clear-cut determination of the presence of life—or of its absence—on Mars looked like an almost impossible task to me. Ambiguous results seemed practically foreordained. Any scientific test for life must involve detecting the induced growth or metabolic interaction of Martian microbes in a controlled experiment. Mars's soil, and any putative microbial life within it, would be scooped up from the Martian surface robotically and then combined with special nutrients, carefully packaged within the Viking spacecraft. Any biochemical products produced in that interaction would then be measured. Distinct interaction could indicate the presence of life. But what attitude should we adopt if there were no interaction—simply the neutral response of the nutrients to a passive, sterile soil? How could we be sure the right nutrient had been used? Hypothetical Martian life might interact biochemically with nutrients differently from the way we anticipated. Would other locations on Mars perhaps be more favorable to life? Sagan and the Nobel Prize winner Joshua Lederberg already speculated that life might be concentrated in moist, favorable "oases."

Fortunately, a second key instrument package in Viking's payload would look for organic chemicals rather than directly searching for life. Developing this instrument cost an unprecedented amount of time, money, and grief but in the end proved well worth it. It was a fully automated and miniaturized gas chromatograph / mass spectrometer (GCMS). The GCMS, a primary laboratory tool of the organic chemist and the biochemist, is a machine designed to separate organic compounds in the gas chromatograph phase and then to identify them precisely in the mass spectrometer section of the instrument.

Viking's GCMS could detect a few parts in one billion of nearly any organic compound in a soil sample. With such sensitivity, even the most barren soils on Earth, such as those from the dry valleys of Antarctica or the Altiplano of Peru, reveal a complex organic chemical signature. Any microbial life, even if in a dormant phase, exists in equilibrium with nonliving carbon compounds (mostly the remains of previous generations of microorganisms). Organic chemicals are an inevitable part of its environment. Thus, Viking's GCMS would reveal the organic chemical signature of microbes in the Martian soil, regardless of what they liked to eat or whether they might be in a dormant phase.

The scientific potential of the GCMS complemented that of three specific biological experiments aboard Viking. The *gas-exchange* experiment would expose Martian soil to a broth containing nutrients that terrestrial organisms commonly choose to eat and metabolize. In contrast, the *labeled-release* experiment would offer a more basic, if less tasty, menu of fundamental organic compounds. Both experiments would combine water vapor and liquid water with Martian soil.

The third experiment, developed by my Caltech colleague Norm Horowitz, elegantly steered clear of any terrestrial assumptions about Mars. Horowitz had learned his environmental lessons well from the early Mariner flights. In his experiment only two assumptions were made about Martian life: first, that carbon dioxide gas, the principal constituent of Mars's atmosphere, must somehow be involved in the metabolic processes of any Martian microorganisms (as it usually is on Earth); second, that the radiant energy from the Sun (which Horowitz simulated with a Xenon lamp) would provide the chemical energy for any Martian life. Horowitz specifically avoided introducing any water vapor or moisture into the Martian soil while testing for Martian microbial metabolism.

Viking did not reach Mars's surface on July 4, 1976. Engineering concerns about the safety for landing correctly took precedence over the symbolism of the bicentennial. Indeed, the selection of a landing site proved much more difficult than expected, despite comprehensive ground-based radar information, earlier Mariner photography, and high-resolution Viking orbiter photography. The Viking leadership finally chose the touchdown target, and on July 20, *Viking 1* plummeted toward the windswept, rocky surface of Chryse Plain, "downstream" from ancient water-carved channels.

It was four in the morning in Pasadena, but the mission operations area of JPL was jammed. Dignitaries and working engineers alike crowded around the video monitors, listening intently as a calm male voice on the public-address system ticked off the milestones of *Viking 1*'s descent through the atmosphere. "Parachute deployment . . . Retros firing . . . TOUCHDOWN!" The voice lost its calmness as the announcer, and I and everyone

BRIEFING PRESIDENT GERALD FORD • Shortly after *Viking 1*'s successful first landing on Mars, the President received the news directly from JPL. *From left,* William Pickering, recently retired director of JPL, Alan Lovelace, deputy administrator of NASA, James Fletcher, NASA administrator, James Martin, Viking project manager, Thomas Paine, Fletcher's predecessor as NASA administrator, and John Naugle, chief of NASA's unmanned space efforts. *NASA / JPL photo*

else in the place, burst with excitement. When *Viking 1* plopped onto Mars, its radio signals continued to flow clearly through to us as it carried out a prearranged robotic health check. Everything was working fine. Viking had evaded the fatal embrace of Mars's surface that had terminated Russia's *Mars 3* landing mission four years earlier.

Quickly, *Viking 1*'s camera lenses blinked open and began relaying, line by line, a sharp panorama of the barren, rocky, and windswept landscape in which the Volkswagen-sized lander had plunked down.

No bushes, no grasses, no footprints or other indications of life relieved the barrenness of this geologically fascinating terrain. But microbial life might be waiting in the soil. The robotic arm soon started initial soil testing and sampling.

Finally, after days of gouging the Martian soil with Viking's remote-controlled arm and of preliminary environmental measurements, the broths of the labeled-release and gas-release experiments received Martian soil samples to test. It was late in the evening in Pasadena when those biological instrument readings finally were broadcast back by Viking. Frank Colella, JPL's press liaison, reached me at home with a stunning report.

"Bruce, they've found something," he said, with barely contained excitement. "You know, there are only two possibilities," he continued, savoring a press agent's dream of a lifetime. "Either there *is* or there *isn't* life on Mars." Then he described how the first radio indications showed a definite reaction with the tasty broth cooked up by Viking's scientists to tempt the appetites of Martian microbes.

But the chemical reaction between Viking's "gourmet" soups and the Martian soil samples proved too strong to be true indications of microbes. Both the gas-exchange experiment and the labeled-release experiments showed changes much greater than would be expected for plausible microbial life. In fact, even oxygen gas was liberated in the reaction, as well as carbon dioxide. Something was breaking up the solvent molecules in the broth solutions aboard Viking.

The Viking scientists quickly realized that the nutrient soup was reacting chemically with the Martian soil itself. They had discovered a chemical rather than a biochemical reaction. This was eventually demonstrated by Horowitz's experiment. It, too, showed peculiarities, but of much lower magnitude, because it did not involve the use of water. As most planetary geologists recognized, the *present* minerals in the Martian soil probably had never encountered liquid water before being submerged in Viking's experiments (regardless of what may have happened on Mars's surface billions of years ago).

Finally, definitive results arrived from the gas chromatograph / mass spectrometer. Despite the most careful searching with soil samples from the *Viking 1* and the *Viking 2* sites, not a single organic molecule was detected by the supersensitive GCMS. Mars's soil is far more sterile than any environment on Earth. It would serve as an excellent standard for sterile controls in terrestrial biological laboratories.

Why should this sterility be? After all, the Martian atmosphere is made up of carbon dioxide gas. Why is all the carbon held in that oxidized gaseous state and none in organic compounds in the soil? Also, carbon-rich meteorites commonly carry a small percentage of organic material to Mars, as they do to Earth, where they end up in museums and are labeled "carbonaceous chondrites."

A preliminary indication of unusual surface conditions emerged in 1969 when *Mariners 6* and *7* discovered that the ultraviolet radiation from the Sun, efficiently absorbed by Earth's ozone layer, strikes Mars practically without atmospheric intervention. Furthermore, the tiny amount of water vapor in Mars's atmosphere is photochemically converted to minute quantities of ozone by some of that ultraviolet light. Noting these facts prior to the Viking landing, some scientists had speculated that the Sun's ultraviolet radiation on Mars might interact directly with mineral grains, or indirectly through the ozone, to produce "superoxidized" surfaces of those minerals, or caustic vapors. Such substances are not common on Earth, because water everywhere inhibits the development of such unusually oxidized materials (although eye-stinging smog is one counterexample). But on Mars such superoxidized materials might be accumulating everywhere at the top of the soil—so the speculations went.

Viking's GCMS results proved those speculations insightful. Whenever organic material is delivered to the soil of Mars—for example, by the impact of an organic-rich meteorite—it is rapidly burned (that is, oxidized) by the reactive Martian soil. Thus, it is converted to carbon dioxide gas and in turn blends indistinguishably into the regular atmosphere of Mars. The mineral grains or vapors responsible for this process are then quickly recharged back to their superoxidized state by more ultraviolet radiation.

Mars's surface is therefore self-sterilizing to a degree totally unexpected before Viking. For this reason, the absence of organic compounds in the Viking tests is a strong indication of the absence of life over all of that desolate planet's surface. Organic material—that is, carbon-based compounds—simply cannot survive in the Martian soil.* Neither microbes nor their organic chemical traces are chemically stable there. Viking gave us a more definitive answer than anyone had anticipated. Mars very probably has been lifeless for at least the last several billion years.

Even as the Viking landers vainly tried to culture the sterile Martian soil, scientists and mission planners at JPL were dreaming and scheming about the future. How about a Viking lander on wheels, to move about and repeatedly test the soil in search of a Martian "oasis"? There followed more-thoughtful suggestions about new ways to explore Mars itself, not just about chasing putative microbes. Rovers were conceived to traverse tens or even hundreds of miles across the Martian surface, transmitting detailed images of the mysterious terrain. Complex schemes were proposed to bring rocks from Mars to Earth for analysis. Even the concept of an unmanned Mars airplane made its way onto the JPL drawing boards.

But launch vehicles remained NASA's obsession, far outstripping concern for new exploration. The Titan / Centaur, the key to Viking's success

---

*Just how deep this extremely oxidizing zone reaches is a key question for the automated exploration of the surface in the mid-1990s.

(and that of Voyager to follow), was already targeted for extinction. NA-SA's strategy was clear: force all future U.S. planetary missions onto the Space Shuttle—a vehicle thoroughly unsuitable for the purpose.

So humankind's burst of exploratory energy toward its neighbor Mars subsided quietly, for lack of a rocket and a sponsor—and because Mars proved to be lifeless. The flip side of high expectations for life on Mars can be intense public disappointment at its absence. The Viking landers and orbiters sent back radio signals for years, then gradually fell silent one by one in the late 1970s. *Viking 2*'s lander held out until 1980 before the infirmities of old age silenced the last earthly voice on Mars. Since then, Mars has been undisturbed by terrestrial probing.

Can our search for life on Mars be put on a scorecard? How do those passionate debates look now? My constant concern about taking the great leap to direct biological tests turned out to be justified. Early, and inexpensive, soil chemistry tests would have shown the surface to be hostile to water and organic compounds. They would have confirmed that life in any form that we could recognize or even imagine is not possible on Mars. The expensive direct biological tests of Viking would have been avoided.

Nevertheless, Sagan's imaginative scientific papers turned up some important and lasting leads. For example, Carl's fourth book, really an expanded pamphlet, entitled *Planetary Exploration,* was published in 1970.[8] I was asked to review it for the scientific journal *Icarus.* My review was harsh because I disapproved of his speculations about glaciation on Mars and of his notion that Mars's polar regions could provide clues to Earth's glaciation. It seemed obvious to me then that glaciers, as we know them on Earth, could not exist on Mars. I felt Carl was unnecessarily prolonging the tired old Lowell legacy.

Yet, after *Mariner 9*'s discovery of the layered polar terrains, and our unraveling of the astronomical changes in Mars's orbit and orientation that caused them, I became deeply interested in the polar history recorded on Mars. I am convinced the poles of Mars do hold clues to Earth's glaciation. The oceans on Earth carry heat across latitudes and mediate our global temperatures. Mars, though, has no oceanic moderation. Therefore, small variations in the amount of sunlight falling on Mars's polar regions can have large climatic effects, as is recorded in the alternating layers of dust and ice accumulated in those regions. But what was happening on Mars about two to three million years ago, when Earth's present Ice Age, the Pleistocene period, began?

We know the causes of many climatic changes *during* Earth's ice ages, but we do not know what caused the onset of the Pleistocene or of other, more ancient ice ages. Has the Sun's output of light varied? Could the

EXPLORING MARS TOGETHER • Carl Sagan resides permanently in Ithaca, New York, where he is a professor at Cornell University. However, while the Viking search for life on Mars was under way, in 1976 and 1977, Sagan lived in Pasadena in order to concentrate on that exciting endeavor. During that and similar episodes he and the author worked closely together—on founding the Planetary Society, for example. In this 1976 photo taken in my JPL office, we posed among maps of Mars and other paraphernalia. *NASA / JPL photo*

passage of an obscuring interstellar cloud have cooled Earth? Alternatively, did Earth's crustal plates, slipping and sliding about on the surface, perhaps produce a dramatic and unstable shift in the evaporation and circulation of the oceans that profoundly altered global climate? Or did some gigantic volcanic episode keep the atmosphere dusty so long that Earth's climate changed?

All of these ideas about Earth's glaciation are ad hoc. They lack firm, conclusive evidence. Furthermore, owing to the complex interdependence of oceans, biosphere, ice sheets, and atmosphere, it may *never* be possible to isolate the principal cause of Earth's ice ages solely from information preserved here. That is why Mars's extremely sensitive response to slight variations in sunlight is so important to us. Any external change that might have affected our climate should have been greatly amplified on oceanless Mars, and its signature should be clearly recorded in the layered terrains.

Following such reasoning, I have come to advocate automated, and eventually human, missions to sample the polar terrains of Mars, just as scientists now collect and scrutinize cores from Antarctica and from the ocean bottoms in order to unravel Earth's climatic history. The objective on Mars would be to identify and sample layers, including those formed about two to three million years ago. Were there any anomalies in Mars's climate at that time, as revealed in variations of the dust, mineral particles, and ice of those layers? If the sun's radiation did indeed fluctuate on Earth to initiate glaciation, it must also have fluctuated on Mars. That fluctuation should be manifest in Mars's layered terrains. If the Mars layers do not record a "Pleistocene" episode, glaciation on Earth must have had internal causes.[9] Our search for the origins of ice ages could then be greatly narrowed.

Carl's speculations, which I had strongly resisted in the 1960s, thus included an important insight.

For his part, Carl and a handful of others still cling to a hope that life now exists on the surface of Mars despite Viking's negative results. Virtually all other space scientists dismiss such hopes as unrealistic, and many regard them as a needless obstacle to serious future work on Mars. But nearly all do agree that many billions of years ago Mars had an aqueous environment. Ancient life is not ruled out by Viking's findings. We can search for a *fossil* chemical ecology surviving from those ancient times as organic traces, or even as biogenic calcium carbonate deposits. In principle, such chemical traces of an ancient Martian biosphere might still be preserved on Mars somewhere beneath its incredibly oxidizing outer skin. Eventually, robotic or human explorers will drill in search of such sites, which could tell us whether Mars was once a starting point for life.

Mars is the planet of the future. It will constitute a playing field for the adventuresome members of future generations.

Despite its excesses, Viking was a great thing for science and for the United States. We know more about Mars now than we would have if my cautious, efficient, step-by-step process had been followed. The United States ran out of national momentum for planetary exploration long before the full scientific yield I had anticipated could be harvested. American, and Soviet, visions of what is possible in space were stretched by Viking. The dream was pursued and the answer obtained. Mars *is* lifeless, and public interest in Mars waned for a while.

But Viking dramatized our Earth-centered perspective. We are indeed alone in this Solar System. Earth, exhibiting the only watery surface, is the oasis of life. We do not have distant microbial cousins on Mars or anywhere else plausibly in this Solar System. In our search for living analogs, we must extend our imaginations and, ultimately, our sensors toward other stars. Thus, although Carl and I differed strongly over the probability of life on Mars, we both advocated searching for intelligent life around other stars by listening for faint cosmic radio transmissions.[10]

Looking back to that "Mars and the Mind of Man" panel on the eve of *Mariner 9*'s revelations about Mars, I am struck by how much we labored under an unexamined assumption. The Mariners' planetary discoveries in the sixties and early seventies seduced us into believing that a new era was at hand. We assumed that the United States would continue to reach out to the limits of its exploratory technology. But as the seventies rolled by and the eighties dawned, we learned instead that America had accumulated far more technological capacity than will for space exploration. The country deviated from the pursuit of a grand destiny and turned inward, toward consumerism and individual self-satisfaction.

Our trips to the Moon, Mars, Mercury, and Venus and our search for life on Mars proved to be only bright anomalies in our history, not a baseline for the future. Apollo happened, in some sense, three or four decades too early, an unlikely consequence of unique political, economic, and technological circumstances and inspired political leadership. The search for life on Mars flowed from the same national enthusiasm that powered Apollo. Thus, Viking, Voyager, and the Mariners themselves became a reality decades earlier than they would have if Apollo had not happened.

It would take time for us, and for others, to find the energy and the will to return to Mars.

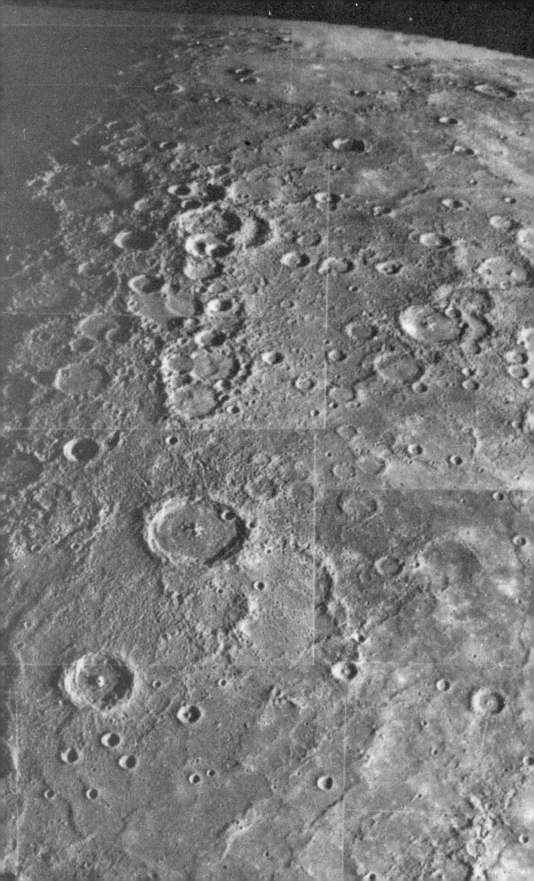

PART TWO

Probing Warmer Worlds

4.1 MERCURY FROM *Mariner 10* • The northern limb of Mercury frames this striking scene, showing craters up to fifty miles (eighty kilometers) in diameter among expansive volcanic plains. Mercury, like the Moon, is atmosphereless and displays similarly crisp photographic details. This image is taken from a large mosaic compiled from over 150 computer-processed *Mariner 10* images. *NASA/JPL photo*

4.2

The Exploration of Venus and Mercury

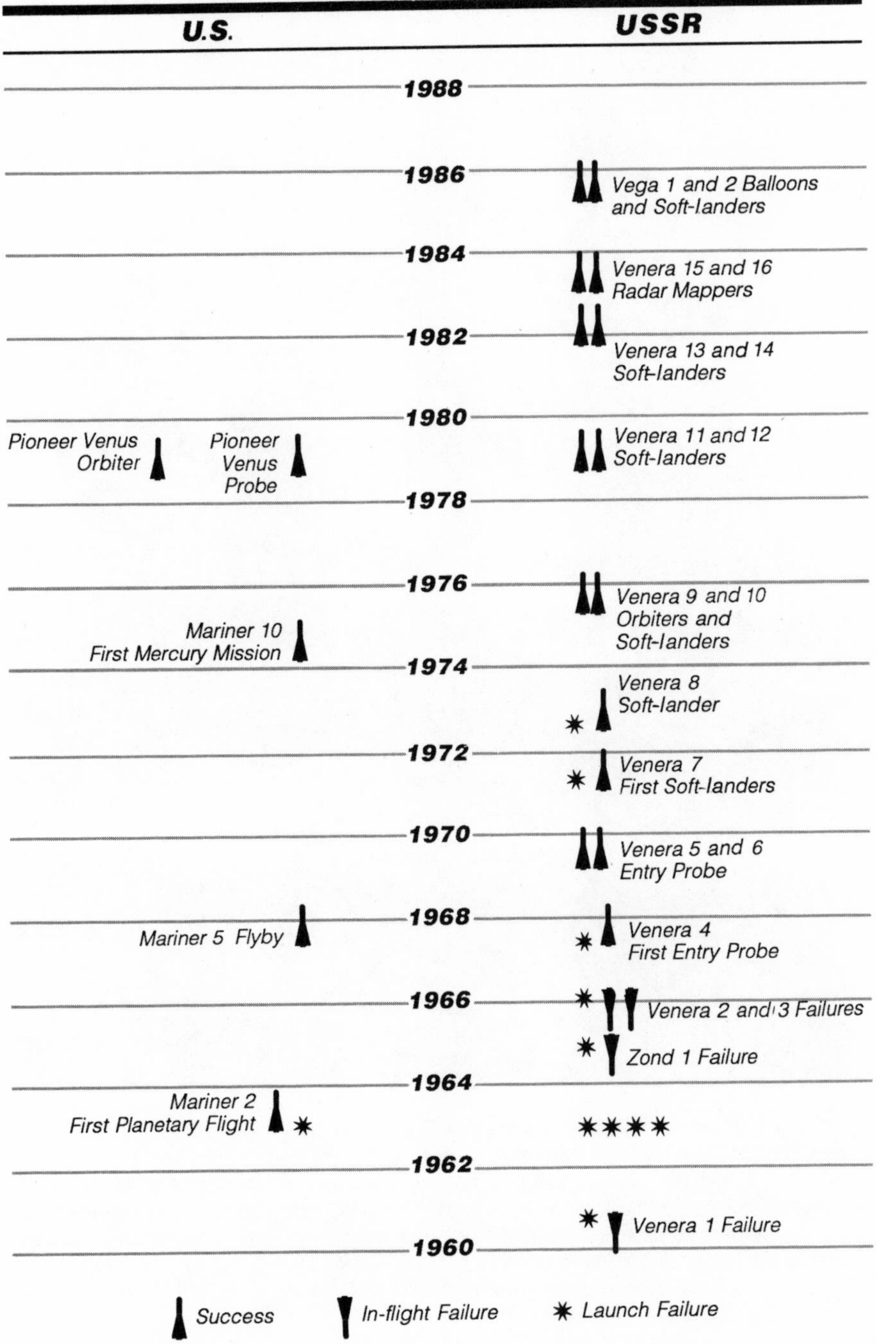

U.S.	USSR

1988

1986 — Vega 1 and 2 Balloons and Soft-landers

1984 — Venera 15 and 16 Radar Mappers

1982 — Venera 13 and 14 Soft-landers

1980 — Venera 11 and 12 Soft-landers

Pioneer Venus Orbiter   Pioneer Venus Probe

1978

1976 — Venera 9 and 10 Orbiters and Soft-landers

Mariner 10 First Mercury Mission

1974 — Venera 8 Soft-lander

1972 — Venera 7 First Soft-landers

1970 — Venera 5 and 6 Entry Probe

Mariner 5 Flyby

1968 — Venera 4 First Entry Probe

1966 — Venera 2 and 3 Failures

Zond 1 Failure

Mariner 2 First Planetary Flight

1964

1962

Venera 1 Failure

1960

▲ *Success*   ▼ *In-flight Failure*   ✳ *Launch Failure*

4.3 THE MAINSTAY OF THE DEEP SPACE NET • This 210-foot (68-meter) antenna in the Mojave Desert of California is one of three identical deep-space tracking systems constructed in the mid-1960s. The others are located near Madrid, Spain, and Canberra, Australia. Together with smaller dishes at the same sites, they have provided nearly continuous coverage of all U.S. deep-space missions. (Spacecraft radio signals received overseas are often relayed back to JPL by geosynchronous satellites.) This worldwide system, termed the Deep Space Net, or DSN, has been progressively improved and managed by JPL from its inception. *NASA/JPL photo*

4. The Elusive Surface of Venus

IT WAS SIX O'CLOCK in the morning on October 19, 1967. Still in my pajamas, I held the kitchen telephone close to my ear. Walter Sullivan was calling from the *New York Times*. "The Russians claim to have landed on Venus, Bruce," he said. "Is that possible?" He went on to read me the Tass release the *Times* received during the night. There could be little doubt. The Soviet Academy of Sciences proudly proclaimed *Venera 4*'s delivery of an atmospheric-entry capsule to Venus and its safe landing on the surface. Man's first robotic visit to the surface of another planet had radioed back crucial measurements of Venus's atmospheric pressure and temperature.

At last a major scientific controversy would be decisively resolved. Was Venus's atmosphere at all comparable to that of Earth, the planet it resembled closely in size and mass? Or, as many American specialists thought, was Venus's atmosphere forbiddingly thick and dense and its surface hot enough to melt lead?

It had been known for centuries that Venus had a permanently cloudy atmosphere. In 1761 the Russian astronomer Lomonosov noted an anomalous brightening of Venus just before and after it crossed the Sun's path. Lomonosov had speculated that such brightenings were diffuse images of the Sun refracted through the thick atmospheric layers on Venus, which acted like a natural lens. So, in 1769, Captain James Cook journeyed to Tahiti specially to observe Venus crossing the disk of the Sun, which was best observed from a tropical site.

By the 1950s radio telescopes were examining the radio emissions of the cosmos, and in 1956 a Naval Research Laboratory astronomer, Cornell Mayer, made an unexpected discovery. About three times more radio noise poured out from Venus than would be expected to accompany the planet's solar heating and thermal reradiation. A calibration error? No. Other observers soon obtained similar results. Could vast thunderstorms be rumbling deep within Venus's atmosphere and producing a deafening background of radio "static"? Or was there an extraordinarily dense ionosphere surrounding our sister planet, powering a stream of natural radio emission?

Neither theory was true, argued a brilliant young Ph.D. student at the University of Chicago named Carl Sagan, who graduated in 1960, after completing his thesis on the radiation balance of Venus. Venus's surface must be extraordinarily hot, but why? Then the MIT radio astronomer Alan Barrett calculated the radio emissions from Venus, which indicated an enormously thick and cloudy atmosphere composed mostly of carbon dioxide. We have a local corollary. Just a tiny amount of carbon dioxide (0.03 percent) is sufficient to retain enough of the Sun's radiant heat to keep our planet habitably warm—the famous "greenhouse effect." Without this carbon dioxide, Earth would be a lifeless, icy desert.

But Venus differs in extraordinary degree, Barrett and others argued. Its atmosphere retains so much of the heat reradiated from the sunlit surface and clouds that a prodigious surface temperature results. The "excessive" radio noise observed at Venus, he concluded, is simply the natural radio noise given off by an unnaturally hot surface. (Arkady Kuzmin, the Russian visitor at Caltech who followed our Mariner Mars research so intently in 1965 [chapter 1], used Caltech's high-resolution radio telescope to test this theory of a hot Venus surface more precisely.)

The controversy had continued into 1962, when Mariner 2, the world's first successful planetary probe, flew by Venus on December 10, carrying instruments intended to detect a Venusian magnetic field and any associated radiation belts. (See figure 4.4.) It also carried a miniature radio telescope to pursue at close range the question of radio brightness. *Mariner 2* laid to rest two issues: Venus had neither a magnetic field nor radiation belts.

These results were impressive, especially because *Mariner 2* almost failed to reach Venus in 1962. The ever increasing sunlight encountered by the spacecraft as it traveled from Earth toward Venus gradually raised the internal temperature of its instruments and electronics beyond design ranges. (See figure 4.4.) The delicate and complex calibration of the miniature radio telescope was compromised, and the radio brightness issue was not completely resolved. The ionospheric hypothesis was indeed eliminated, but *Mariner 2* still could not prove decisively that Venus's radio signals emanated from a hot surface. Four days after the flyby of Venus, *Mariner 2*—the first flight to another planet, and the first American first in space—expired of robotic sunstroke.

This was an early American effort, but the Soviet Union stayed at it. Beginning in February 1961, it attempted launches every nineteen months—the length of the cycle when Venus and Earth come closest to each other. By the time *Venera 4* finally made the journey, in 1967, the Soviets had tried without success at least eleven times. (See tables 4.1 and 4.2.) No wonder the Soviet Academy of Sciences had proudly trumpeted that *Venera 4*'s entry probe had determined that the pressure at the surface of Venus was eighteen times that of Earth and the temperature a steamy 280 degrees centigrade. Venus was clearly more hostile to life than Earth, but not as outlandishly hot as the uninhabitable oven predicted by Barrett.

The Soviet engineers tried to make sure that the *Venera 4* capsule would survive on mysterious Venus. The capsule was designed to float if it encountered Earth-like oceans. Even the bizarre possibility of oceans of *oil* had been anticipated, a wild speculation thrown out some years earlier by the English cosmologist Fred Hoyle. Good design or no, however, *Venera 4* strangely ceased to transmit right at the surface. Was it blown over by winds, coming to rest with its radio antenna pointed away from Earth?

Venera 4 was not the only visitor to Venus in October 1967. An

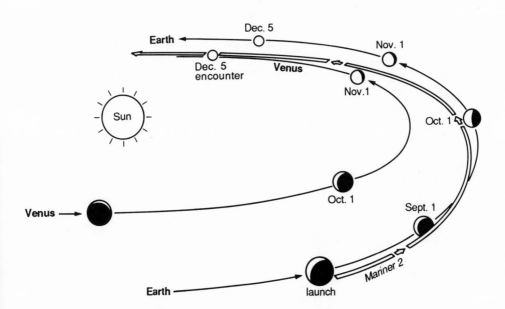

4.4 First flight to Venus • *Mariner 2* was launched in late August 1962 and traveled the ribbon-like path indicated here to encounter Venus on December 5, 1962. Mercury, which orbits between Venus and the Sun with a period of eighty-eight days, is not shown in this diagram.

TABLE *4.1*

SOVIET EXPLORATION OF VENUS

Launch	Spacecraft name	Spacecraft weight (kg)	Mission objective	Results
1961				
Feb. 4	Tyazheliy Sputnik 4	640?	Flyby	Launch failure
Feb. 14	Venera 1	644	Flyby	Flight failure; passed Venus at 100,000 km
1962				
Aug. 25		890?	Flyby and probe?	Launch failure
Sept. 1		890?	Flyby and probe?	Launch failure
Sept. 12		890?	Flyby and probe?	Launch failure
Nov. 11	Kosmos 21	890?	Test	Launch failure
1964				
Mar. 27	Kosmos 27	890?	Flyby and probe?	Launch failure
Apr. 2	Zond 1	890?	Flyby and probe?	Flight failure; passed Venus at 100,000 km
1965				
Nov. 12	Venera 2	963	Flyby and probe	Flight failure; passed Venus at 24,000 km
Nov. 16	Venera 3	960	Flyby and probe	Flight failure; struck Venus 450 km from visible center
Nov. 23	Kosmos 96	960?	Flyby and probe	Launch failure
1967				
June 12	Venera 4	1,106	Flyby and probe	SUCCESS—returned direct readings of atmosphere to 25 km altitude
June 17	Kosmos 167	1,100?	Flyby and probe	Launch failure
1969				
Jan. 5	Venera 5	1,130	Flyby and lander	SUCCESS—returned direct readings of atmosphere to near surface
Jan. 10	Venera 6	1,130	Flyby and lander	SUCCESS—returned direct readings of atmosphere to near surface

Launch	Spacecraft name	Spacecraft weight (kg)	Mission objective	Results
			1970	
Aug. 17	*Venera 7*	1,180	Flyby and soft-lander	SUCCESS—sent back data from atmosphere and surface of Venus
Aug. 22	*Kosmos 359*	1,180	Flyby and soft-lander	Launch failure
			1972	
Mar. 27	*Venera 8*	1,180?	Flyby and soft-lander	SUCCESS—atmospheric data & soil analysis returned
Mar. 31	*Kosmos 482*	1,180?	Flyby and soft-lander	Launch failure
			1975	
June 8	*Venera 9*	4,936	Orbiter and soft-lander	SUCCESS—returned surface pictures & other data
June 14	*Venera 10*	5,033	Orbiter and soft-lander	SUCCESS—returned surface pictures & other data
			1978	
Sept. 9	*Venera 11*	3,940?	Flyby and soft-lander	SUCCESS—returned data from surface
Sept. 14	*Venera 12*	3,940?	Flyby and soft-lander	SUCCESS—returned data from surface
			1981	
Oct. 20	*Venera 13*	3,940?	Flyby and soft-lander	SUCCESS—returned color pictures from surface
Nov. 4	*Venera 14*	3,940?	Flyby and soft-lander	SUCCESS—returned color pictures from surface
			1983	
June 2	*Venera 15*	3,940?	Radar orbiter	SUCCESS—mapped north polar regions
June 7	*Venera 16*	3,940?	Radar orbiter	SUCCESS—mapped north polar regions
			1984	
Dec. 15	*VEGA 1*	4,000?	Venus balloon and lander and Halley's comet flyby	SUCCESS—balloon yielded new data
Dec. 21	*VEGA 2*	4,000?	Venus balloon and lander and Halley's comet flyby	SUCCESS—balloon yielded new data

TABLE *4.2*

U.S. EXPLORATION OF VENUS AND MARS

Launch	Spacecraft name	Spacecraft weight (kg)	Mission objective	Results
1962				
July 22	*Mariner 1*	202	Flyby	Launch failure
Aug. 27	*Mariner 2*	203	Flyby	SUCCESS—passed Venus at 34,853 km
1967				
June 14	*Mariner 5*	245	Flyby	SUCCESS—returned data, passed Venus at 4,094 km
1973				
Nov. 3	*Mariner 10*	504	Flyby of Venus/Mercury	SUCCESS—returned orbital data and pictures at Venus; first flyby of Mercury
1978				
May 20	*Pioneer Venus 1*	549	Orbiter	SUCCESS—returned abundant orbital data and pictures for several years
Aug. 8	*Pioneer Venus 2*	904	Probes	SUCCESS—five vehicles returned direct data from atmosphere

American spacecraft, *Mariner 5,* was approaching Venus, too, flying by just two days after *Venera 4. Mariner 5* was a low-cost conversion of the backup Mars spacecraft for *Mariner 4* in 1965. In those first, tentative years of planetary exploration, it was not enough for JPL simply to prepare two identical spacecraft per planetary mission. A third complete spacecraft, a spare, was built as well. As *Mariner 4* sped toward Mars, that spare was carefully "flown" in a test chamber back at JPL. Before any command was radioed to *Mariner 4,* it was first rehearsed and certified on the spare. Such painstaking care and conservatism made possible America's early planetary successes—no in-flight spacecraft failures and only three launch failures (*Mariner 1* to Venus and *Mariners 3* and *8* to Mars)—in eighteen American tries.

But what to do with a spare spacecraft after that *Mariner 4* flyby of Mars in 1965? Fly it to Venus, of course, and probe that mysterious atmosphere, using the same elegant radio occultation trick pioneered by *Mariner 4* at Mars. So, in less than two years, that spare spacecraft, now renamed *Mariner 5,* was reconfigured to fly *toward* Venus and the Sun—

not away from the Sun, like *Mariner 4*. A second specialized radio system was added to probe Venus's atmosphere and ionosphere further. On June 14, 1967, this rig was launched on an Atlas / Agena. It was by far the least expensive planetary mission in history.

Bob Leighton and I, however, were not entirely happy with the Venus mission's priorities. The Mars mission was astounding because of the unexpected features in the images Mariner beamed back to Earth. The Venus mission, by contrast, scrapped the television camera and tape recorder. Thus, we felt, a great opportunity would be lost if small gaps in the clouds of Venus perhaps allowed a glimpse at the surface. In any case, faint and mysterious atmospheric markings had been observed photographically from Earth, in invisible ultraviolet light. We had proposed that NASA add an ultraviolet filter to *Mariner 4*'s camera system, for *Mariner 5* to provide the first orbital close-up of our next-door planetary neighbor.

But it was not to be. *Mariner 5* would have to make its mark on scientific history strictly through the use of radio waves.

These radio waves had an interesting trip. Very much the way today's air traffic radar systems probe our crowded air space, continuously transmitted radio tones from the huge antennae (see figure 4.3) of the Deep Space Net (DSN) were received on *Mariner 5* and then amplified. Then they were continuously retransmitted by *Mariner 5*'s radio to the same DSN antennae. Thus, a continuous "two-way loop" linked Earth and *Mariner 5*, locating the craft in space to within a fraction of a mile. When the spacecraft flew behind Venus, the continuous radio link with Earth would gradually be occulted and distorted by the thick atmosphere. Those observed distortions in *Mariner 5*'s radio signal denoted the pressure and temperature profiles of the upper atmosphere of Venus (the actual estimates being obtained through computer modeling). This approach had paid off handsomely in exploring Mars in 1965 with *Mariner 4*.

Another type of radio wave experiment supplied the answer to a key question: Where was the surface of Venus in relation to the precisely located *Mariner 5?* Remember, one "sees" this planet's surface only through the radio waves. Shortly after World War II, radar systems of growing power first bounced echoes off the Moon. By May 1961 JPL's Goldstone radar system was able to reach out and touch even distant Venus. Suddenly, the distance from Earth to Venus could be measured to an accuracy of a few miles. Such extremely accurate measurements of range led to a new fix on Venus's orbital motion about the Sun and also fixed the diameter of this solid planet, forever shrouded in cloud, at 7,523 miles (12,104 kilometers), just a shade less than that of Earth.

JPL's computers knew where the surface of Venus was, and the DSN tracking located *Mariner 5* precisely, so the computers could keep track of both.

Thanks to tiny *Mariner 5* and the giant American ground-based radars and precise radio tracking systems, it was now possible to closely reconstruct how the pressure of the Venus atmosphere increased in moving toward Venus's surface. The *Mariner 5 / DSN* team could go further. Well-known physical laws describe how atmospheric pressure must increase downward in a planetary atmosphere. Thus, by extrapolating, one could use the *Mariner 5* observation to determine the actual surface pressure on Venus. It was simply a matter of extending the upper atmospheric profile all the way down to the depth of the radar-determined surface.

On October 18, 1967, the *Venera 4* capsule acquired in situ measurements of the Venusian atmospheric pressure as it sank to the surface. Two days later *Mariner 5* confirmed and augmented the Russian information. This was the first, and quite unplanned, joint planetary investigation. *Venera 4*'s design, mission, and purpose had been kept secret by the USSR practically until the first Tass announcement. American scientists and engineers involved with *Mariner 5* heard of *Venera 4* with as much surprise as I did when awakened by Walter Sullivan.

Then came another surprise. *Mariner 5* did not, in fact, find the same Venus described by *Venera 4*. Intensive computer analysis at JPL of the bending and fading of *Mariner 5*'s radio signals did indeed produce a steady pressure-versus-altitude profile quite similar to *Venera 4*'s first measurements, at the top of Venus's atmosphere. And Russian and U.S. data sets continued to yield comparable estimates of pressure through the upper atmosphere, but not through the lower. The U.S. radio tracking showed that *Mariner 5* was probing atmospheric layers much higher above the solid radar reflection surface than the Soviets reported. It became obvious that the Russians had made a mistake. Their entry capsule had stopped transmitting some sixteen miles (26 kilometers) above the surface of Venus. By the time *Venera 4* finally fell to the surface, it was long dead.

Why was the Soviet Academy of Sciences fooled by the signals of its own spacecraft? The *Venera 4* capsule probably had imploded near its own design limits—near the a priori (and unrealistically low) estimates of Venus's surface pressure available to its designers. Because of their expectation of the "thinner" atmosphere, the Soviet scientists and engineers probably assumed that *Venera 4*'s demise corresponded to the capsule's reaching the surface. Instead, they encountered an atmosphere that is incredibly thick, with a surface pressure nearly a hundred times that of Earth. (And it is hot enough there to melt the solder in the instruments.)

A year later, when Arkady Kuzmin, the roving Russian astronomer, was in Tucson, Arizona, for a conference on planetary atmospheres, we asked him how he reconciled the discrepancies in the two sets of spacecraft data.

"Venus must have great high plateaus—like the Pamirs," * he replied. "Venera must have landed on such a high place."

"A plateau sixteen miles (25 kilometers) high?" we asked incredulously. "Isn't that rather improbable?"

"Let me tell you about 'improbable things,' " Kuzmin said. "What would be the chances that the first German bomb dropped on Leningrad would kill the only elephant in the Leningrad zoo? Yet that's exactly what happened!"

After hearing this facetious view of mathematical probability theory, we dropped further questioning. Clearly, the Soviet Academy was trapped by its premature claim that its spacecraft had reached the surface of Venus. No official representative like Kuzmin would back away from it.

For two years thereafter, Soviet planetary scientists wrote about Venus as if it were a world from *Alice in Wonderland*. Sometimes they were forced to run their atmospheric model profiles of pressure versus altitude down into the range of negative altitude values (which would place the surface of Venus miles below the crust). They knew that *Venera 4* had not reached the surface, but they were trapped by an overbearing national pride. In 1969 the *Venera 5* and *6* landers survived to greater depths in Venus's atmosphere. The true surface was finally reached by *Venera 7* in 1970 and by *Venera 8* in 1972. Quietly, the negative altitudes disappeared from Soviet scientific papers. The Soviets' prolonged success at Venus finally buried their earlier embarrassment.

* A high, mountainous plateau in Soviet Asia.

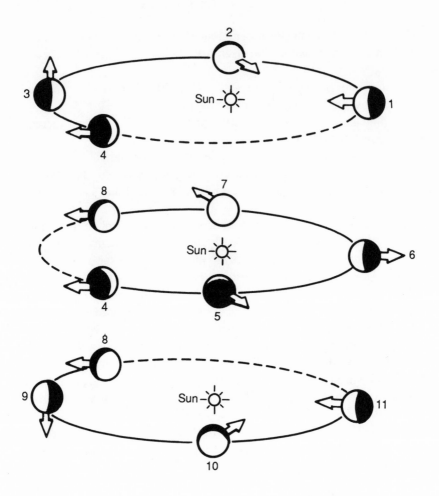

5.1 THE ROTATION OF MERCURY • Mercury spins about its axis of rotation precisely three times while orbiting the Sun exactly twice. This coupling between Mercury's day (58.7 Earth days) and its year (88 Earth days) is illustrated here, the arrow denoting a specific site on Mercury. By following the rotation and orbital positions numbered 1 through 11, we can see how the planet returns to the same orientation after two complete revolutions.

5. One Chance for Mercury

"YOU MEAN you want to go to Mercury to learn about the *Moon*?" cried the Nobel Prize winner Richard Feynman, incredulously.

It was 1968, and I had just tried out some new ideas about Mercury on my colleagues at Caltech's weekly physics colloquium. Diminutive Mercury, with only 38 percent the diameter of Earth and 53 percent that of Mars, reflects sunlight and radar waves, largely the way the Moon does. After Earth, however, Mercury is the densest planet, very much denser than the Moon. Its strong gravitational field reaches out to perturb the orbits of even some asteroids, betraying a large mass packed into a small, spherical body. Mercury, like Earth, must thus be far richer in the heavy planet-forming element iron than the other planets or moons are. But there is a puzzle. If the materials that make up Mercury's surface were likewise so iron rich, the reflection of sunlight and radar waves would not resemble that of the Moon, which is covered with silicate mineral debris and practically no free iron. Mercury, like its cousins Mars and Venus, presented us with contradictions from the beginning. (See figure 5.2.)

A unifying hypothesis was needed. Suppose that all the iron on Mercury long ago condensed into a central core like Earth's. In that case, the iron core alone would be three-fourths of the diameter of the whole planet, slightly bigger than the entire Moon. Puzzling possibilities like these led Gerard Kuiper, the leading American planetary astronomer of the postwar period, to speculate that the Sun might have erupted early in the history of the Solar System. Intense outpourings of searing hot gases could then have stripped away the atmosphere and perhaps even some of the outer rock layers of Mercury. Furthermore, Kuiper argued, the huge basins of the Moon that hold the maria (dark plains) could be giant scars recording collisions from an extinct family of Earth-circling satellites that once accompanied the Moon. Such an origin of lunar basins from Earth-circling moonlets would be unique to the Earth-Moon system.

If the surface of Mercury really resembles the Moon's, I reasoned, the search there for dark plains like those of the maria of the Moon could be important to unraveling the Moon's history as well as Mercury's. We could test Kuiper's theory of the Moon by going to Mercury because of its pre-

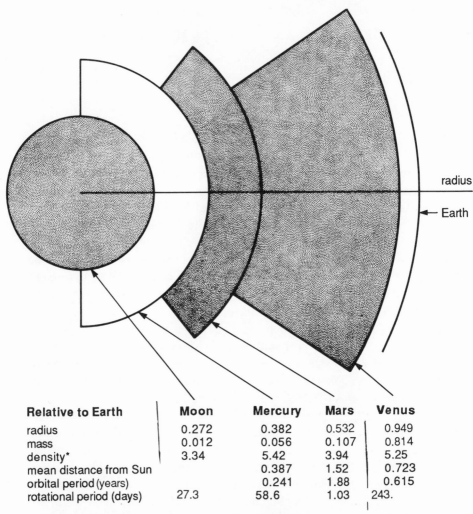

Relative to Earth	Moon	Mercury	Mars	Venus
radius	0.272	0.382	0.532	0.949
mass	0.012	0.056	0.107	0.814
density*	3.34	5.42	3.94	5.25
mean distance from Sun		0.387	1.52	0.723
orbital period (years)		0.241	1.88	0.615
rotational period (days)	27.3	58.6	1.03	243.

*Grams per centimeter cubed; Earth's density is 5.52 g/cm^3.

5.2 THE TERRESTRIAL PLANETS • The diameter, mass, and other properties of the rocky bodies that make up the inner Solar System are compared here. The Moon, although a satellite and not a planet orbitally, is usually included in such comparisons because it is large enough to exhibit global differentiation and deformation.

diction that mare-filled basins would not be found on the solitary planet Mercury. It was this tentative excursion into interplanetary geological comparisons that aroused Feynman in his front-row seat at the Caltech colloquium.

In fact, the geological nature and history of lonely Mercury that *Mariner 10* revealed six years later would stun us all. But in 1968 we were still trying to extract the reality of Mercury from the uncritical scientific mythology that had overgrown the subject, like jungle vines hiding an ancient Mayan temple.

Since the end of the nineteenth century, textbooks had reported that Mercury rotated synchronously about the Sun, just as the Moon rotates, showing only one face to the Earth. If that were so, Mercury's rotation on its own axis (its day) would be the same length as its revolution about the Sun (its year), eighty-eight Earth days. Mercury is much closer to the Sun than Earth is, barely a third as far away, so the planet's sunny side would always be in daylight and be incredibly hot. Conversely, Mercury's other side would orbit in perpetual darkness and be the coldest place in the Solar System. Here was a planet that only a geologist could love—and pursue.

In 1961 a leading European authority wrote that Mercury's synchronous rotation about the Sun had been demonstrated to an accuracy of at least one part in ten thousand.[1] That same year radio astronomers in Michigan finally managed to measure the radio noise emitted from Mercury. It was a difficult observation because Mercury is always close to the Sun. To get a maximum signal, it was necessary to observe Mercury when its apparent size was greatest, which is when it approaches closest to Earth. But the part of Mercury viewed then, when Mercury lies between Earth and the Sun, is mostly the frigid, nighttime side. The radio astronomers were surprised to find that Mercury, like Venus, was emitting excess radio noise even though the supposedly frigid nighttime hemisphere was being viewed. Their observation proved that the nighttime side of Mercury is *not* the coldest place in the Solar System after all.

Why is it warm? some of us wondered. Does Mercury perhaps have a modest atmosphere like that of Mars, an atmosphere that warms the nighttime side with heat carried from the hot, daytime side? But the heat-retaining carbon dioxide gas of Mars and Venus could not be the cause at Mercury, because its spectral signature was missing from sunlight reflected from Mercury's surface. Mercury's atmosphere would have to be composed of an unusual gas, perhaps nonreactive argon gas (which freezes only at extremely low temperatures). Colorless argon occurs naturally as an invisible trace in Earth's atmosphere and artificially in some gaudy "neon" lights.

In order to pursue these questions about Mercury, I became an airborne commuter between Pasadena and the Palomar Observatory in July 1965, just as *Mariner 4* encountered Mars. During the long development of *Mariner 4*'s television system, I had also vigorously pursued a separate program of telescopic observations of the planets in the wavelengths far beyond the visible—the thermal infrared. In 1961 Caltech's Jim Westphal and I first measured the heat radiated from an object *outside* the Solar System (the star Alpha Orionis, popularly known as Betelgeuse), and in the process we helped create the field of infrared astronomy.[2] This part of the spectrum is an exciting one for the study of planets because the heat reradiated from planetary surfaces can reveal much about the physical state of those surfaces.

Mercury posed a particular challenge. I realized that the equipment Jim Westphal and I had developed and used on the 200-inch telescope at Mt. Palomar to observe Venus, Mars, and Jupiter should also be capable of measuring the heat emitted from Mercury's nighttime surface. But the observation was just barely possible, because, from Earth, one must always observe Mercury close to the Sun, especially if one wants to view the planet when it appears with a reasonable angular diameter in the sky. The same geometric alignment in 1962 caused the Michigan radio astronomers to view the unexpectedly warm nighttime hemisphere.

I finally managed to get observing time on the 200-inch telescope around sunset, when the planet's image would be only 17° away from the Sun. (This was apparently the first time the 200-inch instrument had ever been pointed so close to the Sun.) This critical time for Mercury observations on this telescope happened to correspond to the last four days leading up to *Mariner 4*'s encounter with Mars. The drive from Pasadena to Palomar took three and a half hours each way. What to do? Simple. Commute by air. There was a small deserted airstrip near Mt. Palomar. I chartered a plane and pilot to shuttle me every day between Pasadena and Palomar. By day I sat in increasingly tense meetings preparing from the first spacecraft encounter with Mars. At night I used the largest telescope in the world, with what was then the most advanced infrared instrumentation, to observe the surface heat emitted from the nighttime side of Mercury. With the din of the Beechcraft's motor loud in my ears in the late afternoon, I would peruse the *Astronomical Almanac* to determine the correct right ascensions and declinations for Mercury. The next morning, on the flight back, I would go over the printouts describing the detailed sequence for the imminent *Mariner 4* encounter. What a time to be alive! What remarkable good fortune to be able to participate simultaneously in the best planetary studies in space and on the ground!

Others were also looking at Mercury in 1965. The giant radar antenna in Arecibo, Puerto Rico, cradled in a natural amphitheater, could view Mercury that year. Radar astronomers there not only detected echoes from Mercury and refined its distance from Earth but also precisely measured the distortion of the reflected radar "tone" caused by the rotation of the planet. Because a radar beam from Earth illuminates the entire hemisphere of Mercury facing Earth, the reflected radar signal from Mercury is made up of waves reflected not just from the point nearest Earth but from other nearby terrains east and west as well. Waves reflected from those sites west of the central point have a slight extra rotational velocity toward Earth, and those east of it have one away from the Earth. The result of this planetary Doppler effect is to broaden slightly the overall spectrum of the received signal.

The broadening of Mercury's radar signal told the Arecibo astronomers that Mercury's rotation about its own axis was faster than that required simply to maintain one face permanently toward the Sun. The rotational period of Mercury about its own axis was not eighty-eight Earth days, the period of revolution of Mercury about the sun, but about fifty-nine Earth days.

I first learned of this discovery from the Sunday newspaper, shortly after I pursued my airborne commuting to Palomar in search of the signature of Mercury's elusive atmosphere. "Of course," I said to myself, "there is no mysterious atmosphere carrying dayside heat around to a perpetually sunless hemisphere. Every part of Mercury's surface is warmed by the Sun every fifty-nine days. That's why the nighttime soil is still warm enough to emit detectable radio waves. That's just the way the Moon is, with its twenty-eight-day cycle of night and day. All of us studying Mercury should have been suspicious of its 'synchronous' rotation when the 'anomalous' radio observation in 1962 indicated that the backside was warmed by something. But none of us asked the obvious question."

While I was finding intellectual satisfaction (and some self-criticism) in resolving for myself the mystery of the radio brightness, Giuseppe ("Beppe") Colombo, a distinguished Italian specialist in celestial mechanics who taught at Galileo's own University of Padua, was going further and asking, "Why fifty-nine days?" Colombo quickly grasped the significance of the ratio of Mercury's newly discovered fifty-nine Earth days to its year of eighty-eight Earth days. Fifty-nine is to eighty-eight almost in the exact ratio of two to three. He saw that there must be a connection between the periods of Mercury's day and of its year. Before the end of 1965 he published a seminal paper proposing that Mercury spins in a locked rotation rate about the Sun, so that it rotates on its own axis three times for every two revolutions it makes about the Sun. (See figure 5.1.) Other radar studies quickly confirmed Columbo's theory.

It is no wonder that astronomers had long been led to believe that Mercury is in synchronous rotation about the Sun. Because Mercury is so difficult to observe from Earth, they had concluded wrongly that one side always faced the Sun and the other never was heated at all. A sampling bias can easily creep into the data, because uniform coverage of the entire surface is difficult to build up. Most of all, this episode showed how scientists can sometimes ignore data points that do not "fit" if they think they already know the answer.

Mercury was proving to be more deceptive than I had expected. (Even so, Willy Fowler, my Caltech colleague and a Nobel Prize winner, continued to goad me by saying, "It's just another hot rock.") But how to get to Mercury? It orbits so much closer to the Sun than Earth does that it would take a much larger and more expensive rocket than that needed to go to "nearby" Mars.

Ray Bradbury had pointed us toward the answer:

> ... I send my rockets forth between my ears,
> Hoping an inch of Will is worth a pound of years.

To explore Mercury, we had somehow to find a way to substitute brain power for rocket power. Fortunately, visionary members of the British Interplanetary Society had been worrying about such questions even before Sputnik startled the world in 1957. They showed that brain power was worth a lot of rocket power if a planetary spacecraft could use the extra gravity pull it encountered during a close flyby of a planet. Under the right circumstances a close passage of Venus, for example, could impart to a spacecraft just enough extra acceleration to speed it along to more distant Mercury.

When I first heard of these three-body "swing-by" trajectories, the idea sounded to me like humbug—like some sort of perpetual-motion machine. But it involves no violation of physical laws. Both the spacecraft and the planet are orbiting the Sun. As the spacecraft approaches the first planet, it is pulled by both the Sun and that planet. The spacecraft is perceptibly accelerated relative to the Sun, and the massive planet is imperceptibly slowed.

In October 1962 the UCLA graduate student Michael Minovich, after working several summers at JPL on arcane trajectories, discovered that in 1970 and again in 1973 Venus swing-by trajectories to Mercury would exist. A Mariner-class spacecraft, propelled by an Atlas / Centaur from

Earth to Venus, could then continue on "for free" to a rendezvous with mysterious Mercury.

But not easily. There is only one point adjacent to Venus—almost a science fiction "gateway"—that leads directly to Mercury. Each one-mile miss in hitting that invisible moving target translates into a thousand-mile error at Mercury. The voyage to Mercury, economy class, would require unprecedented accuracy in space navigation.

I found out about these launch opportunities in 1970 and 1973 through the JPL grapevine in 1966 and immediately urged that we exploit them. Here was a chance to explore a completely unknown planetary surface. Only one generation of Earthlings would ever do that, and I could help make it happen.

So I listened keenly in early 1968 to a CIA expert assessing the possibility of a *Soviet* mission to Mercury. I was at CIA headquarters in Langley, Virginia, along with other members of the Space Science and Technology Panel of the President's Science Advisory Committee (PSAC, pronounced "pea-sack"). PSAC had been created by President Eisenhower in the mid-1950s when high-technology intelligence and space issues increasingly demanded presidential attention. From time to time our panel was given secret information concerning the space capabilities and intentions of the Soviets.

Swiftly ushered through the lofty CIA lobby, we were escorted into a small and disappointingly ordinary conference room, not at all like the Hollywood version of high-intelligence digs. Using casually prepared transparencies laid on an ancient projector, the CIA expert summarized past Soviet automated missions to the Moon, Mars, and Venus, including many unannounced launch failures. He then speculated about future Russian intentions, among them a possible Mercury flyby.

The Russians had propelled a spacecraft toward Venus every nineteen months since 1961 (see table 4.1). Finally, just the year before our meeting, *Venera 4* succeeded in directly probing Venus's dense atmosphere for the first time (see chapter 4). Certainly, they would continue to probe Venus at each nineteen-month interval. And now, I feared, they had only to modify the 1970 Venera flyby bus in some way to handle the greater intensity of sunlight at Mercury, and they would be there.

But the CIA chart also posited a possible Soviet mission to Mercury before then, in 1969. How could that be? Had JPL missed a window? Growing uncomfortable under my persistent questioning, the expert pulled out more-extensive data prepared for the CIA by a contractor. Here, for each possible trajectory, were listed the distances from the center of Venus to the critical aiming point that led on to Mercury. Yes, the data regarding 1970 and 1973 were familiar. But the distance listed for 1969 was much closer to Venus—too close, in fact. The distance from the center of the planet to the aiming point was less than the planet's radius. A spacecraft heading toward Mercury in 1969 would have to fly *through* Venus! That was obviously why the 1969 "opportunity" had been rejected outright by Minovich and others at JPL.

How could a top technical agency of the U.S. government have made such a mistake? The answer is simple: except for the Apollo program, peaceful space competition with the Soviets simply was not very important to the CIA. Thus, a second-rate person had been permitted to do a second-rate job.

The CIA does work for the White House, however, and it reflects the priorities of the President and his advisers. Even as our briefing at CIA headquarters continued, the electoral wars to determine who would succeed the embattled President Lyndon Johnson were heating up. In 1969 that new President would gain the political windfall of Apollo's success. He would also inherit the responsibility for establishing a new goal—or at least a new direction—for civilian space efforts. The precious political, technical, and human assets accumulated by a successful implementing of Kennedy's manned lunar enterprise should not be allowed to dissipate.

But Richard M. Nixon never evinced a positive vision for America in space. Nor did he care much for independent-minded technical advisers. By 1972, less than four years after Nixon took office, his original science adviser, Lee DuBridge, had been dismissed, and PSAC and most other sources of domestic advice were on their way out. Domestic policy issues lay firmly in the grasp of his political operatives, H. R. Haldeman and John Ehrlichman. Instead of goals for future accomplishment in space, America got a gimmicky aerospace project—the Space Shuttle—justified by fallacious and inadequately reviewed technical and economic arguments.

By contrast, NASA in 1968 still pursued America's *future* in space, even if the President's men were overwhelmed by Vietnam and other earthly problems of the here and now. Consequently, a study of the 1973 Venus-Mercury opportunity got under way quietly at JPL in late 1967, and in 1968 NASA formed a science advisory group to help JPL define and develop a Mariner-class mission to Mercury and Venus in 1973. I was named imaging-team leader. Mariner / Venus / Mercury (MVM) would be at the center of my life for the next six years.

The first meeting to discuss our newly formed project to fly to Venus and Mercury was sobering. "MVM will be *Mariner Mars 6 and 7* with no changes,'' Bill Cunningham announced, somewhat defensively. Bill, a respected NASA headquarters official, had been NASA's man at JPL for the *Mariner 6* and *7* Mars flyby missions just nearing completion. Now, from the podium of JPL's main meeting room, he was setting Spartan standards for the newly selected MVM scientific participants. Minimum cost and no frills. In response to this lean requirement, JPL's director, Dr. William Pickering, clinched NASA's approval for this mission by committing JPL to do it for a bargain-basement price of $98 million. Scientific support for the mission was reasonably strong, but only if it would not detract from the mainline Mars efforts or from the emerging Jupiter initiative.

In 1968 the Space Science Board (SSB) of the National Academy of Sciences endorsed NASA's proposal for a 1973 flyby of Mercury and Venus. But the board's report strongly suggested that only a single launch would be needed: "Now that technological advances have made failures infrequent, whether at launch or during the mission, ... planetary exploration is no longer a primitive or risky act."

History did not support the SSB's optimism about space systems. *Mariners 1* and *3* had failed at launch in 1962 and 1964. *Mariner 7* experienced a nearly catastrophic flight failure at Mars in 1969. Two years later *Mariner 8,* one of the twin spacecraft for orbiting Mars, plunged into the Atlantic Ocean. What the SSB really meant, it seemed to me, was that more risk was acceptable for the Mercury probe because the scientific results were more problematical to them.

As I listened to Bill Cunningham, I thought about the solar radiation at Mercury—ten times more intense than at the Moon. No wonder the Soviets never modified a spacecraft to go on from Venus to Mercury. We quickly learned that *Mariner 10* would need a gold-plated sunshade to shield most of the spacecraft from the scorching sunlight. Even its solar panels had to be made tiltable to reduce the intense solar heating at Mercury's orbit. The mission also required unprecedented accuracy in aiming and communications.

Was this "*Mariner Mars 6* and *7* without changes"?

"Come on, Bill," I argued repeatedly. "Flying by Mercury just isn't the same as flying by Mars. And, if you want MVM to search for a magnetic field as well, we'll have to fly over the darkened *nighttime* hemisphere of Mercury. But then, without sunlight, no pictures!"

The test for a magnetic field on Mercury was very important. The origin of Earth's magnetic field is one of the most intractable problems in geophysics. There is, for one thing, no way to get to it, since the magnetic field originates in Earth's core, forever beyond the reach of direct measurement. Physical conditions in the core cannot be simulated completely in any laboratory. Thus, explanations for Earth's field must be based upon extrapolation and speculation. Somehow, Earth's core is a self-sustaining electrical dynamo, powered ultimately by Earth's internal heat interacting with its global rotation. Internal heat drives slow convection within the electrically active metallic core. Vast electrical currents must accompany the rotationally modified convection motions. Those currents then produce the strong, invisible magnetic field that orients compasses all over the world.

Mariner 2's key experiment on its historic flight in 1962 was the search for a magnetic field at Venus. None was found, even though Venus is nearly as large as Earth and probably has a similarly iron-rich and molten core. However, Venus rotates much more slowly than Earth—possibly too slowly to maintain a dynamo inside—or perhaps it lacks convection in its metal core. Two and a half years later, *Mariner 4* searched for a magnetic field at Mars, but no Earth-like field turned up there either. The rotation of Mars is almost exactly the same as Earth's. However, smaller, less iron-rich Mars probably lacks a currently convecting iron core, necessary to power a natural dynamo.

Mercury could have a much larger iron core than Mars because it is very dense and must be full of iron. But its core probably cooled long ago and, by conventional wisdom, should also not still be convecting heat from its deep interior. Furthermore, Mercury rotates very slowly, although not quite so slowly as Venus. Thus, it did not seem likely to the MVM team that Mercury would exhibit a significant magnetic field to MVM's sensors. But that important possibility could not be ruled out. These terrestrial planets are all we have for comparison with our own.

So we had to find a new approach. We would crowd the blackboard with sketches of the differing geometries at the Mercury and the Mars encounters (see figure 5.3). But Bill was determined to prevent cost increases through expensive or complex scientific enhancements. Washington had spoken. A cheap flight to Mercury was the only way we would get there. And, indeed, in the fall of 1969 the House Space Committee deleted MVM entirely from the NASA authorization bill. The committee's chairman, Joseph Karth of Minnesota, wanted NASA to give greater emphasis to new satellites to study Earth. "Practical" uses of space sounded better to his constituents in the Midwest.

In the end, after a fight, the Mercury-Venus flyby was grudgingly restored in a House-Senate compromise, but along with it came an extra $10 million for new Earth satellites to appease Karth. The Apollo honeymoon was ending, even as Neil Armstrong and his fellow visitors to the Moon made their triumphal journey around the world.

Gradually, Bill Cunningham and JPL's Gene Giberson, the new project manager, began to show interest in a novel solution proposed by our imaging team to solve the conflict between photographing the daytime side and also passing near to the nighttime side to search for a magnetic field. We devised a way for MVM to get good pictures (from the sunlit surface) and still carry out a good test for a magnetic field (which would require a flyby on the nighttime side).

The trick was optical (see figure 5.3). If the focal length of the high-resolution TV camera design we had inherited from viewing Mars with *Mariners 6* and *7* were increased by, say, a factor of three, good high-resolution pictures of the afternoon side of Mercury could be acquired from far out on the approach trajectory, well before the closest approach. MVM would become a supertelephoto camera. MVM could then continue on above the dark, nighttime side of Mercury for its closest encounter and test directly for a magnetic field. Then, more high-resolution pictures of the morning side of Mercury could be taken on the departing trajectory.

"But won't the camera weigh more, Bruce, if the focal length is three times longer?" Bill asked.

"We'll make sure it isn't any longer overall and doesn't weigh appre-

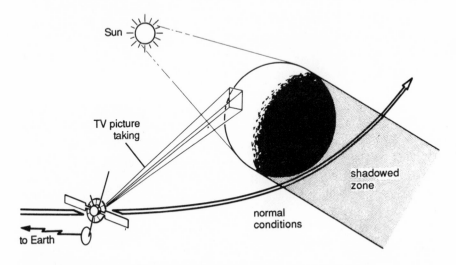

5.3 IMAGING MERCURY • The illumination of Mercury as encountered by *Mariner 10* was unfavorable for close-up imaging, necessitating the development of new high-resolution optics for the television cameras. Compare this with the more favorable Mars viewing shown in figure 1.5, p. 39.

ciably more,'' I answered. ''We'll do that by increasing the magnification of the secondary mirror and by keeping the overall dimensions about the same as those of *Mariners 6* and *7.*'' My friend John Casani, now spacecraft manager for MVM, was at times like this a mighty spokesman for economy—whether in money or in weight. He chimed in, ''But Bruce, your team is asking for *two* large long-focal-length cameras. On *Mariners 6* and *7* there was only one. The other was a small wide-angle lens.''

Fortunately, the newly selected TV team of MVM included most of those who had developed the optics of *Mariners 6* and *7,* among them the space-photography pioneer Merton Davies. In replying to John Casani, Mert Davies and I were ably supported by Ed Danielson, a JPL optical engineer and scientist who had helped us greatly with *Mariners 6* and *7.* Now he was JPL's key man on the MVM imaging team. Together, we pointed out that the wide-angle lens used at Mars could never be effectively used at Mercury. MVM's closest approach to Mercury (when Mercury's image could finally fill the field of view of the wide-angle camera) occurred at night, when the surface was dark. The wide-angle-lens approach was therefore useless for MVM. But a dual electronic camera system was already in place, inherited by MVM from the earlier Mars mission. It was not practical now to throw away one electronic camera simply because its wide-angle lens was not useful. So, we argued, if we simply replaced the wide-angle lens of *Mariners 6* and *7* with a duplicate high-resolution one, the two long-focal-length (MVM) cameras could deliver up to twice the number of useful pictures. John balanced the alternatives and ultimately came to our support. At last we got the go-ahead from Giberson and Cunningham to develop twin high-resolution cameras.

Now came the hard part—building the optics. The optical-design experts we brought in emphasized the difficulty in getting such a large amount of magnification into such a short overall length. But that was the optical compromise necessary. Furthermore, JPL knew a lot about vidicon cameras, the ''retinas'' of these supersensitive digital television cameras, but not much about designing and building high-resolution optics. Our design called for much higher optical resolution than JPL had achieved in deep space before.

It was a tough problem, and the MVM project fell further and further behind schedule in developing this new optical system. The JPL engineers and their contractors were so committed to success that they found it difficult to acknowledge imminent failure. It looked as if I would have to wear the ''black hat'' and tell the project leaders we had to back down.

But, just as time was running out, the critical optical assembly finally performed satisfactorily. We were able to continue. In the end the MVM camera system provided an optical resolution of four arc seconds—good enough to let one read the classified ads of a newspaper held up a quarter of a mile away.

At about this time MVM got another boost from the giant brain of Beppe Colombo. I barely knew this short, balding man, with one of the most engaging smiles in the world, when he showed up at an MVM science conference at Caltech in February 1970. Afterward, he came up to speak to me.

"Dr. Murray, Dr. Murray," he said, "before I return to Italy, there is something I must ask you. What would be the orbital period of the space-craft about the Sun after the Mercury encounter? Can the spacecraft be made to come back?"

"Come *back?*"

"Yes, the spacecraft could return to Mercury."

"Are you sure?"

"Why don't you check?"

He was right. After flying by Mercury, MVM would orbit the Sun with a period of revolution of 176 days, exactly twice that of Mercury's 88 days. With small maneuvers the spacecraft could be made to return to Mercury's orbit every two Mercurian years at the same time Mercury itself was there. *Mariner 10* could fly by Mercury several more times (although under the same surface-lighting conditions). Once again, brain power had enhanced rocket power, as Beppe Colombo's fine mind grasped a profound reality that had escaped the rest of us. (See figure 5.4.)

Extending MVM's life for another year in space would strain the space-craft's capabilities but could yield much greater amounts of data and more viewing opportunities. Here was another challenge to Gene Giberson and his $98 million, fixed-price mission.

Despite the problems in building the TV optics, I felt that our novel camera was going to work. The real challenge was how we would radio back, from fifty million miles away, the huge numbers of TV pictures the camera could acquire.

The Deep Space Net (DSN) engineers at JPL, the finest in the world, wanted a space communications system to have enormous signal margins. With great performance margins (excess capability over the minimum necessary), degradations in components or bad luck could be handled— and the engineers would still look good.

Whereas the engineers pushed for a small but very sure number of TV pictures, we explorers wanted the largest-possible yield—more surface coverage, higher surface resolution. The best approach for us was rapid transmission of large amounts of imaging data. That means taking it all, including the noise in the signal, like the "salt-and-pepper" flickering you see on a home TV screen while switching channels. In fact, such background noise does not interfere much with a viewer's ability to recognize the content of a scene in the regular TV programming. It would be the same with the interpreting of pictures of Mars or Mercury, we argued. We

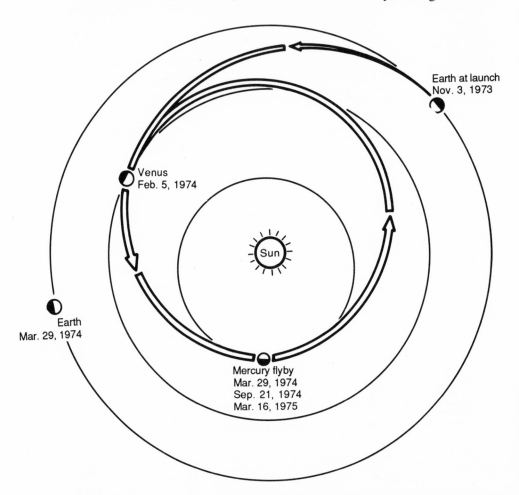

5.4 To Mercury, again and again • *Mariner 10* ended up in an orbit about the Sun with a period of 176 days, exactly twice that of Mercury. As a consequence the spacecraft was able to make two additional encounters with Mercury during its useful lifetime. However, because Mercury presents the same surface under the same lighting conditions every two years (see figure 5.1, p. 92), *Mariner 10* could study only the same hemisphere each time.

insisted on designing the MVM communications for the highest rate of pictures, even with some noise under *nominal* conditions. The battle over communications capability between the DSN engineers and the scientists started all over again.

In 1965 the pictures from *Mariner 4* had come back at eight bits of information per second—a rate roughly comparable to that of Morse code transmission in the early days of railroad telegraphers. (A bit is the basic unit of information, such as the 0 or 1 of a binary number. Each alphabetic character on a computer, for instance, is represented by a "word" composed of eight bits, for example, eight individual 0's or 1's.) As a consequence, it took most of a week to return the five million bits of data that *Mariner 4* gathered about Mars. That is approximately the amount of data in a single picture taken by the Viking orbiter in 1976 and radioed back in *five minutes*.

The DSN engineers were probably justified in being so conservative for *Mariner 4* in 1965, because pictures had never been transmitted from planetary distances before that time. (Those were the days of the Great Galactic Ghoul.) For the Mariner Mars flybys four years later, the DSN engineers seemed to be very supportive of us when they announced a thirty-two-fold increase in the data rate. This time 256 bits per second would be available—a fine increase, I thought, but surely no threat to their safety margin. I noted that now NASA had bigger antennae on the ground and on the spacecraft, more-sensitive receivers were on the ground, and more radiated power flowed from the spacecraft. That meant the transmission capacity had increased greatly since the days of *Mariner 4*.

So I badgered the DSN officials until we finally struck a deal: this compromise between the TV team and the DSN would be termed an "engineering experiment" in which communications at 16,000 bits per second would be tried. That rate was a 2,000-fold increase over *Mariner 4* in 1965, and a 60-fold increase over the nominal rate of 256 bits per second. In fact, in 1969 *Mariners 6* and *7* used the "engineering experiment" to return 200 times the volume of picture data sent by *Mariner 4*.

But there had been no discernible noise in those *Mariner 6* and *7* pictures either. Now, in thinking through the MVM mission, I realized that we could get even more pictures back from Mercury if only we could avoid storing the picture data temporarily on board the spacecraft with a tape recorder. That bottleneck was slowing us down.

The vidicon sensor of the television camera poured out its electronic signal covering each TV frame every forty-eight seconds, and that analog electrical representation of the picture was then digitized, picture element by picture element, into 8-bit words. This electronic process resulted in a signal that poured out of the electronic camera at a rate of 117,600 bits per second. But the maximum rate of data readout for MVM transmission to Earth was to be only 16,000 bits per second, just as when *Mariners 6* and *7* were at Mars. So, once again, only selected frames of picture data from the camera would be read into a tape recorder; most of the electronic pictures taken of Mercury by the camera would never be recorded and thus would be lost. The tape recorder full of the highly selected frames would be played back later at the much slower rate of 16,000 bits per second that the radio system could handle. The total picture return, as a result, would ultimately be limited by the capacity of reliable tape recorders that could be carried on the spacecraft, even if the communications link back to Earth could be opened up, as I suspected it could.

To remove this bottleneck, we needed to be able to send the digitized data directly from the camera to the radio system, and then have the radio send back data from Mercury at a rate of 117,600 bits per second. Our goal was nothing less than real-time television from Mercury, just like home TV. But the DSN engineers insisted that a radio link from the spacecraft to Earth of 117,600 bits per second was not feasible.

We chipped away at the problem relentlessly. But MVM communications improvements over *Mariners 6* and *7* remained slow and small—a slightly larger antenna on the spacecraft, a somewhat better receiver on the ground. We still fell short by at least a factor of four in the communications rate we needed for real-time TV transmission, by the DSN signal standards.

The DSN officials were not exactly stonewalling, but it is fair to say that they did maintain a military-like discipline and control over their employees. Open communication with that tightly run organization was not easy. I had not managed to promote a full and detailed discussion with anybody in the DSN about possible additional safety margins that might be buried inside the DSN calculations. Then I got a break. I found an engineer, a specialist in communications performance, who was so detached from the day-to-day worries of his organization that he naively invited me to look at an internal memo he had written, analyzing the communications performance of the *Mariner 6* and *7* Mars flybys.

And there it was—clear evidence of additional safety margins buried in the *Mariner 6* and *7* transmissions. Additional safety margins likewise had to be hidden in their projections for MVM. Clutching the evidence I needed to negotiate with the project leaders and with the DSN officials for real-time transmissions by MVM, I whipped up to Project Manager Gene Giberson's JPL office. He and his staff agreed and threw their weight behind us in the debate with the DSN. We won. MVM would couple the highest-optical-resolution camera yet developed for a planetary spacecraft, with by far the highest rate of data communications ever tried in deep space, 117,600 bits per second. Unprecedented scientific riches would be the prize.

If we could make it all work.

6.1 VENUS IN ULTRAVIOLET LIGHT • *Mariner 10* captured this portrait of Earth's nearest planetary neighbor in February 1974. The dozens of individual frames that were specially processed by JPL computers to produce this remarkable mosaic were all shuttered through the same filter, which blocked out visible light but allowed the invisible light beyond the blue to fall onto the vidicon detector. When seen in visible light, Venus is featureless. These ultraviolet features are believed to be produced by variations in the properties of cloud particles high in Venus's stratosphere. Their patterns chronicle global circulation on Venus. *NASA/JPL photo*

6. Perilous Flight of Discovery

THE FLORIDA AIR was heavy and warm even on a November night. Floodlights illuminated the gleaming white rocket pointed skyward. A delicate vapor trail from the oxygen vent at the top of the gantry tower at Pad 36B provided the only motion in the scene. The Atlas / Centaur was truly a twentieth-century cathedral, painstakingly built by thousands of disciplined supplicants. Inside the conically shaped shroud at the very top lay *Mariner 10*, folded up like a jack-in-the-box ready to spring fully into life once it was released in the sunlight on the other side of Earth. The twin high-resolution cameras—and my hopes and anxieties—were there too.

My wife Suzanne and I had slipped past the nominal safety barriers surrounding the launch pad to spend a few moments together in silent reverie before the final countdown. It was the second marriage for each of us, and this occasion was full of intensely shared feelings. Mariner / Venus / Mercury (MVM) had been a constant part of our life together. Now it was time for MVM to take on real life. *Mariner 10* was to be born—or stillborn. Just two years earlier, at this same launch pad at the Kennedy Space Center, in Florida, *Mariner 8* had sat on top of an identical Atlas / Centaur. It, too, gleamed in the sunlight awaiting launch. But the Centaur stage failed, and *Mariner 8* died. Two weeks later, our hopes were saved as its twin, *Mariner 9*, lived to reveal a breathtaking new Mars.

No twin had been authorized for *Mariner 10*. I had testified before the House Space Committee in March of 1972 that a dual-launch mission was well worth the extra cost. But the $98 million budget cap stayed on.

However, Gene Giberson, the JPL project manager, is very resourceful. Somehow, within the fixed price of $98 million, a backup MVM spacecraft had been assembled and tested. It now sat just half a mile away from Pad 36B, ready for use if necessary. Furthermore, Gene and NASA's

Bill Cunningham had persuaded the users of the next Atlas / Centaur scheduled for launch from this pad to *lend* us their launch vehicle in the event that *Mariner 10* ended up in the Atlantic Ocean. So we had an economy-class backup for the launch itself.

But *Mariner 10*'s launch was flawless. The dark, humid night was shattered by roaring sound and blinding light as the Atlas / Centaur slowly lifted off the launch pad. These modern cathedrals literally fly toward heaven.

Then smiles, handshakes, and embraces as knotted nerves slowly released, followed by ceremonial champagne, smuggled into the control center after control of the spacecraft shifted to JPL. Then fatigue-drugged sleep in a Cocoa Beach motel.

My fellow TV team members, Ed Danielson and Mert Davies, back at JPL, were not sleeping but working that night, along with others of the MVM team. Ed and Mert monitored the birth and health of our TV cameras and electronics after the spacecraft got established. Functioning beautifully, *Mariner 10* quickly found the Sun by elegant, ballet-like twirls in space, spreading its solar panels to be warmed by life-giving sunlight. Sun-generated electrical power flowed at once into the rechargeable batteries. Time was of the essence. Key heating circuits needed to be energized immediately to protect delicate instruments, especially our compactly folded optics, from the extreme cold of space.

In space there are always design compromises. In this case the critical high-magnification secondary mirror, which had almost defeated JPL's determined efforts of implementation, had to be mounted nearly at the outermost, and potentially coldest, place on the telescope in order to keep the telescope as compact as possible. A robust electrical heating element looped around it there to keep the entire telescope reasonably warm and, in any case, well above minus forty degrees centigrade. If the telescope ever approached temperatures that low, it would shoot out of focus.

Suddenly, Davies and Danielson received alarming news. The critical heater on the TV camera did not come on as commanded. The temperature readings radioed back from the critical location continued to drop—from zero degrees centigrade to minus five. Frantically, JPL controllers radioed alternative command sequences. To no effect. The key electrical circuit powering the heaters had failed completely.

It was 2:00 A.M. in Pasadena when Mert and Ed wearily began improvising new command sequences with the spacecraft engineers to order the camera to take test pictures of star fields. Minus ten degrees centigrade, then minus fifteen. As the TV images were radioed back to JPL, they were computer processed immediately, printed and magnified in search of blur among the tiny star images. Camera A looked all right. Camera B showed a small but real degradation. We were right on the edge of failure, fortunately cushioned by a thoughtful mechanical approach to the camera that partly compensated for unplanned thermal changes.

Finally, the temperature started to stabilize around minus twenty degrees centigrade. So far, so good. But we dared not turn off the vidicon tubes as we had originally planned, because they might be fatally cracked at the extreme temperatures. Moreover, they dissipated some of the heat that was helping stabilize the camera temperatures. Could those delicate vidicons last all the way to Mercury with power on? They had been qualified for weeks of usage, not for months.

Did we really have a useful camera system? Soon there would be a dramatic way to find out. Another of Project Manager Gene Giberson's wizard-like accomplishments was to launch *Mariner 10* on November 3, 1973, despite Bill Cunningham's pressure to launch earlier, before some new trouble cropped up. November 3 was the only day when the spacecraft could fly near the Moon on its way to Venus and Mercury. We on the imaging team had urged this lunar flyby in order to calibrate our complex camera against a well-known target, one that was probably similar photographically to Mercury. We also took pictures looking back at Earth's cloudy disk for comparison with Venus.

The lunar images proved excellent. Our cameras were working fine. In addition to enabling us to make a critical test of optical performance, those lunar images provided valuable new geodetic control points stretching across the north polar regions of the Moon. They filled in a gap left by the equatorially oriented Apollo missions more than a decade before. Our TV crisis was over, and Giberson finally released his hold on the backup launch vehicle.

But it proved to be only crisis number one on a flight plagued with serious and, for the most part, unexplained problems. On November 21, just eighteen days after launch, *Mariner 10*'s delicate electronic brain reset itself. This problem recurred intermittently throughout the mission. That electronic glitch came perilously close every time to causing robotic amnesia, like the dreaded "disk failure" on a personal computer.

On Christmas Day 1973 *Mariner 10*'s radio signal abruptly fell to one-fourth its power. Our long struggle for real-time video pictures seemed doomed. Four days later the signal returned, then faded away again after four hours. Eight days after the New Year's parade in Pasadena, the spacecraft automatically switched to its backup power system. Something was wrong with the primary power system. *Mariner 10* was down to one "lung" of power barely two months into its flight. Then, on January 28,

1974, *Mariner 10* nearly expired as a result of a sudden hemorrhaging of part of its vital attitude-control gas. By the time the picture-taking sequence for Venus began, on February 5, 1974, the movable scan platform carrying our cameras had started to stick.

Economy class was proving very hard on the nerves.

At 10:10 A.M. on February 5, 1974, after much anticipation we saw the clouds of Venus close up for the first time. The pictures, taken in ordinary white light, were intended to catch any shadowing because of relief on the top of the cloud deck. But they proved completely feature-less, even duller than *Mariner 4*'s first view of Mars. This time the flat-ness did not surprise us. Some Venus specialists had predicted that the top of Venus's atmosphere was hazy, without well-defined cloud layers and boundaries.

But Venus in the ultraviolet was another matter. Telescopic photo-graphs of Venus sensitive to the invisible ultraviolet radiation (beyond the visible, blue portion of the spectrum) displayed very faint markings. Some telescope observers even thought they could deduce a four-day rotation period from these markings.

In 1962 radar measurements made at JPL's Goldstone site, and else-where, had determined that the surface of Venus rotates very slowly— once every 243.1 days, and backward. The sun rises in the west on Venus. (Furthermore, 243.1 days is almost, but not exactly, the period for Earth and Venus to pass in their orbits. Venus presents nearly the same portion of its surface toward Earth when they are closest. Implausibly, Venus acts as if it were somehow under Earth's gravity control. Another coincidence? Who knows? Beppe Colombo is gone now, and no one has deduced a new and subtle dynamical connection between Earth and Venus.)

In any case, a four-day rotation of the ultraviolet markings would mean that high up in Venus's atmosphere (where the ultraviolet markings form in some unknown manner) the air is moving eighty times faster than at the surface. How strange!

But true. We chronicled the motions of Venus's ultraviolet markings for a week following the encounter. Delicate wavelike patterns and cumu-lus-shaped clouds all shared in a general four-day motion in the equatorial areas. (See figure 6.1.) At higher latitudes the wind speeds were much greater, mapping out a unique pattern of global circulation. *Mariner 10*'s time-lapse "moving pictures" made Venus's atmosphere real and compre-hensible on a global scale for the first time.

Mercury lay ahead, still hidden in obscurity. *Mariner 10* continued to buck like a poorly broken bronco; there were more control oscillations, and more precious attitude-control gas was lost. But we had one good break. The radio antenna suddenly regained full capability, a technical outcome predicted by some of the JPL specialists. A transmission rate of 117,600 bits per second was back.

As the long-awaited first look at Mercury neared, preparations at JPL for the public display of the results quickened. However, one technical issue remained unresolved. Bill Cunningham from NASA wanted eye-catching color pictures of Mercury.

"You can't take color pictures of a black-and-white planet with a black-and-white camera,"* I kept telling him, with growing impatience. Then I would recite again why the TV team was so sure that Mercury would exhibit a dark gray appearance, just like the Moon. Both atmosphereless bodies suffer unfiltered irradiation by the Sun's damaging ultraviolet rays. Mercury should experience at least a comparable loss of intrinsic mineral coloration, as does the Moon, where the solar flux is ten times more intense.

Yet Bill persisted. Shortly before the Mercury encounter, I found out why. NASA Administrator James Fletcher had been badgering Bill about color pictures since late 1973 when he presided over the Jupiter encounter of *Pioneer 10*. The delicate orange shades in the reconstructed† images from *Pioneer 10* attracted great media interest, Fletcher felt. So why not give the public the same thing from Mercury?

As leader of the seemingly recalcitrant TV team, I was summoned to JPL Director William Pickering's office, where Fletcher waited. Persuading this man should be an easy task, I thought. Fletcher had a Ph.D. in physics from Caltech, was a recognized leader in the aerospace industry, had been president of the University of Utah, and had in 1969 been runner-up to Harold Brown to succeed the retiring Lee DuBridge as Caltech's

Mariner 10's TV camera, like most then used in space, snapped each TV image in white light or through a single broad color filter. The resultant black-and-white photographic print that was later re-created on Earth displayed tonal variations proportional to the variation in the incident intensity only in white light or in the single hue. Such an image contains no color information. If an object, like Jupiter or Mars, is strongly enough colored, the relative intensities across it when imaged in, say, red, green, and blue will show marked differences in tonal patterns. Such images can then be combined numerically in a computer to generate a "color TV" representation of that colored planetary scene. At the end of this process, Mars appeared reddish brown for Viking, and Jupiter appeared orangish brown and white for Voyager. Mercury and the Moon, however, would appear only gray in a properly controlled color TV.

†*Pioneer 10* actually recorded Jupiter images through only two filters. Three colors are needed for color photography. In order to create color photographs for public use, the third color (blue) was initially added "in the darkroom," in sufficient amount that the Pioneer pictures resembled to the eye the colors of Jupiter through Earth-based telescopes.

president. A man so accomplished technically would no doubt quickly grasp why drab, gray Mercury could never look like gaudy Jupiter, regardless of any computer wizardry we might devise.

Fletcher is tall and fair, low-key and genial. He possessed an engaging style, more ecclesiastical than executive, as befits his devout Mormon background. We had a pleasant chat, except for one detail—Fletcher did not seem satisfied with my explanation of why grayish Mercury must always produce grayish pictures. Finally, he asked, "Bruce, can't you just twist the knobs, somehow?"

I was dismayed. Did he really not understand color imaging, despite his Caltech Ph.D.? Or was he urging me deliberately to produce a distorted "color" picture to enhance public interest?

Mariner 10's portraits of Mercury remained black-and-white. And I remained puzzled about what was going on inside the head of that affable, distinguished leader of the American aerospace enterprise. We would interact repeatedly over the next fourteen years, and I would end up as puzzled in 1988 as I was in 1974.

Finally, at a range of two million miles, and 134 days after launch, an image relayed back from our cameras showed details of Mercury's surface never before seen. And more and more details appeared as *Mariner 10* rapidly overtook Mercury in its orbit.

Then came the shock: Mercury *really* looks like the Moon!

Anyone who sees another planet for the first time feels a special excitement and sense of wonder. You savor the alien image, the magic of the moment, before you begin systematic analysis. It's like sipping a great hundred-year-old brandy.

A sip of Mercury turned out to be like a sip of the Moon. Mercury in fact resembled the Moon in astonishing detail. Great circular basins lay amid smooth plains, appearing very much like the lunar maria amid the heavily cratered uplands.

"What will Feynman think about *this?*" I asked myself in the privacy of the imaging-team room at JPL.

In fact, our cameras had revealed a planetary surface that not only mimicked most features found on the Moon but also recorded the same sequence of events (see figure 4.1). The surface of the Moon (and of Mars, as revealed by *Mariner 9*) records an early, cataclysmic bombardment that ended abruptly long ago. Our planetary neighbors have had a quiet life for the most part, but when it was rough for them, it was very rough. *Mariner 10*'s pictures proved that this ancient cataclysm of bombardment was Solar system–wide and very probably contemporaneous. A multitude of comets

SAVORING THE FIRST LOOK AT MERCURY • As a half-illuminated image of Mercury relayed back from *Mariner 10* is displayed on the TV monitor, spirits run high at JPL in March 1974. *From left,* N. W. ("Bill") Cunningham, the NASA program manager for *Mariner 10,* the author (who was the leader of the imaging team), Harris M. ("Bud") Schurmeier, the manager of the Voyager project to study Jupiter and Saturn, and Walker E. ("Gene") Giberson, the *Mariner 10* project manager. *NASA / JPL photo*

or asteroids hurtled through the Solar System some four billion years ago, wreaking indescribable havoc. *Mariner 10* revealed that the "End of Heavy Bombardment," the cessation of the impacting period, is the primary geological marker horizon of our Solar System.

The similarities of impact cratering on the Moon, Mars, and Mercury extend beyond the presence of giant basins. All three exhibit smooth volcanic plains that fill up these ancient basins. The plains are in each case peppered with uneroded impact craters, some as large as twelve miles (twenty kilometers) in diameter. On each planet the smaller the size of the craters, the larger their number. There are about four times as many six-mile (ten-kilometer) craters as twelve-mile (twenty-kilometer) ones, about four times more three-mile (five-kilometer) ones, and so on down to the point where craters overlap, obliterating one another.

Two years earlier, in 1972, *Mariner 9*'s pictures had shown that the sizes and distribution of these smaller craters on the oldest Martian plains were very similar to those on the oldest lunar maria, determined to be nearly four billion years old by means of rocks returned with the Apollo astronauts. *If* the rate of impact on Mars and the Moon had been the same, both surfaces would have had to be the same age, too. Thus, the Martian basalts would also be three to four billion years old. The chanels and other unusual aqueous features would be even older, because those Mars basalts generally fill in and overlap the earlier aqueous features. Moreover, the water features on Mars must have formed there practically at the birth of the planet. Mars could be inferred to have been completely dry on a global scale for billions of years.

But this was before *Mariner 10*'s discovery that Mercury also showed this same surface history. So life-on-Mars enthusiasts argued instead for a younger aqueous era on Mars. A prevalent idea concerning the impacting of ancient objects on the terrestrial planets had been that those impactors originated in the asteroid belt between Mars and Jupiter. (Some meteorites now striking Earth can be tracked backward to such origins.) If this old idea were correct, Mars would have been bombarded perhaps ten times more frequently than the Moon (because Mars orbits very close to the asteroid belt) and maybe a hundred times more frequently than distant Mercury. By that line of reasoning, even though the accumulated crater abundance that *Mariner 9* found on Mars was similar to the Moon's, it was judged a coincidence. Those who believed in life on Mars reasoned that the topographic features on Mars could be ten times younger, perhaps only 300–400 million years old instead of 3–4 billion. In that case, the aqueous period on Mars might have lasted billions of years and ended only yesterday, geologically speaking. Plenty of time would have been available for life to form and to adapt to harsher conditions.

This theory of a "young" Martian surface fell to pieces with *Mariner 10*. Our pictures showed that Mercury exhibits about the same abundance and sequence of postmare craters as do both Mars and the Moon. Thus, the rate of impact and the sequence of events did not depend on proximity to the asteroid belt. Instead, the rain of smaller impacting objects has been similar throughout the Solar System, just as with the cataclysmic bombardment that formed the huge basins in the first place. Perhaps comets originating at great distances from the terrestrial planets played a larger role than asteroids. In any case, the smooth plains of both Mercury and Mars must be about as old as the lunar ones—three to four billion years. The global aqueous phase on Mars must have occurred in a geological flash, followed by billions of years of hostile conditions for life.

So there, Dr. Feynman! Comparative planetology makes sense. We *do* go to Mercury to learn about the Moon—and unexpectedly learn about Mars as well.

The life of a planetary explorer is spent reacting to the unexpected. On Sunday, April 1, 1974, Ed Danielson rushed into our TV analysis room at JPL with news of an apparent Mercury satellite, reported by the ultraviolet team. Their report was pounced on by the press corps at JPL's Von Karman auditorium, a press corps struggling to compete with the unfolding news of the Watergate affair, which was pushing Nixon steadily toward resignation from the presidency. But Danielson and I were puzzled by the satellite report. Large moons of Mercury had been ruled out long before by surveys with Earth-based telescopes. And we had seen nothing so far with the *Mariner 10* television, although a major TV search for such objects was still programmed to take place. How could an ultraviolet *spectrometer*—not even an imaging instrument—chance upon the faint signal from a small Mercurian satellite?

A few hours later the *Mariner 10* navigation team figured out that by chance a relatively bright star in the cosmic background was visible from the spacecraft in a direction near Mercury, close to where the putative satellite had been reported. After a little more checking, the "satellite" report faded away. Mercury has no satellites. The press called it Mariner's April Fool's joke.

Instead, the real scientific news was that Mercury has a diminutive, Earth-like magnetic field, complete with a small Van Allen radiation belt. *Mariner 10* revealed that, notwithstanding its small size and fifty-nine-day rotation, Mercury ranks next to Earth and distant Jupiter and Uranus in terms of intrinsic magnetic field. Planetary exploration had finally discovered another still active Earth-like planet on which to test theories of Earth's magnetism. This discovery of a Mercurian magnetic field required us to alter our analogies even further. Mercury is like the Moon on the outside and like Earth on the inside. But an Earth-like convecting interior has not crumpled the outer skin of Mercury into mountains, as it has on Earth. This was extraordinary news indeed. ("Hot rock" it may be, Dr. Fowler, but one filled with mystery, wonder, and significance for us Earthlings.)

Mariner 10 had been a great explorer, but it still was not tamed. Right after the Mercury encounter, a major electrical short developed. Twenty percent of the spacecraft's power suddenly sparked into an electrical malfunction. The spacecraft began to heat up from the power dissipation. Deputy Project Manager John Casani's dark-ringed eyes looked like those of a punch-drunk prizefighter.

Gradually, John regained control, but *Mariner 10* was a very sick spacecraft. Next, the tape recorder failed in August 1974, just as Nixon left the White House forever. Now real-time TV was not simply an option;

it was our only way to get pictures from the second encounter, thereby bypassing the failed tape recorder. Somehow, the JPL engineers managed to nurse *Mariner 10* along the course around the Sun once more. On September 21, 1974, it passed over the sunlit southern polar regions on a perfect imaging trajectory. Two thousand more images of the planet flowed back, bridging across the south polar regions the preencounter and postencounter views from our first visit to Mercury.

As the spacecraft finished its second encounter, the crucial nitrogen gas that was fired in tiny bursts to stabilize the spacecraft was practically used up. More ingenuity. The large high-gain antenna was commanded to move in a delicately coordinated dance with the tilting solar panels. As a consequence, the solar pressure on the spacecraft was balanced so evenly that hardly any nitrogen control gas was used up to stabilize *Mariner 10* during the third revolution about the Sun. On March 16, 1975, the ailing *Mariner 10* flew by Mercury for a third time, passing a mere 125 miles (200 kilometers) above its nighttime hemisphere and confirming even more strongly the earlier indications of a magnetic field.

This pass was meant to be an important one for us, to yield our highest-resolution images. But more bad news hit just before encounter, this time on the ground. An experimental component in JPL's 210-foot antenna in Australia, needed to provide the critical margin of signal strength, began to fail. The noise in the faint signal received from *Mariner 10* immediately increased. The real-time TV images became garbled. The tape recorder, of course, was gone. Otherwise, we could have used it to buffer the rate down for slower transmission.

One trick still remained, though. We had built in an electronic means to send only the center quarter of a strip of each frame in real time, at one-fifth the nominal rate of real-time data (22,050 bits per second, instead of 117,600). Our last ace in the hole worked. We were able to acquire important strips of the highest-resolution images, in spite of the failure of both tape recorder and ground-tracking equipment.

Finally, like a robotic John Henry, our heroic spacecraft expired. On March 24, 1975, just 506 days after birth on that heavy Florida night, the last of *Mariner 10*'s attitude-control gas seeped away. The spacecraft started slowly tumbling out of control and was turned off.

Our risk taking in optics and communications had paid off handsomely. *Mariner 10* returned 12,000 pictures to Earth. This included more than 2,400 useful and nonredundant pictures of Mercury (thirteen times the combined production of *Mariners 6* and *7* at Mars in 1969) and 3,400 useful pictures of Venus (eighteen times the combined production of *Mariners 6* and *7*). The economy-class *Mariner 10* flyby of Venus and Mercury was one of the most productive space-science experiments ever carried out.

Ed Danielson and Mert Davies joined the Voyager imaging team. With their leadership Voyager ultimately adopted the *Mariner 10* optical design and real-time communications and used them to view Jupiter, Saturn, and Uranus. Indeed, Voyager transmitted pictures back from Jupiter initially at almost our same, magic rate of 117,600 bits per second (actually, 115,000 because of some slight redesign of the camera electronics). Thus, the MVM camera design that was to be *"Mariners 6 and 7 with no changes"* continues to discover new worlds even now through Voyager.

Then came the systematic scientific follow-up. *Mariner 10*'s mapping of the surface of Mercury gave scientists an unprecedented opportunity to assign logical and tasteful names to all of a planet's geographical and topographical features at once. Earlier spacecraft missions to the Moon and Mars had to cope with a fanciful hodgepodge of names applied over a century of observing with telescopes. But Mercury is just a tiny blur through a telescope, so no official naming had occurred.

However, I was dismayed at Mert Davies's status report on the naming process. Mert represented the *Mariner 10* TV team on the "IAU Task Group on Nomenclature for Planet Mercury of the Working Group on Planetary System Nomenclature."

"The nomenclature committee can't be serious about naming craters on Mercury, an airless planet, after *birds!*" I exploded.

"The other major candidate is cities," replied Mert evenly. "Authors, my favorite, is a distant third."

Mert Davies is an unusually even-tempered individual, but this birds business was too much even for him. The international scientists involved were caught up in a maze of illogical precedents and petty national concerns. Fortunately, the chairman of this task group had postponed a final decision favoring either birds or cities, but time was running out. Never ask what's in a name when NASA is getting ready to print the first maps. The answer is—a lot. Names of some kind had to be chosen.

Help arrived from an unexpected quarter. Carl Sagan was not involved in any way with *Mariner 10*. But Carl has an unusually refined sense of the global and of the future. He is also a constant adversary of provincialism and chauvinism, be it based on race, gender, national origin, or occupation. So he wrote as follows to the members of the committee:

In recent discussions I have discovered the drift of the last meetings of the IAU Mercury nomenclature group. I hope I may be excused if I make a few remarks on the question.

As I understand it, the present debate is whether to name major features on Mercury after cities or after birds. My starting point is to imagine the solar system after it has been entirely named and see the impression which various nomenclature committees will have left for the following millennia. If the present trend continues, we will find major features on the moon, Mars (and to a small extent, Mercury and elsewhere) named after scientists—largely European and American astronomers. If the present inclination of the nonmenclature committee is followed, I suppose we will find other solar system objects sporting the names of fish, minerals, butterflies, spiders and salamanders. I have nothing whatever against salamanders; but from the perspective of a millennium hence, it will be thought interesting that no features larger than a kilometer or so across have been named after Shakespeare, Dostoyevsky, Mozart, Dante, Bach, Vermeer, Homer, Hiroshige, Cervantes, Locke, Bacon, Hume, Wittgenstein, Aquinas, Tacitus, Michelangelo, Beethoven, Keats, Shelley, Lorca, Leeuwenhoek, Claude Bernard, Mark Twain, Thucydides, Bertrand Russell, John Donne, Dvorak, William Blake, Tolstoy, Stravinsky, van Gogh, Pavlov, Freud, Kafka, Franz Boas, Phidias, Ruth Benedict, Schliemann, Gregor Mendel, Thomas Hunt Morgan, Herodotus, Rabindranath Tagore, and so on. Such a list can obviously be extended significantly. Each person will have others he would consider essential to such a list...

The net result will be the impression that we esteem birds and salamanders very highly, that we cherish the memory of those scientists close to our own discipline, but that we care little or nothing for the greatest poets, musicians, historians, anthropologists, sculptors, archaeologists, philosophers, and painters in human history. I cannot believe that this is the intention of anyone who views the situation from a long perspective. ... It is important that we make sure that the end result will be an unprovincial distribution of nationalities, epochs and occupations—a distribution that our great grandchildren can be proud of. How about poets and prose writers on Mercury?

After a year more of bitter wrangling over possible compromises (including the possibility of demon names), the IAU decreed that craters on Mercury would be named after great human contributors to the humanities and arts—drama, prose, poetry, painting, sculpture, architecture, and music.

Names of prominent craters on Mercury now include Abu-Nuwas, Bach, Beethoven, Cervantes, Chopin, Dostoyevsky, Goya, Hiroshige, Homer, Li Po, Melville, Michelangelo, Proust, Renoir, Rodin, Shakespeare, Sophocles, Mark Twain, Tolstoy, van Gogh, Wagner, Wren, Yeats, and Zola.

My final encounter with *Mariner 10* was in August 1980. "Boring, boring," was the clear message on my fourteen-year-old daughter's face as we wandered through the National Air and Space Museum. Exposing her to Washington, D.C., had been fun, but I shouldn't overdo it. "Okay, we'll leave in just a minute,'' I temporized. Suddenly, I stopped and stared. There were those twin lenses. That backup version of the *Mariner 10* spacecraft and our cameras were hanging right there, in the center of the Air and Space Museum. I drank deeply of the memories. Could a mere five years have passed since the last Mercury encounter? It seemed decades.

The next time I visited the Smithsonian, *Mariner 10* was gone, replaced by a big Space Shuttle display.

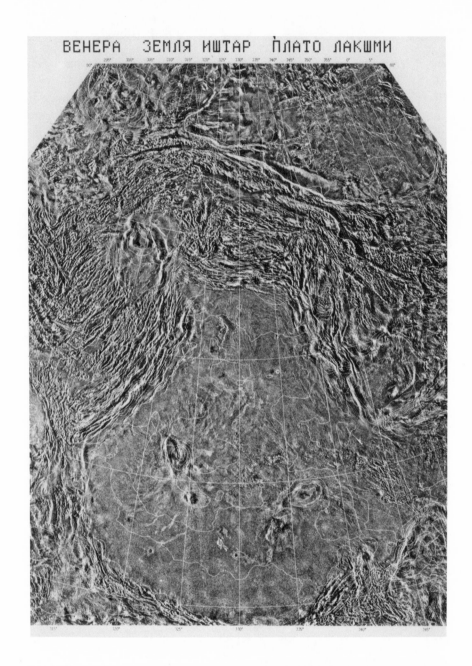

ВЕНЕРА ЗЕМЛЯ ИШТАР ПЛАТО ЛАКШМИ

7.1 A SOVIET RADAR MOSAIC OF VENUS'S SURFACE • In 1983 the Soviet radar-mapping spacecraft *Veneras 15* and *16* radioed to Earth the high-resolution radar data on which a stunning series of regional "photomaps" is based. The example shown here in much reduced form covers the Ishtar region, ranging from about 60° to 80° north latitude and encompassing over sixty degrees of longitude, from about 300° west to the zero meridian. Altitude contours in one-kilometer intervals are also displayed. The USSR has released the mosaics to the U.S. scientific community as part of U.S.-USSR exchanges of scientific data. *NASA/JPL photo*

7. Spell Venus
V-e-n-e-r-a

"I DON'T SEE HOW the Russians could pull off a competitive radar mission at Venus before us, do you, Bob?" It was late 1979, and Bob Parks, JPL's head of flight projects, agreed with me in the privacy of my office as we pondered Soviet intentions and capabilities. There had been rumors about a Soviet radar mission to Venus as early as 1983. Still, we did not think that the Soviet Union could map Venus's surface terrain in the early 1980s with radar from an orbiting spacecraft. We at JPL certainly intended to do it.

In 1978 JPL had pioneered the use of a complex, high-resolution radar to observe Earth from space. That year's mission of the SEASAT satellite was a low-cost demonstration of novel radar and radio sensing instruments to monitor Earth's oceans from orbit, day or night, cloudy or clear. It brought much practical information about the oceans, including data on sea state, surface winds, and ice conditions. It also radioed back stunning radar "pictures" of land areas. Exciting geological applications of imaging radar on spacecraft jumped out at us. As a result, new efforts took shape to map Earth in this new way with the Space Shuttle during the 1980s. The key to SEASAT's radar success had been adapting for civilian space use the sophisticated military radar intelligence technology long used on airplanes and termed side-looking radar and, more precisely, synthetic aperture radar (SAR).

Of most immediate interest to us was the application of SAR radar to Venus. Here was a planetary surface almost as large as the area of Earth's land and water combined, virtually unexplored. Venus's dense cloud layer hid that surface from orbital cameras like those used at Mars and Mercury. In 1975, however, Soviet scientists had managed to peek under the clouds. Facsimile images of Venus's surface immediately surrounding the landing sites of *Veneras 9* and *10* were radioed back to Earth, despite intense heat and high pressure.

These Russian closeups of Venus were surprising. I had presumed that its surface was buried under a uniform blanket of soil and dust. Chemical weathering should be intense in such a hot and acid environment. And there could scarcely be scouring winds at the bottom of an ocean of carbon dioxide. My expectations notwithstanding, the *Venera* pictures showed bare rock surfaces, sometimes broken by cracks and sharp edges. To a geologist those pictures meant that Venus's surface is, like Earth's, active geologically and full of surprises. Unknown processes of topographic renewal evidently manage to outstrip degradation and burial.

Earlier hints of unusual global-scale topographic features had emerged through radar imaging of Venus from Earth. Following up on the first radar distance measurements, JPL scientists in the early 1960s, using the powerful Deep Space Net transmitters and antennae, named areas of conspicuous radar reflection simply α, β, and γ, not knowing what they were sensing. More powerful investigations, from the giant dish at Arecibo, Puerto Rico, later added the "Maxwell" mountains and intriguing outlines of new kinds of features. But these Earth-based observations were limited to the area facing Earth and to a surface resolution that did not reveal diagnostic geological features.

Then, in 1978, the large-scale, global topography of Venus fell into sharper focus, thanks to a radar height-measuring instrument carried aboard an American spacecraft orbiting the planet. That orbiter and a companion package of exquisitely instrumented atmospheric probes constituted the Pioneer / Venus (PV) mission. PV was developed by the Hughes Aircraft Company for the Ames Research Center of NASA.

Launched on two Atlas / Centaurs, PV was the last American planetary mission before the Shuttle era—or since. The features revealed by the PV radar altimeter showed that Venus apparently has a tectonically active crust, like Earth and Mars. But Venus's overall geographical patterns were unfamiliar. For example, the characteristic topographic features signifying boundaries between continents and oceans, and other plate-tectonic features on Earth, seem to be absent on Venus. Instead, strange irregular forms cluster in highland and lowland groups. The PV "images" contained intriguing, large circular features as well. Giant scars left from ancient impacts? Or enormous volcanic calderas like those on Mars? Many questions could not be answered until high-resolution radar imaging from an orbiting spacecraft became available.

This is what JPL had in mind with the Venus Orbiting Imaging Radar (VOIR) concept, an extension of the SEASAT technology. Initially, VOIR was targeted for the 1983 Venus opportunity. NASA's obsession with the Shuttle and with low-Earth orbit, however, soon demoted VOIR to the 1984 opportunity. (Venus opportunities occur every nineteen months.) That was why, in late 1979, Bob Parks and I were in my office reviewing all we knew about Soviet space capabilities. It had been great for science and for America that JPL's *Mariners* first mapped the surfaces of Mars and Mercury. Now JPL and America could play a similarly important role for Venus's surface and vastly extend PV's first global glimpse. But this would happen only if we capitalized promptly on SEASAT, before the Russians developed a similar capability.

The SEASAT experience had taught us that imaging with radar from orbit is much more difficult than snapping pictures in ordinary, visible light with an electronic camera. The *Mariner 10* camera lenses were only about ten inches (twenty-five centimeters) in diameter. SEASAT's extendable radar antenna was thirty feet (ten meters) long. Furthermore, the transmission, reception, and decoding of SEASAT's radar signal was far more complex than the simple reading out of electrical signals from the earlier Mariner video cameras had been. The advantage—and the challenge—of imaging with radar lies in measuring the phase of the radio waves as well as their intensity; it is somewhat like making holographic pictures with a laser. In addition, Venus radar imaging necessitates a much higher rate of data transmission to Earth than did *Mariner 10*'s imaging of Venus's clouds. We felt that the technological alternative, the computer processing of the signal automatically on board the spacecraft, was too complicated for use at Venus in the early 1980s.

A particularly difficult problem with space-borne radar, compared with a passive camera like that on *Mariner 10,* is the enormous electrical power a transmitter requires to bounce detectable radar reflections off the planet's surface. SEASAT carried unique and immense solar arrays that had to be continuously pointed at the Sun in order to provide that extra electrical power for its radar. A new and (as we discovered too late) untested rotating electrical connection was required on SEASAT where the large electrical current from the movable solar panels passed into the spacecraft bus. That connection on SEASAT short-circuited after three months in orbit and brought the mission to a catastrophic end. Fortunately, the SEASAT data harvest was already so rich that a new field, satellite oceanography, was born nonetheless.

Thus, the provision of electrical power for a radar mapper at Venus would be another big challenge. In fact, the problem of power for Earth-orbiting radar satellites had bedeviled the Russians as well. For many years the USSR had routinely used a low-resolution system of radar ocean mapping. Its purpose was to find and target U.S. aircraft carriers at sea and to pass that information to the guidance systems of air-to-sea missiles carried aboard lurking Soviet bombers. RORSATs, as these satellites are called in the West, used a uranium-filled nuclear reactor to provide the large amounts of electrical power for their radar. Solar panels were not adequate. At the end of each mission, the RORSAT's nuclear reactor was typically rocketed up to a safe, long-lived orbit. But in January 1978 the kick stage failed on *Kosmos* 954. Fragments of the uranium reactor came down in northern Canada, resulting in a great international outcry.

If it took all of that to complete a low-resolution radar mission about Earth, how could the Russians carry out a SEASAT-type mission around Venus ahead of JPL—the only institution in the world in 1979 with SEASAT orbital experience?

In 1980 VOIR suffered further slippage in NASA's planning. It would have to wait until 1986, the next later time when Earth and Venus approached closely. Was there a chance we might end up playing second fiddle to the Russians, after all? Then, in 1981, Reagan's black-hatted budget boss, David Stockman, forced the mission to further scale down its objectives and pace. The launch date was deferred yet another nineteen months, to 1988. VOIR was going backward. Its planned launch date had been postponed nearly five years in only three years' time. The scaled-down mission was renamed Venus Radar Mapper (VRM), then was re-christened "Magellan" to identify it with that great explorer.

But Magellan, the man, is remembered because he was first in his achievement, the leader of the first ship to circumnavigate the globe. Magellan, the spacecraft, lost that distinction in October 1983, when two Soviet spacecraft, *Veneras 15* and *16,* went into orbit about Venus. Bob Parks and I, and probably the CIA, had completely underestimated the growing radar capability and aggressiveness of the Soviets. With SAR radars *Veneras 15* and *16* efficiently mapped the north polar regions of Venus. (Mapping the whole planet was evidently far too large a task.) Clear radar picture mosaics revealed heretofore unknown surface features—giant parallel ridges and steep slopes of a size not found elsewhere in the Solar System. (See figure 7.1.)

After the *Challenger* crash Magellan's outlook did not improve. It was deferred again, to the 1990 opportunity,* which would be six and a half

---

*Magellan is scheduled to be launched in April 1989, the first planetary mission to be launched from the Shuttle, and the first launched by the United States since 1978. However, there will be a further delay. Magellan will be placed on a needlessly long oribital path to "kill" an extra six months before approaching the 1990 Venus opportunity. Normally, a Venus probe would be launched only four months before the intended arrival, that is, in

years after the radar discoveries of *Veneras 15* and *16* and thirteen years after SEASAT had first pioneered the key technology. Magellan is still an important scientific prospect. It should provide higher-resolution radar pictures than those of *Veneras 15* and *16*. And it will yield much greater geographical coverage, nearly the entire planet's surface.

Meanwhile, in the dispassionate mirror of space, *Veneras 15* and *16* dramatized the growing Soviet exploratory capabilities, boldly and purposefully deployed. Magellan, by contrast, has come to symbolize a powerful nation adrift, rudderless in space, its former technological leadership in space slipping away.

---

fall 1989. However, that is when another long-delayed planetary mission is scheduled. The Shuttle is so marginal as a planetary launcher that NASA opted to schedule Magellan for the earlier launch to avoid two planetary launches within a few months of each other. Previous U.S. planetary launches on expendable vehicles were routinely planned to occur within a three-week launch window. It is current Soviet practice to launch two planetary probes within one week.

PART THREE

Voyager and the Grandest Tour Ever

8.1 THE VOYAGER SPACECRAFT • The structure of the identical spacecraft *Voyagers 1* and *2* is dominated by (1) the 11.5-foot (3.7-meter) high-gain antenna, (2) the boom and scan platform carrying the television camera and other instruments (*bottom*), (3) the long, thin, slanting extendable boom that separates the magnetic sensors (off this picture) from the main body of the spacecraft, where spurious magnetic fields are generated, and (4) the plutonium batteries, one of which juts directly up from the spacecraft and which provide a steady supply of electrical power for space travel beyond the useful range for solar-generated electrical power. *NASA/JPL photo*

8.2

Exploring the Outer Solar System

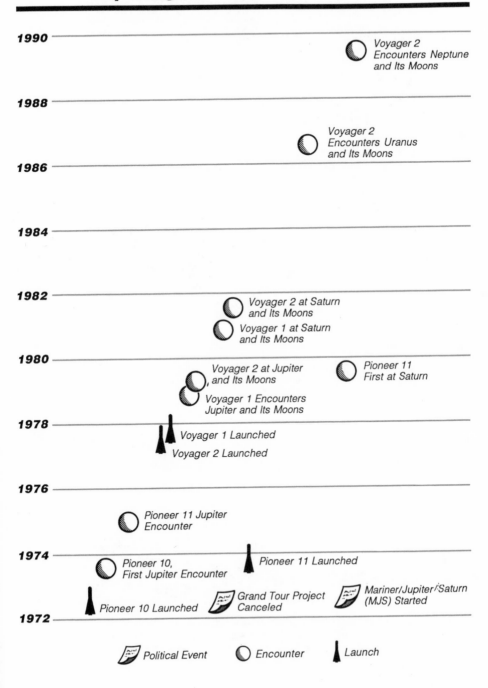

1990
Voyager 2
Encounters Neptune
and Its Moons

1988

Voyager 2
Encounters Uranus
and Its Moons

1986

1984

1982
Voyager 2 at Saturn
and Its Moons
Voyager 1 at Saturn
and Its Moons

1980
Voyager 2 at Jupiter
and Its Moons
Pioneer 11
First at Saturn
Voyager 1 Encounters
Jupiter and Its Moons

1978
Voyager 1 Launched
Voyager 2 Launched

1976

Pioneer 11 Jupiter
Encounter

1974
Pioneer 10,
First Jupiter Encounter
Pioneer 11 Launched
Pioneer 10 Launched
Grand Tour Project
Canceled
Mariner/Jupiter/Saturn
(MJS) Started

1972

Political Event   Encounter   Launch

8.3 TITAN/CENTAUR LAUNCHES *Voyager 2* • The two solid-fuel rocket boosters and the main (liquid-fuel) engine all fire to lift off the Centaur stage and the encapsulated *Voyager 2* spacecraft from the Kennedy Space Center on August 20, 1977. A diagram of the *Titan/Centaur* is shown in figure 2.2, on p. 52. *NASA/JPL photo*

8. Catching the Wave of the Century

IMAGINE "STAR TREK" IN REVERSE. An alien starship exploring our portion of the Milky Way trains high-powered sensors on a still distant star of no apparent distinction. As the starship moves nearer, four cold, uninhabited planets come into view, orbiting majestically about the ball of burning hydrogen we call Sun. A few hours later (the starship is moving at nearly the speed of light), four rocky spheres are sighted orbiting much nearer the star than the four big planets. One of those is our home, the water-covered ball where we live, dream, and die. The starship aliens (or, more likely, their robotic surrogates) might soon streak on in search of a more interesting stellar environment, rather than expend fuel and time in a close-up inspection of insignificant rocky debris, orbiting a rather average star.

Those four conspicuous planets—Jupiter, Saturn, Uranus, and Neptune—are very large indeed. They constitute 99.8 percent of the aggregate planetary mass orbiting the Sun. Viewed from interstellar space, these gas giants *are* the planetary system. Earth, however, is luckier. Tucked into a narrow zone of optimal solar warming, it has a moist surface, and because of that moisture life has evolved on it.

In ancient and medieval times it was fashionable to speak of the harmony of the spheres. The users of that phrase could not know how true that was. Jupiter, Saturn, Uranus, and Neptune *are* harmonious. Every century and three-quarters a magical alignment occurs among them. During this short-lived alignment a properly designed spacecraft launched to Jupiter from Earth can, with the right choice of close trajectory at Jupiter, break the Sun's gravitational pull and fly on to Saturn, Uranus, and Neptune—the Grand Tour (see figure 8.4)—and then, past Neptune, disappear forever into interstellar space. When that once-in-this-century opportunity occurred, in 1977, the United States was ready to ride that cosmic "wave" and make stunning discoveries about our giant planetary cousins.

James Van Allen, of the University of Iowa, was one of the world's first space scientists. In 1958 he discovered the invisible, globe-girdling belts of lethal charged particles that bear his name. In the spring of 1970 he sat at the head of our conference table, chairing a scientific group brought together by NASA to explore the scientific opportunities offered by the Grand Tour trajectory discovered in the early 1960s. Of the fifteen specialists meeting that day at JPL, I was the lone geologist, and the only member of the team especially interested in rocky worlds. My role rapidly became clear—to evangelize not just for good imaging but also for serious and close observation of the many satellites, the moons of these gas giants.

From Earth the moons of Jupiter and Saturn *look* insignificant, being no more than smudges on the photographic plates. Jupiter and Saturn dominate the telescopic views, physically and intellectually, presenting enormous challenges to the practitioners of physics. Jupiter's magnetic field challenges even the Sun's. Intense radiation trapped within huge belts by that field generates the Solar System's strongest radio signals. Jupiter's Great Red Spot, a giant vortex visible even from Earth, has persisted mysteriously over three centuries. Larger than Earth, that oval eye rules a colorful kaleidoscope of atmospheric swirls and bands, begging for theo-

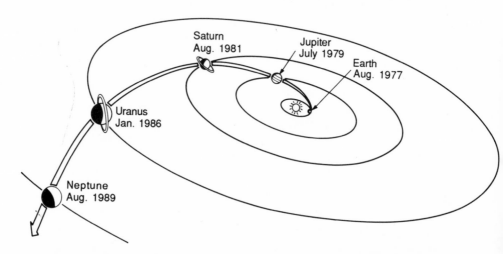

8.4 THE GRAND TOUR • The alignments and motions of Earth and the outer planets very rarely coincide such that a spacecraft can be launched from Earth to Jupiter, be perturbed by Jupiter's gravity to move on to Saturn, and so forth all the way to Neptune. This trajectory, called the Grand Tour, is the path that *Voyager 2* has followed on its voyage of discovery. The last time the planets were aligned similarly, Thomas Jefferson was President of the United States.

retical interpretation. To be able to fathom the rings of Saturn would alone have justified the voyage. We were only a little wiser about those rings than Galileo was when he first described them.

In 1970, as we met at JPL, the first spaceships intended to scout Jupiter were already under construction only forty miles away at TRW (the large aerospace firm Thompson Ramo Wooldrige). Called *Pioneers 10* and *11,* they were launched in 1972 and 1973, respectively, to sample Jupiter's environment—its electrical currents, magnetism, and energetic particles (see table 8.1). Completely controlled by ground commands, the Pioneers spun passively in space like giant tops. These two spinning spacecraft were sturdily built to scout the gateway to the outer planets. They were mechanically and electrically simple and therefore relatively inexpensive. However, as every amateur photographer knows, imaging (and other remote sensing) is best carried out from a rock-steady and carefully pointed platform. *Pioneers 10* and *11* did pack a small light-scanning sensor, from which a few images of Jupiter and Saturn were built up line by line by the spacecraft's spinning motion during the flyby. But no useful photography of Jupiter's or Saturn's moons was possible with such rudimentary equipment.

Pioneer 10 blazed a historic trail through the unknown hazards of the asteroid belt and extended humankind's reach to Jupiter on December 3, 1973, encountering far more damaging radiation than had been expected. *Pioneer 11* reached Jupiter a year later, then blazed the trail past Saturn.

TABLE **8.1**

U.S. MISSIONS TO THE OUTER PLANETS

Launch	Spacecraft name	Spacecraft weight (kg)	Mission objective	Results
		1972		
Mar. 2	*Pioneer 10*	259	Jupiter flyby	Encountered Jupiter, Dec. 1973
		1973		
Apr. 5	*Pioneer 11*	259	Jupiter flyby	Encountered Jupiter, Dec. 3, 1974; Saturn, Sept. 1979
		1977		
Sept. 8	*Voyager 1*	800	Jupiter/Saturn flyby	Encountered Jupiter, Mar. 8, 1979; Saturn, Nov. 12, 1980
Aug. 20	*Voyager 2*	800	Jupiter/Saturn flyby (on Grand Tour trajectory)	Encountered Jupiter, July 9, 1979; Saturn, Aug. 25, 1981; Uranus, Jan. 26, 1986; Neptune, Aug. 25, 1989

In September 1979 it certified the safety of the narrow zone adjacent to Saturn's rings that leads on to Uranus and Neptune (just as Mariner / Venus / Mercury had found the tiny target adjacent to Venus that led on to Mercury). The success of *Pioneer 11* silenced fears that unseen ring particles lurking there would pepper the unwary robotic explorer like a natural antisatellite weapon.

In 1970 those accomplishments of *Pioneers 10* and *11* were still years away. Intense planning was taking place for the Grand Tour, for visiting all four giant planets in one flight. The Pioneer flights were overtures to the big show. For that next, giant step, new plutonium cells would be used to power the most intelligent robot ever conceived. A breakthrough in artificial intelligence (as it is now called) was essential for sophisticated exploration light hours from Earth. The state-of-the-art brain used on the Grand Tour would have to be as robust as the control mechanism of a nuclear weapon in order to survive Jupiter's murderous radiation. That was the biggest problem, the hidden surcharge for the free ride to Saturn and on to Uranus and to Neptune.

As the months of 1970 passed, meetings proliferated. By October the small, dreary JPL conference room, crammed with engineering diagrams and plans, had become all too familiar. I was running one of Van Allen's subcommittees—the Data Handling Subcommittee for the Grand Tour Mission Definition Phase. Jupiter is half a billion miles away. Radioing back thousands of pictures and other data to Earth would challenge every link in the communications chain—the spacecraft, the Deep Space net, and the ground computers. But it could be done. The Grand Tour was no economy-class mission like the flights to Venus and Mercury. Expensive new technology was to be its hallmark. I experienced déjà vu, recalling nearly a decade of previous meetings for Mars and Mercury missions at which we had labored over the same technical issues.

Suddenly, a revelation hit me. Three successive missions to Mars, plus the upcoming flyby of Venus and Mercury, were enough for me. Other people could convert the promise of the Grand Tour into reality. After MVM, I would not return again to the "trenches" but would move on to something new.

Meanwhile, the Grand Tour seemed to have a sound basis for governmental approval, for it offered the following:

• Great scientific promise for countless physicists, astronomers, meteorologists, and geologists (although official scientific support still reflected a concern that the Grand Tour might suck up a disproportionate share of space resources)

- The cutting edge of space technology to challenge JPL once again
- The prospect of historic American achievement, clearly beyond Soviet capability, certain to elicit worldwide acclaim
- A billion-dollar, high-profile program for NASA, well suited to pick up the budget slack after Viking

NASA promoted the Grand Tour concept strongly. In April 1971 it charged nearly a hundred diverse scientists to put scientific flesh on the concept Van Allen's group had laid out.

Over the next seven months two optimal trajectories were identified. The first involved a blast-off in August 1977 toward Jupiter, where the spacecraft was to catch the free trip to Saturn, Uranus, and Neptune. A few weeks later, the second spacecraft would be launched, also to Jupiter and Saturn. But the Grand Tour gateway to Uranus and Neptune would already have closed. Tiny Pluto would therefore become Number 2's final destination. Fifty-four separate scientific objectives for the Grand Tour were spelled out, but only eight even mentioned the satellites, and only two did so uniquely. For most of the scientists involved, those dozens of little worlds orbiting the gas giants remained insignificant smudges on photographic plates. Geographical discovery—the progenitor of so much natural science on Earth—was not fashionable with the laboratory physicists, chemists, and biologists who dominated American (and Soviet) science.

But Bud Schurmeier, JPL's new project manager, understood the importance of photographic exploration. Schurmeier had put JPL and America into the lunar race in 1964 when his Ranger lunar-impact probes radioed back the first close-up pictures of the Moon. This was a crucial preparatory step for Apollo. In 1969 he led the work on *Mariners 6* and *7* that achieved a hundredfold gain over tiny *Mariner 4* in the return of pictures from Mars. With Schurmeier, JPL was clearly giving the Grand Tour its best effort. The exploration of even those unknown moons would now receive a sympathetic hearing.

Events moved rapidly in late 1971 and early 1972.

- On August 3, 1971, Dave Scott, the test pilot Chuck Yeager's protégé, and his *Apollo 15* crew took viewers all over the world on a roving television ride across the rocky and forbidding lunar landscape.
- On November 12, 1971, *Mariner 9* (launched in March of that year) eased into Mars orbit to wait out the dust storm before astounding us with images of mountains, canyons, and valleys larger than any on earth.
- On January 5, 1972, President Nixon announced the beginning of a new era in space with the development of the Space Shuttle.

• On January 11, 1972, NASA notified JPL that the Grand Tour project had been killed for budgetary reasons.

The Apollo momentum that had powered a generation of virtuoso American planetary first looks was fast dissipating. The Grand Tour simply reached the starting gate too late for a new planetary endeavor of such cost.

JPL responded to this situation by formulating a cheaper, less ambitious alternative. Within days Bud Schurmeier and his engineering team transfused a decade of hard-won experience in space exploration into an attractive new engineering concept for a rich comparative examination of Jupiter and Saturn and of their moons. It was modestly termed Mariner / Jupiter / Saturn. MJS matched the most promising scientific objectives at Jupiter and Saturn with capabilities already largely proven by previous Mariners, by *Pioneers 10* and *11,* and by JPL's large new Viking orbiter development then under way.

A greatly enhanced electronic brain, however, remained to be developed because MJS needed an exceptionally good memory and great intelligence. But Schurmeier and his project were more than five years away from launch, almost twice as long a lead time as earlier Mariner and Pioneer developments had enjoyed.

An immediate challenge loomed. MJS would have to be sold to the Nixon administration practically overnight. Just how fast, I learned in February 1972, in a telephone call from the Old Executive Office Building, adjacent to the White House.

"There are questions around here about Mariner / Jupiter / Saturn, Bruce. Any comments?" was the cryptic opening from an old friend, Russell Drew.

He was alerting me from within Nixon's secretive staff deliberations that MJS needed more justification in Washington. Drew, a naval aviator, with a Ph.D. in electrical engineering, and an active-duty captain in the Navy, joined the science adviser's staff in 1966, swelling the civilian-garbed brigade of military officers that inconspicuously strengthens the White House's staff.

By February 1972 and our off-the-record phone conversation, the President's Science Advisory Council (PSAC), which Russ had ably served, had been reined in. (It would be eliminated altogether in Janaury 1973.) But Drew still cultivated his old PSAC network, quietly funneling outside expert opinion into the budgetary deliberations of the executive branch. I told him that the Mariner-class investigation of Jupiter and Saturn was more promising scientifically than any new mission I could think of, trying to neutralize the damaging scientific opposition that had helped kill the

Grand Tour. The Space Science Board (SSB) of the National Academy of Sciences had sensed another expensive Viking-scale mission that would threaten other space science projects.

"It could be incredibly visual and popular," I said. "And once the Uranus, Neptune, and Pluto requirements are dropped, the mission is more within reach technically, and much cheaper." Then I pointed Russ toward what I hoped was still a key factor at the Nixon White House, despite the national ebbing of interest following the first Apollo landings: "It's certainly the most cost-effective space competition with the Soviets imaginable."

He knew from the secret PSAC briefings, as I did, that, despite the massive Mars and Venus efforts of the Soviets, Jupiter—let alone Saturn—was way beyond their reach. "Very interesting," chuckled the always cheerful Drew as the telephone clicked off.

Broad support for MJS coalesced rapidly. On February 22, 1972, the Space Science Board endorsed MJS as first-rate science (and much more affordable than the Grand Tour). NASA, Congress, and the Office of Management and Budget (OMB) waived their usual procedures to consider MJS right in the middle of the annual budget cycle. NASA approved JPL's MJS proposal officially on May 18, just a month before brief press reports mentioned that five men had been arrested for breaking into Democratic National Headquarters in the Watergate Office Building.

In fact, congressional and OMB support for MJS was so strong that an additional $7 million was added to the MJS appropriation for scientific and technological enhancements. The new autonomous electronic brain would now be made reprogrammable to an unprecedented degree. Moreover, some flashy new computer tricks would be built into the spacecraft to assure deep-space communications even under drastic conditions. A new, more efficient computer-driven attitude-control system could also be counted on. No one imagined then that these extra "mental" capacities, intended to assure that MJS got safely to Jupiter and Saturn, would first almost disable the spacecraft at launch and subsequently enable it to explore Uranus and Neptune.

At that time neither Uranus nor Neptune figured seriously in MJS's plans. Far from the Sun, Uranus receives only one four-hundredth as much sunlight as Earth. It is *really* cold and dark on Uranus. Few of the remote-sensing instruments designed for Voyager, as MJS was called after 1975, were expected to yield much at Uranus, let alone at Neptune, a billion miles farther yet. In addition, more immediate problems were showing up in tests of that brilliant, autonomous electronic brain. JPL was learning the hard way that it is easier to build complex computer programs and systems than to test them adequately for the unforgiving space environment. But technical challenges are the stuff good engineers thrive on. Most important, MJS had gotten started and was on the way to becoming the most productive robot explorer ever built.

Things were going less well for JPL as an institution, however.

NASA, JPL's sole sponsor, lacked the presidential political support that it had enjoyed before the Apollo Moon landings. Since then the agency had steadily declined in national clout and technical capacity. The challenging and uplifting Apollo project had become a glorious, almost mythic, memory for NASA, rather than a gateway to the future. JPL was NASA's only center not staffed by government employees. That made it a natural target for elimination in tough times, a circumstance that added to the "creative tension" inherent in the three-sided relationship between NASA, JPL, and Caltech. Furthermore, JPL had no important role in the post-Apollo focus on the Shuttle and on related uses of astronauts in low-Earth orbit. In fact, JPL seemed to be on everybody's rumored "hit list," awaiting the next big NASA cut. Russ Drew sometimes dropped alarming hints to me of draconian measures under discussion at OMB that would have eliminated JPL.

Nevertheless, such alarmism had seemed to me unwarranted in June of 1975, when I flew to Washington for my first official visit with NASA Administrator James Fletcher. Harold Brown, Caltech's president (and soon to be Jimmy Carter's secretary of defense), was waiting for Fletcher's approval before he would announce publicly that I was to succeed William Pickering as director of JPL.

Affable as always, Fletcher had only one substantive question for me. "Bruce," he said, "how do you feel about JPL working for the Defense Department?"

JPL had done no significant military work since 1958, when it became a charter member of the NASA-led civilian space program. I had no personal problem with future JPL defense work, but what a strange question for the head of the American civilian space program to raise! And Fletcher had nothing at all to say to me about planetary exploration, even though he kept prophesying publicly that the 1980s would be "a golden age of planetary exploration."

The message was clear. NASA, a declining institution, felt it could not support JPL as it had during the halcyon days of the building of America's first satellite, first Moon probes, and first planetary explorers.

Unlike my predecessor Pickering, I would thus have a twofold task as director of JPL—to push U.S. planetary exploration to the hilt and, at the same time, to develop a second governmental role for JPL that would be independent of NASA. In 1976, following the oil-price shocks of 1973, government-supported energy research and development was the best (really the *only*) practical alternative for a high-tech, nonprofit, civilian-oriented place like JPL.

So, on April 1, 1976, my first day as director of JPL, I promoted Bud Schurmeier, with great fanfare and with instructions to carve out a meaningful role for JPL in energy R & D. Neither Bud nor I, as it turned out, would find deep satisfaction in these new strategic responsibilities. The ebbing of the old Apollo spirit was merging with a broader retreat of American self-expectations in the face of failures in Vietnam and at home. America would not blaze new paths in either alternative energy or planetary exploration. But in 1976 we had to try to create the kind of future that JPL, and America, deserved.

Schurmeier's replacement to head Voyager was John Casani, who at last had his own project. He loved every minute of it—at least until August 20, 1988, the date of the first Voyager launch at Cape Canaveral.

The day before that launch Suzanne and I stood on top of Pad 41 at Cape Canaveral, 160 feet up. The great red gantry tower clutched Titan / Centaur Number 5. Flickers of Florida lightning punctuated the sun-and-rain-washed panorama. The lightning was still far away, but if those scattered thunderstorms came within a mile and a half of us, we and everyone else would have to get out of there fast. We were standing on nearly seven hundred tons of high explosives. (See figures 8.3 and 2.2.)

Beneath us, twin 250-ton, solid fuel boosters stood ready to power the Titan 3 / Centaur, the largest existing U.S. launch vehicle, off the pad the next morning. The boosters supported a seventy-foot-long, two-stage Titan rocket originally designed to fling a city-destroying nuclear bomb. Tomorrow's purpose was sublime in contrast. Two liquid propellants tanked separately within the Titan would mix precisely in a delicately sustained spontaneous explosion lasting six and a half minutes. These rocket firings of the Titan would thrust the graceful Centaur stage and its heavy Voyager cargo through Earth's atmosphere and partway into Earth-circling orbit. Then the intelligent Centaur stage would take charge.

As we gingerly descended the winding stairway of the gantry tower on that glorious Florida afternoon, a quick glimpse of a busy engineer working *inside* the Centaur surprised me. It was a reminder that unmanned rockets do indeed fly by the skill and dedication of humans—on the ground. The engineer was checking this giant's intricate computer brain.

At launch, delicate instruments at the very tip of the Centaur stage would sense precisely the rumbling flight motion produced by the noisy solids and the storable liquid Titan rocket. As soon as the second Titan stage burned out and fell away, Centaur's electronic brain would ignite the liquid hydrogen–liquid oxygen mixture of the Centaur high-performance engine.

Then would come the brilliant part. The Centaur's electronic brain would automatically compensate for any shortfall in the propulsion of the earlier stage. It would "fly" the Centaur and its payload precisely into a predetermined low Earth orbit. The Centaur brain would then shut down the engine temporarily while the rocket, with the remaining fuel and the attached planetary spacecraft, would coast along in that orbit for the tens of minutes required to reach a precise location for that day's planetary departure. At that point, out over the Atlantic, liquid hydrogen–liquid oxygen would be reignited by the brain in the cold silence of space to enable the craft to break free of Earth's gravity and to build up most of the final velocity that would propel it to another planet. Then the final burst of velocity to reach Jupiter would be supplied by yet another rocket stage, fastened directly to the Voyager spacecraft.

Throughout these eventful minutes, Centaur's brain would measure thrust and calculate flight path. At the exact second when the correct velocity was reached, the brain would shut off the Centaur rocket engine forever. Explosives would then separate the precious spacecraft from the now useless Centaur.

This is what is meant by a smart machine. Inside it, working intently, was the even smarter engineer who understood it all.

We continued down the gradually widening gantry tower in the yellowing afternoon, past dormant liquid and solid explosives ready to send *Voyager 2* into history the next morning. The lightning drifted harmlessly seaward into a darkening sky.

Why launch *Voyager 2* before *Voyager 1?* The decision had to do with keeping open the possibility of carrying out part of the old Grand Tour dream. *Voyager 2,* if boosted by the maximum performance from the Titan / Centaur, could just barely catch the old Grand Tour trajectory. In that way the "Uranus option" from Grand Tour days could be maintained. Two weeks later *Voyager 1* would leave on an easier and much faster trajectory, to visit Jupiter and Saturn only. *Voyager 1* would make up time and reach Jupiter four months ahead of *Voyager 2* and then go on to arrive at Saturn nine months earlier. (That's why it was called *Voyager 1.*) The nine-month separation between the arrivals at Saturn ensured that if *Voyager 1* failed in its Saturn objectives, *Voyager 2* could still be retargeted to achieve them—but at the expense of any subsequent Uranus or Neptune encounter.

At 10:29 A.M., August 20, 1977, the blast from the twin solids and the Titan core shattered the clear blue Florida morning. Exactly as planned, the Titan / Centaur powered *Voyager 2* into Earth orbit, and thence on a path toward the outer planets.

Voyager 2's gryroscopes and electronic brain were alive during the Titan / Centaur launch, monitoring the sequence of events in order to take control upon separation. But here the unexpected happened: *Voyager 2*'s brain experienced robotic "vertigo." In its confusion, it helplessly switched to backup sensors, presuming its "senses" to be defective. Still no relief from its disorientation. Mercifully, the panicky robot brain remained disconnected from Voyager's powerful thrusters, so it did not cause damage to the launch. The Centaur attitude-control system—under its normally behaving brain—stayed in charge, suffering no "vertigo" and, as planned, electronically correcting the disequilibrium of Voyager's brain just before separation.

From the control center John Casani and his tense engineers helplessly watched (though mostly they listened, because there were not enough monitors available to us in Florida) the antics of *Voyager 2*'s disoriented brain. One hour and eleven minutes after lift-off, *Voyager 2* fired for forty-five seconds its own, special solid rocket to provide the final push it needed to get to Jupiter.

One and a half minutes after Voyager's key rocket burn ended, a ten-foot arm holding the television camera and other remote-sensing instruments was unlatched and deployed as planned. Then, more trouble. Voyager's anxious brain once again sensed an emergency. This time it switched thrusters and actuated valves to control the tiny bursts of the gas used to stabilize its orientation. Voyager's robotic "alter ego" (its executive program) then challenged portions of its own brain in a frantic attempt to correct the orientation failure it sensed. Next, Voyager followed the procedures JPL engineers had installed to cope with the most dreaded emergency for a robot in deep space—spacecraft attitude disorientation.* Voyager shut down most communications with Earth in order to begin its reorientation.

Seventy-nine minutes passed while *Voyager 2* struggled alone and unaided to find the Sun and reestablish a known orientation. Finally, it radioed confirming data. For the moment *Voyager 2* was stable.

It was all work and no celebration that afternoon in the dimly lit High Bay Conference Room, where, just days earlier, a seemingly healthy *Voy-*

---

*In August 1988 the *Phobos 1* spacecraft of the Soviet Union succumbed to such an emergency, and in March 1989, *Phobos 2* evidently met a similar fate.

ager 2 had checked out perfectly. Were the redundant sensors malfunctioning? Was the state-of-the-art brain defective?

The technical discussion in the room was poorly illuminated too. All the new, supersophisticated fault protection in Voyager's electronic brain operated on the now painful presumption that it would be triggered *only* by a hardware failure billions of miles from Earth. In that event Voyager would be unable to establish even emergency communications with its human handlers, who could not help it much at that distance in any case. As a consequence Voyager had been programmed virtually to shut off communications with Earth during such emergencies and to fix itself. But, somehow, these deep-space procedures had been triggered right after the launch.

Now, because of those disrupted communications, we were not receiving the usual flow of engineering-status measurements. We simply lacked enough information to figure out the causes of Voyager's mysterious behavior, even though the spacecraft was so close to Earth that communications normally would have been feasible under any emergency.

Voyager was proving to be far more autonomous than anyone had foreseen or wished.

Casani's frustrated specialists substituted coffee for sleep in that dark and dreary High Bay area. They crowded around the solitary speaker phone while reviewing the skimpy facts with their equally puzzled colleagues back at JPL. Even as they spoke, Voyager experienced yet another spasm of thruster firing, accompanied by frenzied switching of its redundant brain components. What was going on out there? Had Voyager's massive propulsion module, so gently jettisoned after launch, continued as a ghostly companion on Voyager's trajectory? Was it actually bumping into Voyager from time to time?

Ted Kopf, an attitude-control specialist at JPL, had designed some of that intricate computer logic now running rampant on *Voyager 2*. He had come to the Florida launch on his own hook for pleasure, not work. But that was before the emergency. I listened now as Kopf and a brilliant JPL systems engineer, Chris Jones, served up precise clues afforded by *Voyager 2*'s scattered radio transmissions. So detailed were their mental images of that complex Voyager brain that they could reconstruct each stage of its successive anxiety attacks since launch. There had been no hardware problems in the brain—just a slight but serious missetting of computer parameters.

Subsequent detective work shed light on the computer problems first encountered after rocket separation. These were complicated by unexpectedly large vibrations from the unlatching of the instrument boom. There had been no "bumps in the night" from a ghostly companion. Those spacecraft disturbances were simply a complex and overly sensitive reaction by the autonomous Voyager to a familiar space-engineering problem. Small dust particles released by the vibrations from rocket propulsion sometimes drift near a spacecraft. Being close, and being lit by the Sun, they are much brighter than the star images that the spacecraft's optical detector normally tracks. In trying to follow the dust particles, the tracker reorients the spacecraft.

A lot of robotic technology and human expertise had accumulated over the years, ensuring that situations like this would be recognized and compensated for. But this new, superautonomous robot did not have that repertoire of human judgment and experience built into it. So it mistrusted itself but was finally righted by its own logic.

This could not be allowed to happen again. Desperately, corrections were patched onto the computer programs in *Voyager 1*, now waiting to be entombed for launch, even as the analysis of *Voyager 2* stumbled along. A new mechanism to reduce the excessive vibrations from the instrument boom was designed, tested, and installed. Time was running out.

Sixteen days after *Voyager 2*'s heart-stopping launch, on September 5, 1977, *Voyager 1* flew. The last Titan / Centaur in the world shook the ground for miles around Pad 41. NASA had phased out America's most powerful rocket long before a replacement capability, the Shuttle, would be ready. (Indeed, not until the early 1990s will comparable rocket capability be available for American space endeavors, civilian or military—a gap of nearly fifteen years.)

Our attention in the control room was riveted on the continued radio updating describing the execution of key rocket sequences. "Solid booster burnout and separation," a voice over the control room intercom said. "Titan core ignition ... Titan burnout and separation."

But wasn't that sooner than shown on the projection? Yes, there definitely had been an anomaly. In fact, a slight error in the mixture ratio of the two liquid fuels in the Titan had left twelve hundred pounds of them unburned. This was serious. Titan had underperformed, not propelling Centaur and Voyager fast enough. "Centaur ignition," came the words from the public address system. The Centaur's brain, recognizing the Titan performance deficiency, smoothly extended its burn just enough to compensate. After coasting, and then restarting, the Centaur put Voyager and its propulsion module precisely on course—with only a squeaky 3.4 seconds of propulsion left.

"Wow, that was pretty close," I thought at the time. It was only later, over a beer in Cocoa Beach, that I realized that if the two Titan / Centaur rockets for Voyager had been fired in reverse order, if the underperforming one had been used for the more demanding trajectory, *Voyager 2* would not have received enough velocity to catch the Grand Tour trajectory. It would have reached only Jupiter and Saturn. The opportunity of the century would have passed us by (even though the rich Jupiter / Saturn comparison would still have taken place). Pure chance had assigned the unexpectedly low-performing Titan to *Voyager 1* rather than to the more demanding launch of *Voyager 2*.

But in September of 1977 it was hard to take seriously any hypothetical extension to Uranus, two billion miles from Earth. We were not even off to a good start toward Jupiter, a mere five-hundred million miles and eighteen months away.

9.1 APPROACHING JUPITER • Both Voyagers documented Jupiter's rapidly changing colored atmosphere with numerous three-color images for many weeks before encounter. In this black-and-white version of one such image, the Great Red Spot can be seen in the lower left center, along with many adjacent bright features. The Great Red Spot, a permanent feature of Jupiter, is larger in diameter than the entire Earth. *NASA/JPL photo*

9. To Jupiter, Barely

THOSE TROUBLED DEPARTURES from Pad 41 had highlighted JPL's own troubled circumstances. At the other end of my somber flight from Florida back to Pasadena lay threats to its future.

Just six years before, in 1971, when the Grand Tour was still abuilding, JPL's cup had overflowed. *Mariner 9*, like Voyager now, had just been launched from Florida on its way to Mars. *Mariner 9*'s spacecraft specialists moved on smoothly to apply their experience to the next generation of Mars-orbiting spacecraft. Those Viking orbiters, scheduled for launch in 1975, represented a key part of America's grand search for life on Mars. A second group of *Mariner 9* veterans augmented Gene Giberson's erstwhile Mariner / Venus / Mercury band. In 1971 they (with Boeing) were just starting to assemble the component pieces of *Mariner 10*, still two years away from its history-making voyage to Venus and Mercury. A third group joined Bud Schurmeier, readying the great voyage outward, whose exotic handiwork was now coasting fitfully toward Jupiter.

But in 1977 the cup was empty. Not a single new planetary endeavor had been started at JPL since 1972.* Was JPL, this unique national resource for achievement, doomed to grow old without renewal, to drift into irrelevance?

There was one hope left for JPL. In the face of heavy congressional opposition, JPL had sold a new Jupiter orbiter mission, which would also hurl an instrument probe smack into that swirling, colored atmosphere (see chapter 12). JPL's only hope for continuing as a vital institution lay in making that Jupiter orbiter with Probe (JOP) mission so attractive and

---

*Pioneer to Venus, started in 1974, was a low-cost, high-quality endeavor managed by NASA's Ames Research Center and using Hughes Aircraft as prime contractor. Launched in 1978, it was the last U.S. planetary mission before the Shuttle, or since.

credible that NASA and Congress would *have* to back it. The obvious choice to lead this project was the forty-three-year-old John Casani, with two decades of progressive accomplishments and visibility, right up through Voyager's launch.

"But, damn it, Voyager is not running all that smoothly," I mused on the plane back to California. Still, I had seen JPL work itself out of worse problems many times before: that incredible three-week shroud redesign in 1964 before the launching of *Mariner 4;* the crippling battery explosion on *Mariner 7* a week before encounter in 1969; heart stoppers all the way to Mercury with *Mariner 10.* And we had a good year and a half of presumably uneventful cruise flight with Voyager before the first Jupiter encounter.

Ray Heacock succeeded Casani in running Voyager. Ray, like John, had been at JPL for two decades and had personally babied the development of those Voyager spacecraft from their inception. He knew their insides better than anybody else.

But even as the Casani-Heacock transition was taking place through the fall of 1977, those state-of-the-art, autonomous spacecraft were generating problems faster than the struggling project could solve them.

The thrusters on both Voyagers fired inappropriately. Their rocket plumes in some cases deflected off the sides of the spacecraft, an unexpected circumstance that further aggravated the automatic control of the spacecraft. Incorrect commands—so far harmless—had occasionally been sent by a team ill prepared for the enormously complex task of talking to a semi-intelligent and maddeningly literal-minded robot.

Then, in late November of 1977, the first serious hardware failure showed up in a critical subsystem. One of *Voyager 2*'s two duplicate radio transmitters began to degrade. It was switched into a low-power mode to nurse it along. Something was wrong, but there was no way to know exactly what.

On February 23, 1978, when *Voyager 1* had been traveling for six months, its scan platform began to stick. The camera and other crucial remote-sensing instruments were located on that platform in order to be pointed correctly. Would those observations be compromised? More trouble, also poorly understood.

I spoke to Terry Terhune, the deputy director of JPL. "The Voyager project simply isn't getting on top of this mission," I complained. Terhune, a retired Air Force general and space pioneer, admitted he had similar misgivings. The struggling Voyager team needed to be augmented, especially at the top, but doing so would be tricky. In an old, successful organization like JPL, there is always a pecking order. Bypassing it can stir up deep resentments.

"Parks feels committed to Heacock," I said.

"And he is concerned that project management changes now might be disruptive," Terhune added. Bob Parks was the respected head of all JPL space-flight projects.

Yet we both felt the time had come to do something. Past JPL successes would count for little if one or both Voyager spacecraft failed because of human errors. "Putting Parks himself in as project manager would surely get everybody's attention and support," I said. "And even if Heacock reverted back to deputy manager, he'd still be working for the same boss,"

"On the other hand," said Terhune, "Pete Lyman is outstanding in operations." (Dr. Peter Lyman had just earned NASA-wide recognition for his leadership of Viking mission operations.)

"But wouldn't that look like a real demotion for Ray?" I asked with some concern.

JPL was in the business of taking risks successfully. Key risk takers must never be treated like scapegoats. Success, failure, and risk all had to be shared generously if risk taking were to continue to be fashionable. Indeed, JPL veterans sometimes lamented the psychological change in NASA from being a bold, risk-taking organization to being a bureaucratic one. Furthermore, Ray Heacock was high-powered technically, and potentially a great project manager. In retrospect I realized that I had given him full responsibility for Voyager before we knew just how serious the problems were.

With Terhune's support, I overruled JPL's top project man, Parks, on how to respond to Voyager's difficulties. On March 5, 1978, Bob Parks himself was assigned personal command of the Voyager project. Pete Lyman became his deputy for operations. Ray Heacock returned to his earlier post of deputy for spacecraft. A special notice (from me) informed every professional member of the laboratory that Voyager had absolute priority over all other JPL efforts. Casani and the JOP project—and JPL's future—would have to pace their build-up until Voyager was really out of danger. We were declaring an emergency.

But we had not done so soon enough. The now augmented Voyager team continued to battle *Voyager 1*'s problem with its scan platform. Intricate command sequences to move it from one pointing direction to another were carefully transmitted on March 23, 1978, again on April 4, and once more on April 5. After these, *Voyager 1*'s scan platform did indeed seem to have improved. Another crisis was subsiding.

On that same April 5, however, preoccupied with the scan platform, the busy Voyager team forgot to send a routine but critical command to *Voyager 2*. Both Voyager brains were programmed to presume that their "hearing" had gone bad if they did not hear a radio command from Earth for as long as a week. When *Voyager 2* did not receive the routine command that was due by April 5, it followed its instructions and automatically switched to its backup radio system, presuming the primary receiver had failed. Neither *Voyager 2* nor its designer parents had ever anticipated that the ground crew would simply forget.

Then another hardware problem arose: the backup radio was seriously defective. It was "tone deaf." It could not track the rapidly varying tone of Earth's transmitter. That failure revealed yet another electronic trick I had not comprehended previously. These spacecraft radio receivers routinely compensated for the "Doppler" distortion of the radio signal transmitted from Earth. This was necessary because the speed of the spacecraft relative to Earth was always changing, affected by curved orbital paths and planetary motions. A spacecraft's radio tones are no different in this respect from the tone of a horn of a passing train or car, whose pitch goes first up and then down. Consequently, a pure radio tone transmitted from the big dishes of the Deep Space Net (DSN) actually sounded, at the spacecraft, more like a descending wail than a "hum." The tuning filter on voyager's receiver was so narrow and sharp (in order to select and enhance the faint radio signal from Earth) that, without the automatic electronic compensation for changing tone, *Voyager 2* could "hear" no radioed commands at all.

Chiefs and Indians alike instantly overcame their preoccupation with *Voyager 1*'s scan platform to respond to the more immediate *Voyager 2* hearing crisis. Fortunately, the brain aboard *Voyager 2* had been programmed before flight to switch back automatically to the primary receiver and once more to listen for the anticipated Earth signal. In precisely twelve hours the primary receiver switched back. Commands from Earth once again flowed into *Voyager 2*'s isolated brain. The crew in the operations room began to breathe easier.

But after only thirty minutes that primary radio receiver suddenly failed completely. *Voyager 2* was silent again. Its primary receiver was now dead, and the backup was tone-deaf. As a result of component failures on the spacecraft and distraction with *Voyager 1*, JPL had completely lost contact with *Voyager 2*, humankind's first and only potential emissary to the distant kingdoms of Uranus and Neptune. There was no joy in Pasadena that night.

Seven somber days passed before *Voyager 2*'s brain once again followed its prelaunch instructions and automatically switched back to the crippled, "tone deaf" secondary receiver. By then JPL's DSN engineers had worked up a possible solution. They prepared computer tapes that slowly varied the frequency of the radio tone transmitted from Earth in a manner to compensate exactly for the expected Doppler shift. The Earth transmitters of the DSN were taught to "chirp" exactly the right amount to cancel the descending wail.

On April 13, at seven-thirty in the evening, when most Spaniards were getting ready to close their shops, the DSN's white 210-foot dish in a rolling valley fifty miles west of Madrid began the critical transmission to *Voyager 2*. The fifty-three minutes that it took the radio waves to travel to and from the spacecraft dragged by, but JPL's solution worked. Anxiety gave way to relief as *Voyager 2*'s radioed acknowledgment reached Earth.

Like futuristic brain surgeons, JPL engineers fed hour after hour of triple-checked commands into *Voyager 2*'s brain. They were not always successful. A slight change in temperature of the radio receiver on board *Voyager 2* (hardly more than one-tenth of a degree centigrade—the finest markings on a fever thermometer) was sufficient to wreck the improvised scheme.

How to predict such small changes in temperature aboard the space-craft? An elaborate computer model was rapidly developed to account for all the temperature vagaries of the deep-space environment, such as the changing angle of sunlight that fell on the spacecraft and the changing amount of electrical power consumed internally (and given off as heat, necessarily affecting adjacent circuits).

A critical maneuver to correct *Voyager 2*'s course was successfully carried out on May 3. Earth was gradually reestablishing control over *Voyager 2*. On October 3, the spacecraft's memory banks were loaded to the brim with commands that would provide for a bare-minimum-science encounter at both Jupiter and Saturn, should radio contact once again be lost. There was good reason for concern—because the awesome power and electronic interruptions of Jupiter's radiation belt lay ahead.

Gradually, Voyager's human controllers developed into a powerful team. Elaborate computer instructions for each spacecraft flowed through the giant DSN antennae ever more easily—and practically error free. Ambitious commuter command sequences enabling each instrument to exploit Voyager's true capabilities at Jupiter were hotly debated, revised, and then smoothly blended with dozens of other coded instructions.

On December 12 through 14, 1978, *Voyager 1*, now fourteen months old, danced flawlessly through its Jupiter encounter sequence. The once ailing scan platform executed an elaborate series of motions, while the brain shuttered all the instruments and correctly changed data modes and tape recorder sequences. However, Jupiter still lay two and a half months ahead. This was a full-scale dress rehearsal, part of the hard discipline and practice necessary on the ground to ensure success in space.

After a Christmas break for the now confident Voyager team, it was back to business. On January 4, 1979, Voyager began filming Jupiter's photogenic atmosphere. The real-time-TV technique we had pioneered at Venus and Mercury now provided color time-lapse "movies" and other photographic combinations that revealed Jupiter's dynamic atmosphere in awe-inspiring detail. Dr. David Morrison, who had earlier helped save Mercury's features from being named after birds, described the first Jupiter pictures as follows:

Spots chase after each other, meet, whirl around, mingle, and then split up again; filamentary structures curl into spirals that open outward; feathery cloud systems reach out towards neighboring regions; cumulus clouds that look like ostrich plumes may brighten suddenly as they float towards the east; spots stream around the Red Spot or get caught up in its vortical motions—all in an incredible interplay of color texture and eastward and westward flows.[1]

For the millions viewing PBS's "Jupiter Watch" telecasts, ABC's "Nightline" programs, or Carl Sagan's "Cosmos" series or watching the images appearing on screens in Japan, England, Mexico, and South America, Voyager revealed a treasure trove of abstract art. The swirling Red Spot became a cultural symbol overnight. Much of humanity shared the first glimpses of this new world, savoring unfamiliar but surely drawn paintings, like space travelers wandering through a cosmic Louvre.

Other *Voyager 1* instruments sensed a giant electrical conduit flowing through space around Jupiter and interacting with Io, its innermost moon. Sodium atoms glowed in this energetic environment, surrounding Jupiter with a halo of natural sodium-vapor light. This light is powered by five hundred billion watts, comparable to all the electrical power produced in the United States.

One of the biggest surprises was Jupiter's ring. When my longtime associate Ed Danielson announced it, I did not believe him. "There *can't* be a ring at Jupiter," I said. "The tidal forces from the big moons make it theoretically impossible." I was parroting the conventional wisdom regarding planetary rings.

Ed described how he (with strong support from Dr. Toby Owen, of the State University of New York) had managed to get the camera pointed precisely in the plane of Jupiter's equator and to send the commands for the camera exposures to take place just as the spacecraft passed through it.

This would be the only opportunity to detect any faint ring at Jupiter that could have escaped detection from Earth.

Ed had arranged for the camera shutter to remain open for an unprecedented eleven minutes, in order to build up a faint signal. All the while *Voyager 1* was being bombarded by the charged particles of Jupiter's radiation belts. Raw electrical currents flowing on the spacecraft even disturbed the craft's internal clock and damaged some of the instruments. Nevertheless, *Voyager 1* followed Danielson's instructions precisely, providing a big scientific payoff. A faint ring was detected after all; the theoreticians hurried back to the drawing board. Ed is the cleverest cosmic photographer I know.

Then it was on to the Jovian moon Io.

"It's a pepperoni pizza," joked Dr. Laurence Soderblom, as he showed me the freshly printed orange, red, and yellow photograph of Io. Larry had been one of my star Caltech Ph.D. students just a decade earlier. Now he was leading Voyager's close-up imaging of those once ignored "fuzzy spots," the satellites of Jupiter (see figure 9.2). Io, which is the size of Earth's Moon, had been discovered by Galileo himself, along with the three other large moons of Jupiter. To *Voyager 1*'s camera it looked more like a boiling sulfurous caldron escaped from the imagination of Dante than a natural object. Within a week that comparison became apt. *Voyager 1*'s cameras and spectrometers focused on Io recorded gigantic sulfurous

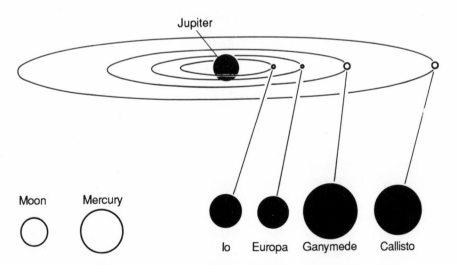

9.2 THE MOONS OF JUPITER • Galileo, with his primitive first telescope, discovered that four large moons orbit Jupiter. *Voyagers 1* and *2* transformed those fuzzy spots into extraordinary worlds. The sizes of the Moon and of Mercury are given for comparison.

WHEN SODERBLOM SPEAKS . . . • Laurence Soderblom, the deputy leader of the Voyager imaging team (and one-time graduate student of the author) was especially articulate in describing to the press the novel attributes of the moons of Jupiter and Saturn. In this candid photo, taken during the Saturn encounter of Voyager in August 1981, Soderblom is surrounded by a galaxy of top space journalists. *From left,* Tom O'Toole, then of the *Washington Post;* George Alexander, then of the *Los Angeles Times;* Larry Eichel of the *Philadelphia Inquirer;* Julian Low, *Pasadena Star News;* Bob Cooke, then of the *Boston Globe;* Mitch Waldrup, *Science* magazine; Bob Cowan, *Christian Science Monitor;* Bob Locke, then of Associated Press; the late Everly Driscoll (*kneeling*); and John Noble Wilford, *New York Times.* Behind Wilford are Ron Clarke and an assistant from Reuters. Bill Hines of the *Chicago Sun-Times* is to Soderblom's left. *Montrose Ledger photo*

volcanos spewing debris a hundred miles above Io's sulfur- and lava-strewn surface. *Voyager 1* had found hell. Only the devil himself was absent.

In frigid contrast, Europa, Io's nearest neighbor, emerged from Voyager's scrutiny as the smoothest object in the Solar System, a giant cosmic egg, complete with faint global cracks (see figure 12.1). Nearly as large as Earth's Moon in diameter, it is very different in composition and structure. Fifty miles beneath its finely cracked, icy crust probably lies a giant ocean, the largest body of liquid water outside of Earth. Such wondrous worlds are belittled by the label "moon" or "satellite."

Next outword from Jupiter is the Mercury-sized moon Ganymede (see figure 12.3). *Voyager 1* discovered there a fossilized surface that uniquely records ancient planetary convection, even surface shear like that of the

San Andreas fault, which sleeps fitfully near Voyager's California birthplace.

The fourth of Galileo's moons, the one most distant from Jupiter, is battered Callisto. *Voyager 1* revealed a cratered surface built up by eons of impacts unrelieved by internal renewal. Collisto's surface may be the most heavily cratered one in the Solar System, recording the bombardment of an ancient flux of asteroid-sized objects passing in the vicinity of Jupiter.

Indeed, *Voyager 1* discovered a miniature solar system surrounding Jupiter. Those four orbiting moons are as unusual and different from one another as four bits of rocky debris orbiting close to our Sun—Mercury, Venus, Earth, and Mars. *Voyager 1* had zoomed in on the four dark smudges near the gaudy telescopic image of Jupiter, revealing worlds and circumstances beyond our imaginations.

10.1 BEYOND SATURN • Beautifully ringed Saturn is displayed in this *Voyager 1* image, taken looking back after encounter. Notice the shadow of the rings on the illuminated surface of Saturn as well as the image of Saturn showing through the partially transparent rings at the terminator. *NASA/JPL photo*

10. A Sublime Rendezvous with Saturn

IT WAS MONDAY MORNING, March 24, 1980. I sat waiting in the Cabinet Room of the White House with four other NASA and JPL officials. Worldwide acclaim for Voyager's accomplishments at Jupiter the year before had won for us the prestigious Goddard Trophy—even before Voyager accomplished the second half of its mission, at Saturn. (Voyager took the trophy an unprecedented second time for *that*). Jimmy Carter, facing a tough reelection campaign, was to make the presentation.

The Cabinet Room surprised me. So small for such a large purpose, I thought. Harry Truman's portrait dominated the sparse formal chamber, and I realized that Carter probably likened himself to that once underrated "people's President."

But my attention was focused on a more immediate question. Would the TV crew get here in time? Frank Colella, JPL's public affairs chief, was trying to work another miracle by rousing a sluggish NASA into emergency action. Only hours earlier I had discovered that Carter's own people in the White House had not thought to publicize this minor but politically useful event, even though he was running for reelection. Presumably, they felt Carter could fritter away half an hour in pleasant conversation without looking for any electoral benefit. There were thus no members of the press on hand, not even a captive video crew—just a lone White House photographer to provide a few personal mementos.

I needed a videotape of this award ceremony for my annual state-of-the-lab talk at JPL. Nearly all five thousand JPLers would watch the talk on closed-circuit TV, or when it was replayed at remote locations. They were all looking to me to provide hope for JPL's future. The recent news had been awful. Shuttle problems had delayed for two years the planned launch of the Galileo mission to Jupiter, JPL's only new planetary project—and JPL's fragile future along with it. No project starts at JPL were included in the NASA budget then winding its way through Congress.

Closer to home for JPLers, interest rates and inflation both approached an unbelievable 20 percent, dramatically shrinking individual purchasing power. Unlike government civil servants, college professors, or unionized employees, the talented people at JPL who had made history with achievement in space from the time of Explorer in 1958 to that of Voyager in 1979 had little job security. Worse, in those inflationary times the nonprofit JPL could not compete with private business in offering such special financial inducements as incentive stock options, favorable pension accumulation, and bonuses. Yet our staff lived within commuting range of the heaviest aerospace concentration in the world. These overheated defense contractors were ravenous for experienced engineers. JPL was running on memories of high adventures and on the prospect of more to come, not on job security.

Suddenly, a side door in the Cabinet Room opened, and a video crew and camera quietly moved in, followed by the sharp-featured, tanned Dr. Frank Press, Carter's science adviser and an old Caltech associate. Then Jimmy Carter himself appeared, flashing an engaging grin.

Carter shook hands with NASA Administrator Dr. Robert Frosch, who would receive the award on behalf of the Voyager team. Neither Frosch nor the next handshaker—Dr. Alan Lovelace, his deputy—had traveled to JPL for either Voyager encounter with Jupiter. I could not understand why. Voyager had been America's principal space winner at the time, yet NASA had virtually ignored it.

Next in line was Tim Mutch, recently arrived at NASA from Brown University to head all of NASA's Office of Space Sciences and Applications. Tim had been the key geologist for the Viking lander's search of the Martian soil and landscape. He really believed in exploration and, in contrast to his NASA bosses, had eloquently represented NASA at the second Jupiter encounter. He remarked, "Those who were fortunate enough to be with the science teams during those weeks will long remember the experience; it was like being in a crow's nest of a ship during landfall and passage through an archipelago of strange islands."

JPL's Bob Parks received the next automatic presidential handshake. Finally, Edward Stone, of Caltech, the project scientist of Voyager, and I completed the JPL representation at this (by now) videotaped ceremony.

Carter displayed technical understanding in his first utterances:

It's a great honor for me as the President of our country to participate this morning in a ceremony delivering the Goddard Memorial Trophy to the NASA team

RECOGNITION FROM JIMMY CARTER • Carter presents the Goddard Trophy (*on table*) for Voyager's achievements at Jupiter. This scene took place on March 24, 1980. Frank Press, Carter's science adviser, watches as Carter shakes hands with Robert Frosch, NASA administrator. JPL's top project official, Bob Parks, is at the far right, next to the author. *NASA / JPL photo*

responsible for the Voyager flight—or odyssey—through the Solar System and beyond. I think the whole world was enraptured on two occasions in 1979 when the Voyager passed by our largest planet, Jupiter, and sent back some absolutely remarkable photographs that probably expanded our knowledge of our largest planet, perhaps even the Solar System, beyond what we had ever known about Jupiter previously in history. Later on this year, I think in November, the same Voyager will go past Saturn, and in 1986 past Uranus. (Is that correct? Well, good, I'm glad I got it right.) We'll be looking with great interest again at this remarkable Voyage.

He followed with the expected pat on the back:

I might say that this team here, the team that's made this flight possible and also had such tremendous success in bringing the images and knowledge so clearly back to Earth, to be shared by scientists and others interested in astronomy and our own Solar System, deserve the highest accolades.

THE PLANETARY EXPLORER TIM MUTCH • Tim Mutch was a pioneer in the exploration of the Moon and Mars, and led the imaging team for the Viking landers. In that capacity he spent many months at JPL working in the mission support area, where this photo was taken in late 1976. His career was cut short by a fatal mountain-climbing accident. The long-lived *Viking 1* lander on Mars was officially renamed by NASA the Thomas A. Mutch Memorial Station, and a special plaque was created to be placed there by some future NASA endeavor. *NASA / JPL photo*

He closed with the obligatory statement about American leadership—even as that leadership was fast disappearing:

I also want to make clear to Dr. Frosch and to other members of this team, that I'm determined as President to make sure that the United States always maintains its leadership in space exploration.

We carefully packed up that precious videotaped drama for its journey back to Pasadena, where it would strengthen the invisible glue of JPL and help keep expectations alive a bit longer.

Bob Parks, rejuvenated by his tactical success in turning Voyager around, had returned to his ranking position on the JPL "general staff" a full year before the ritual with Carter in the Cabinet Room. Ray Heacock had once again become Voyager project manager, this time with the requisite resources and experience, which had been lacking in those uncertain months following the launch from Florida. Under Ray, *Voyager 2*'s encounter at Jupiter in July 1979—four months after that of *Voyager 1*—had gone like clockwork, elegantly supplementing *Voyager 1*'s Jovian findings. New details emerged about the surfaces of Ganymede and especially about the enigmatic, smooth "cosmic egg" Europa. Both Voyager spacecraft now coasted smoothly toward their rendezvous with beautifully ringed Saturn. (See figure 10.1.)

So smoothly, in fact, that I paid little attention until the final JPL project meeting before *Voyager 1*'s Saturn encounter. "We're planning a small burn, the ninth TCM [trajectory-correction maneuver] on November 7," Ray Heacock announced calmly.

A red light flashed in my mind. In order for the onboard rocket to be fired, the spacecraft would have to be rotated on gyros and its radio connection with Earth broken. "Isn't it risky to break communications so close to encounter?" I asked.

Everyone in the room knew that if Voyager encountered any trouble during its reorientation for that ninth firing of its on-board rocket, light-hours away from Earth, the Saturn encounter could be completely lost. It seemed to me too great a risk to take in order to achieve a minor change in the geometry of Voyager's encounter with Saturn's moon Titan. Once reoriented, *Voyager 1* would swing behind Titan and probe its strange atmosphere with radio waves, just as *Mariner 4* once did at Mars and *Mariner 5* at Venus.

"The geometry of the radio occultation of Titan will be significantly improved," Ray replied evenly. This was not the same Ray Heacock who led the shaky Voyager team three years earlier when these spacecraft had surprised everyone with their neurotic anxiety after launch. This man was confident. I had to accede to his judgment.

At 8:23 P.M. (PST) on November 6, commands were beamed toward Saturn from the 210-foot Goldstone antenna, which sat in the remote desert 110 miles east of Pasadena. *Voyager 1*, now relying solely on internal gyroscopic control, rotated a quarter turn away from Earth's direction. Radio communications stopped abruptly as the craft's directional antenna pointed away from Earth. Hydrazine fuel flowed into the combustion chambers of four small thrusters and reacted with their catalysts. The hot gaseous rocket exhaust silently nudged *Voyager 1* imperceptibly "sideways" from its swift flight toward Saturn. Eleven minutes and forty-five seconds later, *Voyager 1*'s brain shut off the rocket engine. Then began the elaborate automatic reorientation of its antenna toward Earth, a tiny, dark disk a billion miles distant. Eight minutes after midnight on November 8,

1980, *Voyager 1*'s minuscule radio signal once again reached that giant electronic ear in the Mojave Desert. Mere static to the untrained, this was sweet music to the ears of the waiting controllers. The ninth maneuver had come off perfectly.

Then, on November 11, precisely as planned, *Voyager 1* swooped within 2,400 miles of Titan, the Mercury-sized moon of Saturn, a body perennially enveloped in a cloudy atmosphere, in which methane gas was known to exist. *Voyager 1*'s close-up sensing discovered that otherwise undetectable nitrogen gas is actually the principal constituent of Titan's atmosphere. Titan's methane (identical to earth's natural-gas fuel and composed of one carbon and four hydrogen atoms, CH_4) is a trace gas, like water vapor on Earth. But Titan's methane is chemically altered by the ultraviolet rays in the sunlight that hits the top of its clouds. The result includes a red-brown smoggy haze, perhaps not altogether different in appearance from that of southern California (but very different in composition).

The smog surrounding Voyager's team in Pasadena forms when California sunshine acts on gasoline-rich automobile exhaust to produce ozone and other oxidizing irritants. At Titan, in contrast, there is no oxygen and the Sun's action produces heavy hydrocarbons, such as ethane (C_2H_6) and propane (C_3H_8). Furthermore, in that reducing, nitrogen-rich environment even a lethal but basic biological building block, hydrogen cyanide (HCN) gas, was discovered.

Titan is a cosmic gas station where future rocket expeditions may refuel before pushing off to really deep space.

At 10:46 A.M. (PST) on November 11, 1980, *Voyager 1*'s radio signal began to attenuate. The radio occultation, which Ray Heacock had insisted on refining with that ninth maneuver, was under way. Within a minute the signal faded entirely. Then came thirteen minutes of silence, and a second minute of reemerging signal, before normal communications resumed.

Those two short intervals of radio probing unlocked the secrets of Titan. Heacock's insistence on that last-minute improvement in the geometry for the radio occultation was completely vindicated. *Mariner 4*'s radio probing revealed a thin Mars atmosphere and *Mariner 5*'s confirmed a very thick Venus atmosphere. Now *Voyager 1*'s radio signal sensed Titan's surface, just barely, beneath an atmosphere denser than many scientists had expected. The amount of bending indicated that Titan's surface pressure is 60 percent greater than Earth's.* The temperature there is an unearthly

---

*Because of Titan's low surface gravity, this surface pressure corresponds to ten times more atmosphere per unit of surface area than exists on Earth.

minus 178 degrees centigrade. Titan is so cold that methane raindrops may fall, and some speculated that solid methane snow occasionlly forms. Refined computer modeling following up on *Voyager 1*'s epochal discoveries suggest an ocean of liquid ethane, perhaps roiling about on Titan like an oil spill on Earth. A thick global layer of organic tar probably forms the "seabed."

All that stands between this cosmic petroleum reserve and a flourishing, life-filled planet is the absence of heat. If Titan were as close to the Sun as Mars is, and were warmed accordingly, we would surely have sent a zoological expedition there—if "Titianians" had not sent one to Earth first! Instead, Titan is the Solar System's only known prebiological planetary preserve, a full-scale natural laboratory steeped in organic clues to Earth's chemical state just prior to that magical moment eons ago when chance events synthesized the first self-replicating molecule.

Throughout October of 1980 a growing abundance of television programming from JPL let the world peer through *Voyager 1*'s telephoto lenses at a steadily approaching image of the rings of Saturn (see fig. 10.1). In those famous globe-girdling sheets of light, viewers could discern first a few new breaks and features; then tens of new rings and gaps between rings; then hundreds of rhythmically ordered rings, bands, and gaps in a dazzling orbital tapestry. Popular interest rose dramatically as encounter approached. Once Ronald Reagan's landslide had buried a hapless Jimmy Carter in the November 1980 election, news of Saturn moved back onto the front pages. Both *Time* and *Newsweek* ran Saturn cover stories. Live programming flowed daily from Pasadena to a network of viewers in countries ranging from Canada to Finland and, especially, Japan. Colorful images of Saturn surrounded by its magnificent rings rapidly became pop art cultural symbols. Everybody was exposed to them.

Well, practically everybody. Bob Frosch had announced his resignation, effective in January 1981, from his position as NASA's administrator. On encounter day, November 12, 1980, Frosch was out of touch, touring rural villages in India on his way to Japan. We sent him a special packet of pictures and news releases because ordinary Japanese citizens knew more than he did about Voyager's Saturn discoveries, thanks to the special Japanese-language live TV programming from JPL. Frosch's deputy, Alan Lovelace, remained rooted in Washington, manning the barricades of a beleaguered NASA battered by cost and schedule overruns of the earthbound Shuttle.

The only spokesman for science and exploration left at NASA headquarters was now gone—and was sorely missed. Tim Mutch had died while climbing in the Himalayas. His engineering deputy, Andrew Stofan, the NASA engineer responsible for integrating the Titan missile with the Centaur upper stage that had powered Voyager into space, now represented NASA at the Saturn encounter.

I felt that Frosch should have been in California, not in India. Frosch's predecessor, Jim Fletcher, who had shifted NASA's priorities away from deep-space exploration, had nonetheless made a special point to be at JPL in March 1974 for *Mariner 10*'s first encounter at Mercury and for the first Viking landing on Mars, on July 20, 1976. He also was on hand at NASA's Ames Research Center to receive *Pioneer 10*'s first reports back from Jupiter, on December 3, 1973.

In an irritable frame of mind, and still buried in the public relations complexities of the Saturn encounter, I was interrupted by a telephone call from a minor White House staff member I knew casually. Did I know of any jobs? Jimmy Carter was leaving the White House, and so was he. There was little I could suggest for a used Carter aide—certainly nothing at JPL. I did grumble a little about NASA's and the White House's lack of interest in the Saturn encounter.

A few hours later, the telephone once again interrupted me. President Carter was on the line. Evidently, my offhand criticism had provoked a White House response. (Normally, presidential phone calls were carefully scheduled and managed by NASA's public affairs office.) Hurriedly, the acting associate administrator for space science and applications, Andy Stofan, was located so I could share the President's call with the ranking NASA official.

"We're honored that you are calling, Mr. President," I said. "The Voyager team will be especially gratified to learn of your personal interest."

"They certainly have done an outstanding job," he said. "The Saturn pictures are fantastic. I watched two hours yesterday as well. Didn't expect to have so much time available," he added, with wry humor.

"How strong are the winds moving the clouds in those Saturn pictures? And are there any auroras there?" asked the Annapolis graduate who had assiduously avoided any political identification with space. No President had better understood space technically, or more poorly understood it politically.

This bittersweet final conversation with Carter triggered some personal memories. Three and a half years earlier on that same telephone on the top floor of JPL's administration building, I had said no to Frank Press when he invited me to become Carter's NASA administrator. It had not, in fact, been a very hard decision.

I knew firsthand how the American space adventure had declined since 1972 and the end of Apollo. True, Americans still wanted to believe in a future in space. And civilian space endeavors in the hands of a Kennedy (or a Reagan) can be marvelous for presidential leadership. But by April 1977, when Press called me, I realized that Carter considered ''space'' to be out of fashion politically, too elitist and too distant from what he considered to be mainstream social issues. He utterly failed to recognize then (or at the Voyager award ceremony in the Cabinet Room three years later) that an effective President will lead by manipulating his two great tools— verbal and visual symbols. Without presidential endorsement, NASA is doomed to make the best deals it can with Congress, the OMB, and special interests. Budget decisions become policy decisions. Mediocrity is the likely outcome.

Thus, it was easy to say no to Press in April 1977. The job could not be done properly without Carter's support. Not only that—if I ''ascended'' to Washington to head NASA, I would probably do so at the cost of my scientific career, leaving myself doomed to a purgatory of administration for the rest of my working life. There would be no going back to personal intellectual pursuits afterward. A high price to pay for four years of frustration.

That was the easy part. It was harder for me to grasp, in 1977, how confused the key NASA officials really were about the Shuttle. Surely, the agency that launched Apollo could see the risks in its Shuttle-only policy. In fact, though, nobody at the top foresaw the magnitude of the Shuttle-focused disaster looming for space exploration, for NASA, and for the United States. In turning Press down, I thus missed the deeper question: Would I—could I—as NASA administrator have gone against the grain of the establishment and focused Carter's attention on the unfolding Shuttle tragedy? Could I have mitigated the outcome, albeit at the cost of professional standing and relationships? Nine years later, on January 28, 1986, that question would well up uninvited in my thoughts, as a numbing silence settled over the ocean off Cape Canaveral.

Carter's November 1980 phone call quickly faded into a new flurry of activity at JPL, where I was trying to capitalize on a wave of public interest in *Voyager 1*'s Saturn encounter. What is more, after *Voyager 1*'s cornucopia of scientific returns from Saturan and Titan, the ''Uranus option'' became the prime objective for *Voyager 2*. Nine years after the cancellation of the Grand Tour scheme, America would once again strive to ride that cosmic wave out from Saturn toward Uranus and even to Neptune. We were still headed outward, to explore new worlds.

By August 1981, *Voyager 2,* though still one-eared and tone-deaf, was

JPL ENCOUNTERS ED MEESE • The presidential counselor Edwin Meese (*right*) pays a flying visit to JPL during *Voyager 2*'s encounter with Saturn in August 1981. The author, gesturing with a clenched fist while extolling the importance of U.S. leadership in space exploration, is observed by NASA Administrator James Beggs (in light suit), while the late Jules Bergman, an ABC News correspondent, holds a microphone in the middle of the group. *NASA / JPL photo*

behaving more or less normally. It was rapidly negotiating the final miles to Saturn. The Shuttle had finally flown in April 1981, once again sparking presidential leadership and space symbolism. Space was back in fashion. As a consequence, *Voyager 2*'s Saturn encounter drew a bigger crowd. The presidential councillor Edwin Meese, key to Reagan's new federal priorities, helicoptered into JPL for a red-carpet, whirlwind tour just before *Voyager 2*'s Saturn encounter. Reagan's new NASA administrator, Jim Beggs, arrived hours later to reestablish NASA's symbolic priority for unmanned space exploration.

For a brief but beautiful moment, one could still fantasize that *Voyager 2* at Saturn might mark a new beginning to the American adventure in space.

11.1 URANUS'S MOON MIRANDA • This small, icy moon, 185 miles (300 kilometers) in diameter, was discovered only in 1948. Just thirty-eight years later *Voyager 2* returned richly detailed images, revealing a perplexing surface unlike any other known in the Solar System. *NASA/JPL photo*

11. Voyager Voyages On, to Uranus and Even Neptune

NINE MONTHS BEFORE *Voyager 2* reached Saturn, members of its brain trust were already plotting how to take the next billion-mile step from Saturn to Uranus. Uranus is twice as far from Earth and from the Sun as Saturn is. If Saturn stretched *Voyager 2*'s communications capability, would Uranus break it? Saturn images could not be radioed to Earth in real time at a rate of 115,000 bits per second, as had been possible at Jupiter. Despite some ground enhancements, 44,800 bits per second was the limit from Saturn.

But the Voyager imaging team had been looking toward Saturn for nearly a decade. With its upgrading of *Mariner 10*'s camera design for Mercury, a new electronic means made it possible to read out the image signal three times more slowly when desired. In this "slow scan" mode, the lower radio transmission rate of 44,800 bits per second from Saturn was adequate for real-time communications because the video signals could still flow directly from the camera to the radio transmitter and on to an attentive Earth.

Once again, high-resolution optics plus real-time TV transmissions had opened the door to scientific riches. As a result, *Voyagers 1* and *2* each returned 15,000 TV images from Saturn. By comparison, if we had still been using the TV communications system on board *Mariner 4* at Mars in 1965, only a single image could have been returned—and that image would have taken a full year to transmit. This represents a technical gain of more than a factor of 15,000 in fifteen years.

But how to get thousands of images back from distant Uranus? First, increase the size of the radio "ears" listening to Voyager's faint voice in California, Spain, and, especially, Australia. (The best viewing of Uranus at that time was from the Southern Hemisphere.) So the great 210-foot dish in the Mojave Desert was electronically linked with two smaller, adjacent dishes. That helped a little. The trick was repeated in Spain and in Canberra, Australia, where identical Deep Space Net (DSN) dishes constitute the other vital legs of the DSN "tripod." In Canberra, in addition, the DSN dishes were linked by a microwave connection to the 210-foot radio astronomy dish located 125 miles away at Parks, Australia. Thus, one of the largest arrays of radio dishes ever used stood ready to amplify the faint radio broadcast from Uranus.

Despite such heroic ground engineering, *Voyager 2*'s signals from Uranus could flow back to Earth only at a meager 21,600 bits per second. This would be just slightly better than our best with *Mariners 6* and *7* to Mars back in 1969, and well below Voyager's 44,800 bits per second at Saturn. We needed more magic to unlock the secrets of Uranus.

That magic for Uranus blossomed out of the past. In 1972, when the Mariner / Jupiter / Saturn (MJS) project arose from the ashes of the canceled Grand Tour mission, $7 million was added to enhance MJS's ability to meet its scientific goals at Jupiter and Saturn. Among the enhancements brought by that farsighted investment was the ability to reprogram the redundant on-board computers. Years later, as the Uranus challenge loomed, team members exploited an early idea about how to use that computer redundancy for improved communications. Why not employ the memory and processor of the backup reprogammable computer to "compress" the TV signals on board Voyager enough to fit the drastically reduced communications rate to Earth. (TV signals can be analyzed and coded by a properly programmed digital computer in a way that reduces the required speed of transmission. This is called compression.) Of course, if a problem arises with the primary computer on board Voyager at Uranus and if the redundant backup is tied up—well, let's not worry about that.

A special computer code soon enabled *Voyager 2* to send back mainly the *differences* in light intensity from adjacent picture elements (pixels) on each line of each image (rather than each full light-intensity value). This technique reduced the communications rate needed by a factor of two and a half.

And then still another trick. That $7 million package of enhancements included the electronic means of transmitting error-free data to Earth. Reed-Solomon coding, as it is termed, in honor of its inventors, is a complex but efficient system. However, only the basic coding hardware in Voyager's brain had been fabricated. More work on the ground and at the spacecraft was necessary if that potential was to be realized. For Uranus a special decoding machine was quickly created at JPL, tested, and made to function perfectly. The necessary computer instructions for Voyager's brain were checked and then radioed up. The whole system was rehearsed repeatedly during the quiet time when the spacecraft was coasting from Saturn to Uranus.

So the communications gap between Saturn and Uranus was closed. We had been lucky, but in 1972 NASA and JPL had also been farsighted. Luck is the residue of good planning. And luck was needed because this trick would not have worked with *Voyager 1*, whose backup computer processor had long since failed. Thus, the TV compression trick for Uranus could not have worked if it had been *Voyager 1* going to Uranus instead of *Voyager 2*. This was another lucky, flip-of-the-coin situation—as with that chance assignment of the low-performing Titan / Centaur, also to *Voyager 1*.

Stretching the communications capacity so elegantly from Saturn to Uranus solved half the picture-taking problem. The Sun at Uranus is only one-quarter as bright as at Saturn and provides less than one four-hundredth of its earthly illumination. Television exposures must be correspondingly longer, just as a camera's shutter speed must be reduced for a dimly lit subject on Earth. *Voyager 2* at Saturn had already provided a marvelously stable platform for long-exposure pictures. There it rocked back and forth between thruster firings very slowly, fourteen times more slowly than the hour-hand movement on a clock. To a hypothetical human eye *Voyager 2* would have seemed perfectly motionless. But taking pictures in the faint Uranus illumination is so challenging that the craft's motion must be four times steadier yet.

Fortunately, engineers had been analyzing the *Voyager 2* thruster performance ever since launch. They suspected that even tinier bursts of its stabilizing hydrazine gas were possible, even though such minuscule firings were not designated in the original design. JPL still had some thrusters left over from *Voyager 2*'s development phase. Engineers rigged them up in the now rarely used vacuum test chamber and found that microbursts of hydrazine were indeed possible.

The engineers carefully devised, tested, and radioed to *Voyager 2* new settings for the attitude-control computer. The neurotic robot that had so unnerved us in August of 1977 now displayed unprecedented sophistication and flexibility in adapting to a mode of operation not even contemplated in its design. *Voyager 2*'s drift rate settled down to one that was sixty times slower than that of a clock's hour hand. At last, *Voyager 2*'s performance at Uranus could approach its return of photographs and data from Saturn.

Yet another challenge loomed ahead at Uranus. *Voyager 2*'s trajectory skimmed only 18,000 miles (29,000 kilometers), just twice Earth's diameter, above the surface of the moon Miranda. Spectacular viewing would be possible by means of Voyager's high-magnification camera. But during such long exposures (two to three seconds), *Voyager 2* would be traveling at 46,000 miles per hour. (See figure 11.2.)

In two seconds the spacecraft would move some twenty miles relative to Miranda, and the image would be badly smeared. The solution to this problem was to pan the camera to keep it pointed precisely at Miranda's surface features. However, to rotate the spacecraft in a panning motion would require programming special gryo-controlled turns in all three of the spacecraft's axes of rotation to compensate precisely for its speedy movement. Furthermore, such robotic panning would require driving the gyros near their maximum rate—hence, more risk, more complexity, and more programming of the attitude-control system. Above all, it would require exquisite timing in the issuing of commands. Light-hours distant from Earth, Voyager must pirouette perfectly on time, far more precisely than any ballerina.

On January 26, 1986, it all worked. Voyager's human and robotic intelligence could display the baffling surface of Miranda fifty thousand times better than the best Earth-based telescope could. Exquisite, if unfamiliar, topographic details emerged from the monitors in the crowded imaging-team room.

Miranda is perplexing. It looks like a planetary conglomerate, a hodge-podge of bizarre fragments from geologically diverse terrains of older moons, long since destroyed in unrecorded catastrophes. Tiny Miranda may in fact be a Rosetta Stone for deciphering the violent history of Uranus, which culminated in the planet's tipped orientation, making Uranus the dynamical black sheep of the planets.

Whereas the Grand Tour mission had initially stressed narrow scientific goals, its Spartan descendant Voyager sang at Uranus a hymn in praise of exploration. "Seek and ye shall find" found startling vindication, hidden behind a tiny photographic smudge two billion miles from home port.

Miranda was only one of many Voyager surprises. Uranus has a magnetic field tipped 60° off the rotation axis of the planet. This is like discovering that Earth's compasses point toward Pasadena, California, rather than toward a point in the high Arctic near Earth's rotational axis. Furthermore, Uranus' atmosphere literally glows with invisible ultraviolet light, energized by electrical particles captured from the solar wind or by strong ionospheric currents racing through its stratosphere.

The rings of Uranus were also a dynamical and puzzling delight. Jupiter's unsuspected ring had invalidated earlier scientific assumptions; Saturn's gravitational fugue of hundreds of rings, bands, and gaps revealed an unimagined diversity of ring phenomena; and now Uranus added its dimension to the saga of planetary rings.

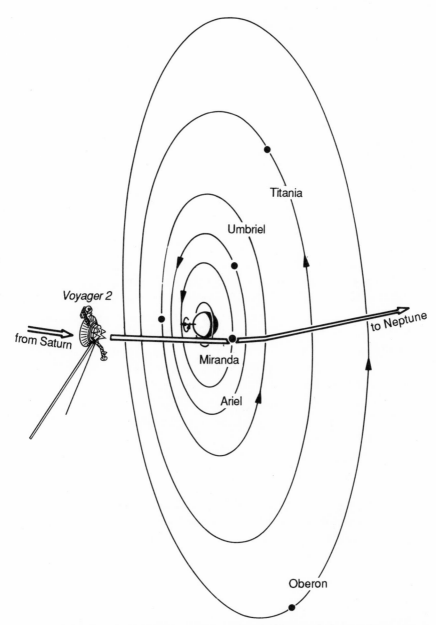

11.2 *Voyager 2* AT URANUS • Uranus differs from all the other planets in that it rotates about an axis pointing nearly in the plane of Uranus's orbit about the Sun, rather than approximately normal to that plane, as do the other planets. Because of this unusual geometry, the orbits of the small moons that lie in the equatorial plane of Uranus presented a bulls-eye appearance to *Voyager 2*. Only one moon could therefore be observed close up—Miranda. To get a sense of the scale of this diagram, note that Oberon, whose orbit is the most distant shown, is about 360,000 miles (583,000 kilometers) from Uranus.

Even as the Voyagers left Florida in 1977, startled astronomers were observing peculiar interruptions of light from stars that passed behind Uranus's disk. Gradually, it became clear that nine very thin, widely separated rings surrounded blue-green Uranus. It was the signature of their occultation of the starlight that astronomers observed on Earth. Caltech's Peter Goldreich and S. D. Tremaine theorized in 1979 that such rings were guarded on either side by "shepherd satellites." In that way they attempted to explain the unexpected narrowness of the rings. One of Voyager's tasks was to look for those Uranian satellite shepherds.

Voyager 2's camera (and another instrument, which repeated the stellar occultation from a close vantage point) revealed two new rings and fine details of the previously discovered rings. Ten new satellites also turned up, but only two shepherds were among them. Remaining hypothetical shepherds may be too small and too dark to have been detected with Voyager's cameras, or perhaps another theory will better explain the facts that *Voyager 2* revealed. Amazingly, some of the rings that encircle Uranus are not as wide as a football field yet are very, very long—as long as 100,000 miles (160,000 kilometers), or four times Earth's circumference.

Voyager 2 had plucked Uranus from the firmament of stars to reveal a miniature solar system full of surprises. On the day after encounter with Uranus—and on the day on which *Challenger* exploded—the *New York Times,* in an editorial entitled "On to Neptune," remarked,

Voyager 2 shows space exploration at its best. The imaginative venture proves that human intelligence can rove the planetary system and fruitfully explore new worlds....

No one can fail to be stirred by Voyager's grand tour of the outer planets. No one, that is, except NASA, which continues to devote the bulk of its $8 billion budget to human sojourns in space, squeezing out funds for unmanned missions. As Voyager 2 approached Uranus, NASA was bestowing a shuttle ride on the representative whose House committee authorizes its budget. As Voyager sped past Uranus, NASA was trying to polish the apple by putting the first schoolteacher into space.

If NASA wants lasting public support for a vigorous space program, the wonder of seeing new worlds will do it a lot more good than soap opera elevated to Earth orbit.

The Voyager team, aging but rejuvenated by its achievement at Uranus, had now earned one last opportunity to write history. Slipping easily through the tiny gravity gateway beside Uranus, *Voyager 2* began its long coast toward an August 1989 rendezvous with Uranus's sibling Neptune. If robotic infirmities could be forestalled for three and a half more years, humankind might at last meet Earth's most distant planetary relative.[2]

To capture that planetary portrait, human brain power and imagination must overcome new challenges of distance and darkness. Neptune is three times farther away from Earth than Saturn is—a staggering three billion miles. Its atmosphere, rings, and moons are lit by sunlight barely one-thousandth as bright as that at Earth. Even more there than at Uranus, all pictures must be time exposures ranging from seconds to minutes in duration. Neptune has its atmosphere-shrouded moon, Triton, just as Saturn has its Titan. And like Titan (and Earth), Triton may have a thick atmosphere composed of nitrogen gas. In some indirect way we may come to better understand our own atmospheric and chemical origins through *Voyager 2*'s exploration of Triton, one of only three nitrogen-covered worlds in the Solar System.

The big engineering payoff this time will be on the ground. In a beautifully lonely and radio-quiet valley near Socorro, New Mexico, stand twenty-seven identical radio telescopes, each twenty-seven yards (twenty-five meters) in diameter. This very large array (VLA), as it is called, was still only a scientific dream when *Voyager 2* burst anxiously into life in 1977. Since then, using ever more exquisite computer processing of the twenty-seven individual signals received simultaneously (from dishes all pointed at exactly the same cosmic target), the world's highest-resolution radio imager grew to maturity. While Voyager zoomed in on the spectacles provided by Jupiter, Saturn, and their unearthly moons, radio astonomers peered through the VLA ever deeper into the science fiction world of black holes and exploding galaxies.

In August 1989 *Voyager 2* and VLA will merge. A microwave link will briefly unite the big DSN antenna at Goldstone, California, and its two smaller companions with the twenty-seven-dish VLA in New Mexico. All thirty dishes will listen in synchronism to *Voyager 2*'s faint, wailing voice as it whispers back the secrets of Neptune. For one, marvelous moment, just as the last decade of this careening century begins, an invisible resonance of precisely tuned radio waves will vibrate across three billion miles. This sound box of the Solar System should provide the greatest live video of our century. Millions of Earthlings will gaze briefly from *Voyager 2*'s bow as it sails majestically past the last significant landfall between here and the stars.

That view of Neptune will take place barely three decades after Sputnik belligerently opened the space age, and little more than three centuries after Galileo's telescope punctured traditional cosmology. In 1989 *Voyager 2* will grasp what was at the very limit of intellectual reach a mere twenty-five years before when the Grand Tour idea flickered tentatively in a human mind.

If ever modern Adam reaches out in search of "God's great hand," substituting brain power for rocket power, this will be it.

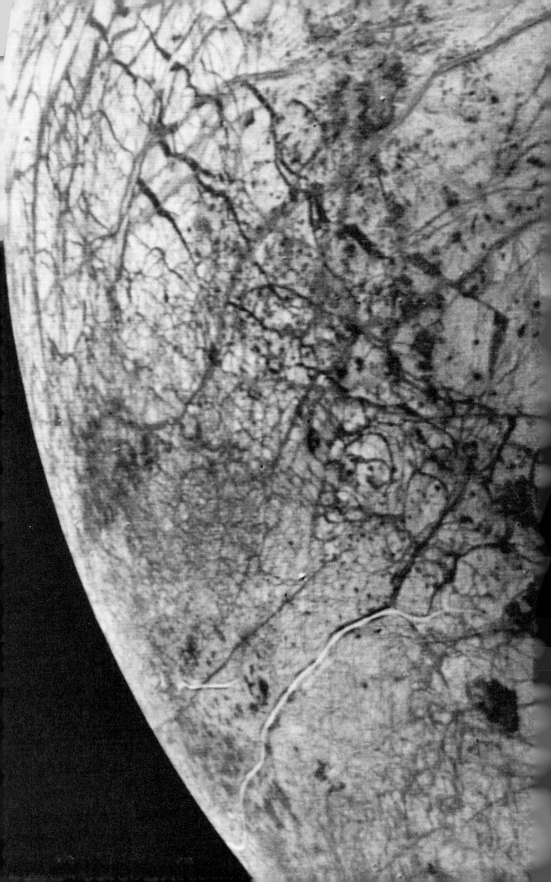

PART FOUR

Lost
in
Space

12.1 JUPITER'S MOON EUROPA • *Voyager 2* provided the best images of the second Galilean moon of Jupiter, from a range of 150,000 miles (240,000 kilometers). Europa's bright, icy surface is almost devoid of impact-crater scars, suggesting that it is being actively resurfaced by some internal process. The thin dark markings may indicate the location of fractures in its icy crust, beneath which a liquid water reservoir is believed to lie buried. Europa's surface is the smoothest yet encountered in the Solar System, with relief less than 300 feet (100 meters). The Galileo orbiter was designed to expand greatly our knowledge of Europa and the other Galilean satellites as well as of Jupiter. *NASA/JPL photo*

12.2

Galileo Jupiter Mission Delays

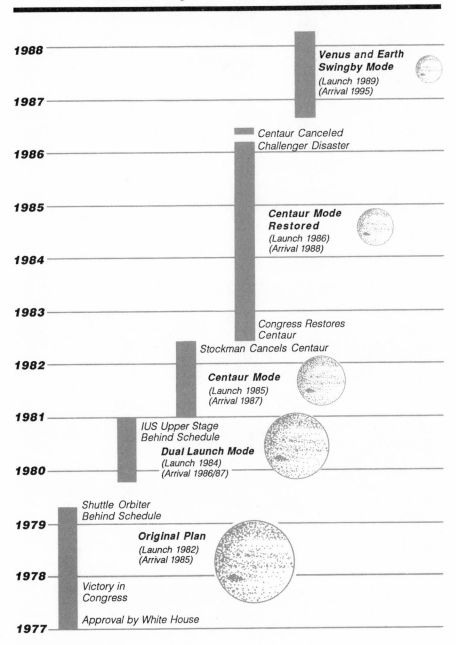

1988

Venus and Earth
Swingby Mode
(Launch 1989)
(Arrival 1995)

1987

Centaur Canceled
Challenger Disaster

1986

1985

Centaur Mode
Restored
(Launch 1986)
(Arrival 1988)

1984

1983

Congress Restores
Centaur
Stockman Cancels Centaur

1982

Centaur Mode
(Launch 1985)
(Arrival 1987)

1981

IUS Upper Stage
Behind Schedule

Dual Launch Mode
(Launch 1984)
(Arrival 1986/87)

1980

Shuttle Orbiter
Behind Schedule

1979

Original Plan
(Launch 1982)
(Arrival 1985)

1978

Victory in
Congress

Approval by White House

1977

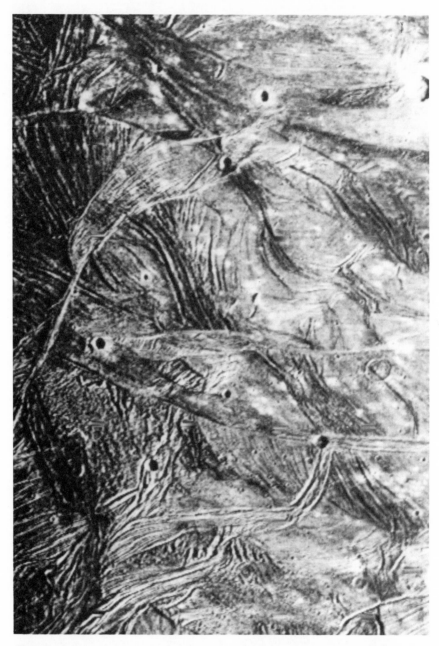

12.3 THE COMPLEX SURFACE OF GANYMEDE • Ganymede's substantial surface (larger than Mercury's) includes relics of many past episodes of internal activity. In this *Voyager 1* image, from a range of 85,000 miles (135,000 kilometers), the very low solar illumination highlights a complex pattern of grooves 3 to 9 miles across (5 to 15 kilometers) and 100 miles (160 kilometers) and more in length. Large, well-preserved impact craters suggest that this surface is billions of years old. The Galileo orbiter mission includes a comprehensive exploration of this strange, planet-sized body circling Jupiter. *NASA / JPL photo*

12. A Triumphant Beginning

MY FIRST YEAR as a university professor turned laboratory director was crammed with new learning experiences.

First came a terrorist's bomb threat, telephoned into JPL just before the July 4, 1976, bicentennial. We evacuated the five thousand employees of the laboratory. No bomb was found, and every one went back to work, a bit squeamish.

Next came the discovery, a few months later, of an unsuspected bedrock outcropping on the lab grounds. Such an outcropping suggested that an earthquake fault ran directly under major JPL buildings. Core drillings and other careful investigations later dispelled initial concern that the fault was active, but a major program of seismic strengthening of JPL's buildings went ahead anyway.

There followed growing agitation for increased representation by minorities and women in the JPL professional staff. We managed to make gradual improvements in both representation and management attitudes.

Topping all of this was the sudden departure in December 1976 of my Caltech boss, Harold Brown, to become President Carter's secretary of defense. I sorely missed his expertise and clout in dealing with the Washington establishment.

There were successes, too. The Viking search for life on Mars in the summer of 1976 attracted the greatest media coverage for a space endeavor since the Apollo landings on the Moon. This fusing of Viking, the bicentennial celebration, and a presidential election year created an ideal environment for drumming up nationwide support for bold new American projects of planetary exploration.

Those purple times offered our best chance to start on a new mission to the planets, now long overdue. That logical next step was an ambitious follow-up to the Pioneer and Voyager flybys, a Jupiter orbiter, just as the *Mariner 9* orbiter brilliantly followed the earlier Mariner flybys. But our proposal offered even more. It combined a powerful orbiter with an atmospheric entry probe (like *Venera 4* at Venus), which would be hurled directly into the swirling, colored atmosphere of the Sun's largest planet—thus its name, Jupiter Orbiter with Probe (JOP).

JPL promoted it in 1976 because America's premier space scientist, James Van Allen, had been a strong and early champion of a Jupiter orbiter mission. After he finished the original Grand Tour studies at JPL in 1971 (see chapter 8), Van Allen played a key role in the *Pioneer 10* and *11* flybys of Jupiter, which captured the first close-up measurements of that giant planet in 1973 and 1974. Next, he worked closely with the staff of the NASA's Ames Research Center at Moffet Field, in Mountain View, California, to advocate a follow-up mission to Jupiter. Another spinning spacecraft, like *Pioneers 10* and *11,* would be placed into a long, looping orbit about Jupiter (rather than simply flying by it). Once there, it would carefully measure the electrical and magnetic fields and the energetic charged particles surrounding Jupiter that had first been probed directly by the *Pioneer 10* and *11* flybys.

Van Allen and the Ames staff saw additional possibilities. Ames was NASA's leading center for the development of heat shields needed to protect payloads during their blazing entries from space into Earth's and other planets' atmospheres. Indeed, by the time *Pioneers 10* and *11* were sampling Jupiter's environment, Ames was already hard at work on a new heat shield intended to shepherd instruments deep into Venus's atmosphere in 1978. Ames had a natural interest in extending its technical capacities to the more challenging task of safely shielding a rapidly entering instrumented probe into Jupiter's atmosphere. The entry velocity at Jupiter is four times greater than that at Venus.

Ultimately, the concept of a spinning orbiter with probe became Pioneer Jupiter Orbiter with Probe (PJOP). In early 1975 a cooperative study between NASA and the (then) European Space Research Organization endorsed PJOP as a relatively low-cost spinning spacecraft to follow up the Pioneer and Voyager flybys of Jupiter. PJOP received solid backing not only from Van Allen and his space-physics colleagues but also from atmospheric physicists eager to get direct measurements of the temperature, pressure, and composition of Jupiter's upper atmosphere.

But even as Van Allen and other scientists polished up a highly focused mission concept that emphasized "science for science's sake" and complemented the broader JPL exploratory missions like Voyager, NASA headquarters finally acknowledged that America's planetary efforts had shrunk so much that most of them should now be concentrated at a single NASA center—JPL. Accordingly, NASA's Langley Research Center, then overseeing the Viking Mars mission, announced it would abandon planetary missions after completing the Mars landing. In the fall of 1975 the Ames Research Center was directed to limit its activities to such special items as the Jupiter entry probe. It had to hand over to JPL the overall development and mission management for the Jupiter orbiter mission, spinning spacecraft or not.

An ominous note was also sounded in late 1975. NASA decreed that the Jupiter orbiter mission would be the first planetary mission to use the Shuttle, in late 1981 or early 1982.

The transfer of PJOP from Ames to JPL became final in 1976 shortly before I, full of enthusiasm and new plans, became director of JPL.

But the PJOP inherited by JPL could hardly serve as a worthy successor to Viking and Voyager. The PJOP approach had virtually no remote-sensing capability—only a simple, very low resolution spinning camera like those on the earlier Pioneers that flew by Jupiter. Van Allen's spinning orbiter would map out superbly the three-dimensional patterns of magnetic and electrical fields surrounding Jupiter, but it would be virtually blind. Neither the spinning orbiter nor the entry probe could show how the planet looked in the visible, infrared, and ultraviolet—much less tell us anything about Jupiter's mysterious moons, which were, in 1976, still only smudges on photographic plates. PJOP was intended to be a narrowly focused enterprise of planetary *science,* not an ambitious voyage of planetary *exploration* that could interest the general public as well.

To the JPL staff, then involved in the final testing of the sophisticated Voyager spacecraft, and to some of NASA's scientific advisers, the question inevitably arose, Why not change PJOP to a three-axis stabilized spacecraft like Voyager, with its rock-steady imaging platform? In fact, why not fly a Voyager-type spacecraft with entry probe, as in the Jupiter orbiter mission?

In addition, a powerful new idea concerning ''free'' navigation about Jupiter had been brewing at JPL. Close traverses of Jupiter's four big moons could provide unlimited gravity-assisted travel throughout the Jovian system, *if* precisely targeted close flybys could be carried out. (See figure 12.4.) ''Voyager with Probe,'' flying exquisitely designed trajectories from moon to moon, promised a mission worthy to be the centerpiece of a renewed American effort in deep space.

But Van Allen and other space physicists felt strongly that a spinning

spacecraft was superior for measuring the electrical and magnetic properties of space. They did not want to abandon the spinning-spacecraft approach
they had been working on for so many years. And NASA could not easily
ignore Van Allen's opinions, especially regarding a mission concept that
he had pioneered.

To JPL, Van Allen's support for a Jupiter orbiter mission carried a high
price. But creative engineers can develop technical solutions even for some
''political'' problems. Why not build a new kind of planetary spacecraft—
one part *stabilized* and one part *spinning?* This seductive suggestion was
welcomed (with more hope than objectivity) by all concerned, including

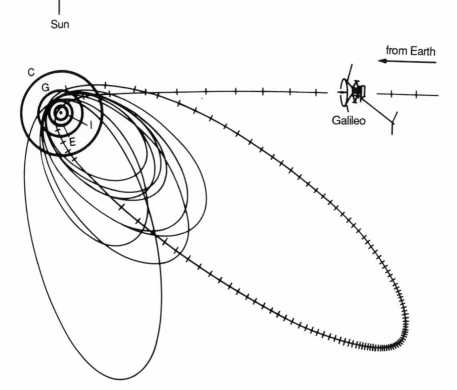

12.4 CELESTIAL SKATEBOARDING • The Galileo orbiter was designed to be lofted from a
near-miss of one big moon of Jupiter to another, in the ultimate application of the "gravity
swing-by" technique. In the diagram, Galileo arrives from the right, fires its braking rocket, and
is captured by Jupiter in an elongated orbit. Tic marks delineate day intervals of motion along
this orbit. Subsequent close passages of the moons perturb Galileo's path in ten more orbits,
permitting eleven close encounters—flybys—of the large moons.

me. Here was a revolutionary idea that would let each of the deadlocked scientific groups have what it wanted.*

NASA headquarters liked the new JPL concept, which now included eleven close flybys with Jupiter's four big moons. In addition, the actual entry point of the probe would be well monitored by time-lapse photography from the orbiter both before and after the instrumented probe's blazing descent into Jupiter's atmosphere. So was born the Jupiter Orbiter with Probe (JOP) concept.

In August 1976, NASA Administrator James Fletcher submitted JOP as a "new start" for the congressional budget cycle of spring 1977. JOP was then endorsed by President Gerald Ford's Office of Management and Budget in the months just before his Nixon-burdened presidential campaign lost to smiling Jimmy Carter's in November 1976.† Carter's OMB quickly reapproved JOP, however, and the proposed project sailed through the House and Senate authorizing committees in the early spring of 1977.

By April 1977 I was feeling pretty good as I reflected on my first year as director of JPL. Popular support for planetary exploration had been highlighted by the Viking mission to Mars, and bipartisan political support seemed to flow from both the White House and Congress. JOP promised to be a high-yield mission for scientists and citizens alike. Perhaps approval of JOP would propel us toward more ambitious adventures—a rendezvous with Halley's comet, a radar portrait of Venus, or even a return to the surface of Mars.

But on Wednesday, May 4, 1977, the Jupiter Orbiter with Probe mission was suddenly ambushed in Congress. Without warning, all JOP funds

---

*I do not think that this impasse and the expensive and weighty compromise that resulted would have developed if NASA had maintained an orderly planetary program in the early 1970s. A Jupiter orbiter mission was obviously the next step in outer-planet exploration after the Voyager flybys. Had NASA in, say, 1974 proposed to the scientific community a "Voyager in Orbit" as a next step for the early 1980s, there would have been, I believe, a positive response, especially if the concept of an atmospheric-entry probe had been added. An orderly and efficient evolution to "Voyager Orbiter with Probe" would probably have ensued, and Van Allen and the Ames staff would have been involved early in the mission design.

Instead, NASA headquarters continued too long to promote rivalry between Ames and JPL and their associated scientific advisers. Scientific and cultural differences were too deeply ingrained for a wise and thoughtful compromise. The dual-spin "compromise" spacecraft, in fact, required substantially more mass than had originally been envisioned, and that in turn gravely burdened the mission when the Shuttle and upper-stage performance failed to materialize.

†Ford's OMB actually approved two new starts for space science in 1976—JOP and the Space Telescope, as well as $10 million for future Mars studies. Had Ford beaten Carter, leadership for space exploration would perhaps have returned to the White House.

were deleted by the House appropriations subcommittee that controlled NASA's funds. This subcommittee also oversaw other independent agencies like the National Science Foundation, the Environmental Protection Agency, the Veterans Administration, and the Department of Housing and Urban Development. Its chairman, the austere Edward P. Boland, hailed from Massachusetts, as did his closest confidant, Tip O'Neill, Speaker of the House. Boland also chaired the House Intelligence Committee.* Therein lay a problem, for Boland was repeatedly assured by intelligence officials that the Space Shuttle was essential for national security. "I have no problem with the Space Shuttle program," he said. "That was the direction in which the Nixon Administration indicated we ought to go, and the *Apollo* program was the direction in which the Kennedy Administration felt we ought to go." Boland supported NASA's Space Shuttle program, even while terming it NASA's "sacred cow" and criticizing NASA's failure to allow for inflation in Shuttle cost estimating.

The flip side of Boland's staunch, if acerbic, support for the Shuttle was his insistence that NASA prioritize its other missions, especially its space astronomy program, into which he lumped all planetary exploration. In 1975 he had temporarily killed the Pioneer / Venus mission. But when the Senate restored those funds, Boland backed off Pioneer / Venus in the ensuing House-Senate conference committee. (Pioneer / Venus was launched on two Atlas / Centaur rockets in 1978.) In 1976 Boland escalated his "prioritizing" campaign and came down hard on his real target—the proposed Space Telescope (ST). He correctly surmised that the Space Telescope, like the Shuttle, would cost far more than NASA had advertised.

But Boland had not reckoned with the avalanche of national support that developed for the Space Telescope. The National Academy of Sciences had in fact given it its highest endorsement. NASA itself had a strong interest in the Space Telescope, the principal scientific use of the Shuttle in the 1980s. Boland's 1976 assault, however, pushed the Space Telescope one year further back in NASA's queue of proposed new projects—to 1977. JOP also had to be approved in 1977, in order for it to be launched during the especially favorable opportunity in December 1981 and January 1982.†

NASA could not focus on everything. In the spring of 1977 it worried

---

*It was the basis of his later role in the "Boland amendment," which set limits to the President's power to bypass Congress in carrying out secret wars.

†To help JOP's performance a gravity assist by Mars was added officially by JPL in November 1977 to make up for the unexpected growth in mass entailed in the dual-spin spacecraft approach. The key Mars / Jupiter alignment was very favorable in 1982, only partly so in 1984, and not at all after that.

about the fate of the Space Telescope, not about JOP, because a further challenge to the Space Telescope loomed in the form of Senator William Proxmire. Long the scourge of NASA,* Proxmire now chaired the Senate subcommittee that voted on appropriations for NASA. His open skepticism created for NASA the nightmare of a Proxmire-Boland alliance to kill the Space Telescope.

But Boland, faced with a determined NASA and with strong National Academy of Sciences support for the Space Telescope, shrewdly shifted at the last minute and struck at JOP. He knew full well that NASA's overall top priority was the Shuttle and that the primary commitment of NASA's Office of Space Science was to the Space Telescope. Thus, in chopping JOP, he could at most expect opposition from the third and fourth layers at NASA headquarters. Of the seven large NASA centers, only two, JPL and the Ames Research Center, were involved. JOP was an easy target. Boland announced,

> But not every project that the scientific community wants can have first priority, and that is why we made the budget priority choice—to provide the Space Telescope—but we denied the Jupiter Orbiter Probe.
>
> Jupiter will be there five or ten or fifteen years from now when this project can be reinstated. In all the twenty years that this Nation has been involved in a major effort—in all those years—the Congress has never, never, made an attempt to deny funding for a major space mission. I think it's time that we did that.

This was no delicate political minuet like the one in 1970 (mentioned in chapter 5) when the Mariner mission to Venus and Mercury was held hostage in Congress by another committee chairman, who forced NASA to kick in more money for his favorite NASA program. No, this was a bare-knuckled power clash. At stake was a symbolic $20 million out of NASA's $4 billion total budget. Overall, Boland's subcommittee controlled $70 billion. Boland did not want compromise. He wanted complete victory—the demise of JOP and the demonstration of his committee's ability to shape "budget priorities."

Most of NASA, JPL quickly discovered, cared more about maintaining Boland's support for the Shuttle and the Space Telescope than about risking an open confrontation with him over JOP.

JOP and JPL suddenly seemed overmatched in a political struggle we did not expect or understand. Our only hope was to gain strong support

---

*Proxmire routinely intimidated governmental agencies with his "Golden Fleece" award. NASA and JPL later "won" his publicity-seeking award for their nascent efforts to start a program to search for extraterrestrial intelligence. As a result, NASA turned tail, and the program has suffered at least a ten-year delay. (See appendix.)

from Senator William Proxmire's Senate appropriations subcommittee. "Imagine *Proxmire* as a white knight," I mused.

But we had friends, who in turn had friends in the Senate. Charles Mathias of Maryland was a special stalwart and was also the ranking minority member of Proxmire's subcommittee. We quickly began to identify out other key players on that subcommittee. How could we get our story to them?

Tuesday, June 15, 1977, was a dramatic day. The full House supported Boland and voted to delete JOP. But Proxmire's subcommittee included JOP in its appropriations bill. It had heard our plea for help.

Four weeks later, at 3:45 P.M. on Tuesday, July 12, Chairman Boland and his chosen House members sat down across the conference table to face Senator Proxmire and his Senate colleagues. Their task was to resolve JOP and thirty-eight other individual differences in the appropriations bill between the House and the Senate.

Time passed slowly in Pasadena as we waited for word to leak out from behind those closed doors under the Capitol rotunda. We had asked our Senate allies to press for a compromise on the first-year, funding—say, $10 million instead of $20 million for JOP—anything, in fact, as long as the mission was authorized.

It was late, even on the West Coast, when we finally learned the result. The conference committee had resolved all but two of the thirty-nine differences. An emotion-laden amendment to the House bill concerning veterans' payments, which had been added from the floor, could not be resolved by the conference committee—nor could the committee reach an agreement on JOP.

"What happens next?" I asked JPL's congressional liaison in Washington over the telephone.

"It goes back to the full House for a special vote."

"When?"

"That's up to Speaker O'Neill and Chairman Boland. It could be just a week."

"You mean we have to try to reach all 435 members of the House and persuade them to openly reject Eddie Boland's position in less than a week?"

"I'm afraid so."

Now we were playing on Eddie Boland's home field, and it was not at all level. In desperation, we tried every political circuit we could reach, along with pleas to the customary supporters of space science like the presidential science adviser Frank Press.

Governor Jerry Brown, of California, had discovered JPL—and its captive press corps—during the Viking missions in 1976. He also queried me frequently on another topic of mutual interest, solar energy. Now he responded vigorously to my mid-July Mayday by making personal phone calls to most of the California congressional delegation. His political support supplemented our scattered contacts in southern California and those that the Ames Research Center (on the south side of San Francisco Bay) could discover in the north.

Many scientists who belong to the Division of Planetary Sciences of the American Astronomical Society correctly perceived Boland's JOP actions as a threat to planetary science. A telephone and letter network spontaneously sprang to life, building its own momentum and reaching some congressmen from home-district constituents. Carl Sagan, James Van Allen, and other prominent scientists spent days making personal visits to members of Congress.

Through good fortune, five thousand "Trekkies" arrived in Philadelphia for their yearly "Star Trek" convention just three days after the conference committee's bombshell. Gene Roddenberry, the originator of the seemingly immortal TV series "Star Trek," was to be their principal speaker. Roddenberry sounded the clarion call, practically exhorting his followers to show their commitment to space exploration by bombarding their congressmen directly with telegrams and phone calls on behalf of JOP.

Howard Simons of the *Washington Post* had covered *Mariner 4*'s dramatic first look at Mars way back in 1965. Now he was managing editor and sympathetic to our predicament. On Monday, July 18, the *Post* led off its editorial page with "On to Jupiter," an incisive analysis of the political realities:

What is involved here is the ability of a subcommittee chairman, who thinks this country is spending too much money on space programs, to veto a worthwhile, already-approved project simply because it is vulnerable. Jupiter Orbiter is a low-cost item, as space missions and most other government projects go. Its start-up costs, which are the ones in difficulty, are $20.8 million. Its established total cost is $280 million. Because it is small and because it lacks the drama of manned space missions, it has little political support. The big items in the $4 billion space budget, like the space shuttle, generate many jobs and much backing from members of Congress. Rep. Boland, who can't curtail the big projects because of that support, picked this one to slice out of next year's budget as a demonstration that the space crowd can't get everything it wants.

The editorial concluded with a crucial endorsement:

This nation ought to use its space capability to gain more basic knowledge of the universe as well as to engage in practical—and exciting—applications of earth-orbiting satellites. Congress accepted that view of our world some years ago, or so we had thought. If it now upholds one subcommittee's veto of this project, it will be turning its back on its own commitment to a space program that makes scientific sense.

To drive the point home to its readership, which included every staffer on the Hill and most members of Congress, the *Post* also ran a news article by its science reporter Tom O'Toole, a longtime admirer of JPL and space exploration. O'Toole detailed how the political goals of Boland were about to destroy a valuable and vulnerable American endeavor.

On Tuesday, July 19, 1977, the full House of Representatives took up the conference committee report. Boland quickly disposed of everything else and staked out his position on JOP. Member after member rose either to voice support for Boland and the deletion of JOP or to differ with him and urge support for JOP. Boland skillfully answered every objection and rejected all compromise.

Finally, the Speaker pro tem announced that time for discussions had expired. A voice vote indicated a stronger "nay" than "aye" vote on the JOP deletion. Boland then called for a quorum. Absent House members rushed to the floor, and the electronic voting machine tallied up the results. Boland lost. JOP won by a vote of 280 to 131.

Nothing like this had ever happened before, nor has it since. To me, this demonstrated the breadth and depth of American political support for planetary exploration. The drought in new efforts to explore the planets must finally be coming to an end. Surely, nothing could stop us now from returning to Jupiter.

After receiving congressional support, JOP was renamed Galileo on January 17, 1978. (This renaming, incidentally, helped remove political associations lingering from our bruising congressional battle.) Galileo was then officially scheduled for launch in January 1982, on the thirtieth planned Shuttle flight (see figure 12.5). But since it was a planetary mission, newly designed upper-stage rockets would have to be carried up in the Shuttle, released in orbit, and then be used to blast Galileo out of Earth orbit and toward distant Jupiter. Galileo would be almost the first user of this new upper stage.

It was April 1978, and the Shuttle's's first flight was only eighteen months away by NASA's schedule.

In boarding a late flight from Los Angeles to Washington, I bumped into John Yardley, NASA's top man for the Shuttle. Along with NASA's deputy administrator, Alan Lovelace, he was returning from a round of meetings with Rockwell International, the Los Angeles–based prime contractor for the whole Space Shuttle effort. Here was a perfect opportunity to air my growing unease over the Shuttle's progress with NASA's Shuttle boss.

"How sure are you about the Shuttle schedule, John?" I asked, polite but anxious. "Galileo has *got* to be launched in the Jupiter window of January 1982." Yardley jotted down a few key dates and smiled back at me tolerantly.

"Hell, Bruce," he said, "we have eighteen months of pad." *

I quietly reminded Yardley of the devastating threat that any Shuttle delay posed to America's *only* new planetary probe. The trusty Atlas / Centaurs, and their successors the Titan / Centaurs, had propelled America to world leadership in robotic exploration of the planets. Now they were museum pieces. NASA had decreed that all new planetary missions be launched with the Shuttle and the new upper stages. America's future in space, robotic and otherwise, had been tied completely to human-piloted Shuttle launches.

This fateful course had begun in January 1972, when NASA Administrator James Fletcher finally sold the Shuttle to President Richard Nixon as America's key to the future in space.

John Yardley, my companion on the airline flight, had been a pioneer in the early manned space flights leading up to Apollo. He was in industry in 1972, leading the McDonnell-Douglas Astronautics team that was competing to win the lucrative prime contract for Shuttle development. But NASA selected Rockwell International instead. Yardley soon left McDonnell-Douglas to join NASA, where he became Fletcher's top Shuttle official—and the Shuttle's supersalesman. John had a favorite analogy: he would emphasize that the planned Shuttle orbiter was rather like the DC-9 jet airliner in wingspan and like the DC-10 in height. Trusting listeners were left to conclude the Shuttle would be a passenger vehicle just like the DC-9 and DC-10.

On May 23, 1977, the newly elected President Jimmy Carter appointed Robert Frosch to replace James Fletcher as NASA administrator. (Fletcher had been appointed by Richard Nixon.) But John Yardley continued as NASA's Shuttle chief, more powerful than ever under the new and less

---

Pad in NASA jargon means "schedule cushion," the amount of extra, unscheduled time available to achieve a particular objective.

experienced NASA management. Indeed, Yardley continued to lead the Shuttle program all the way into President Reagan's administration.*

Back in the early 1970s, Fletcher, Yardley, and Deputy Administrator George Low† knew that the old hands of automated spacecraft, whether in science, commercial communications, or defense, would never voluntarily relinquish Atlas / Centaur and other unmanned launch vehicles. These proven rockets were too reliable. NASA's claims that a *manned* Shuttle could provide lower-cost and more-efficient launches for *unmanned* payloads were viewed very skeptically by those interested primarily in space achievements with automated spacecraft. So how could NASA create for the Shuttle a monopoly on U.S. access to space?

Yardley's Office of Space Transportation controlled NASA's launch vehicle systems. The Office of Space Science in NASA headquarters controlled funds for all missions of space science and planetary exploration. Since it reported to NASA's boss, Fletcher, it was easy, beginning in the mid-1970s, for Fletcher simply to *require* that all new NASA missions of this sort be designed for launch by the Shuttle.

Private companies developing communications satellites were, however, not under NASA control. They could not be forced to choose the Shuttle. So Fletcher and Yardley offered to commercial satellite builders and users a government subsidy in the form of guaranteed low prices for

---

*After the long-delayed first Shuttle flights, in 1981 (three years behind schedule, over budget, and far below design performance), Yardley retired from NASA with honors. Yardley received a resounding endorsement by his peers for his Shuttle leadership, being elected to the prestigious National Academy of Engineering in 1982. That same year he rejoined his old company, McDonnell-Douglas Astronautics, as president.

†Low, an important Apollo manager, lent his credibility as a key technical leader in the early Shuttle days. He left NASA in 1977 to become president of Rensselaer Polytechnic Institute.

12.5 PORTRAIT OF DISASTER • This diagram plots each of NASA's published schedules for upcoming flights. The horizontal axis gives the cumulative number of launches (*Challenger*, shown by a star, was number 25, for example). The vertical axis denotes the year in which each projected flight was supposed to take place (for *Challenger*, that point is plotted opposite the end of January 1986). The date written above each dotted line is when NASA published that particular schedule (*Challenger* was on the October 1985 schedule, for example). The actual flight history of the Shuttle is shown by the solid line. NASA's chronically optimistic forecasting of the Shuttle's launch rate is apparent when one compares successive Shuttle flight schedules (dashed lines with dates) with the actual history (solid line). The *Challenger*'s twenty-fifth flight, in January 1986, took place three and a half years later than NASA had projected in July 1979. The August 1988 projection beginning with *Discovery* puts the fiftieth flight in mid-1991, four years later than the October 1985 projection had put it and eight years later than the July 1979 schedule had indicated. This diagram shows that even after nine years, NASA is still as far from the fiftieth scheduled Shuttle flight as was projected in 1979.

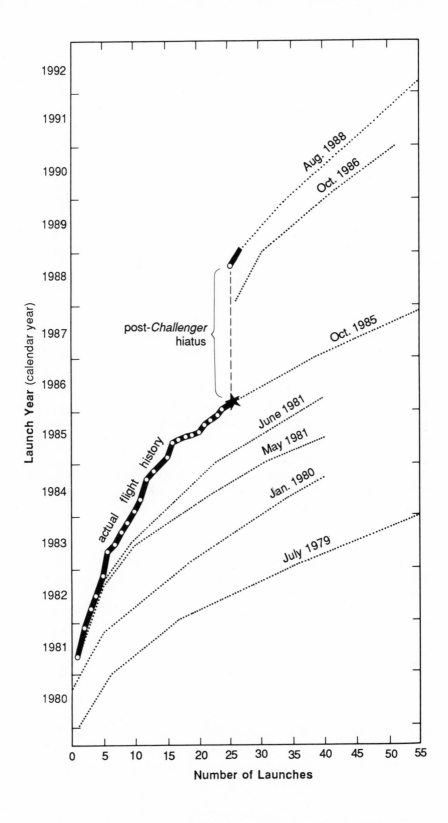

Launch Year (calendar year)

1992

1991

Aug. 1988

1990

Oct. 1986

1989

1988

post-*Challenger*
hiatus

Oct. 1985

1987

1986

June 1981

1985

May 1981

actual flight history

Jan. 1980

1984

1983

July 1979

1982

1981

1980

0   5   10   15   20   25   30   35   40   45   50   55

Number of Launches

future Shuttle launches. These prices were set low enough to discourage the birth and growth of an American private-sector industry for launching satellites with conventional unmanned rockets. Indeed, they were far lower than the likely cost to NASA for launching those commercial satellites on the Shuttle. Thus, big commercial communications satellite companies like Hughes, Ford Aerospace, and RCA, as well as satellite users like Comsat and Satellite Business Systems, had to be content with guaranteed and affordable NASA prices for future transportation into space by the Shuttle.

Many fretted in private, but rarely in public, about NASA's ability to deliver that service reliably and on schedule.

Into this commercial void stepped Europe. France and the European Space Agency shrewdly developed the Ariane expendable-rocket system, specifically tailored for efficient commercial use, employing much of the conventional rocket technology that the United States was abandoning.

NASA still had to deal with the Defense Department, which had its own launching capability and requirements. The DoD was (and is) the largest single user of American launch vehicles. How could the generals and the civilian defense officials be persuaded to abandon their existing fleet of Titans, Deltas, and Atlas / Centaurs? What would bring them to the Shuttle as *the* American launching vehicle? How could they be enticed to let national-security payloads become wholly dependent on the Shuttle? That sales task required even more massive NASA subsidies. All the billions of dollars to develop the Shuttle would come from NASA. And NASA's rates to the DoD for future launches, like those for commercial communications satellites, were set far lower than the probable costs to NASA for providing those services. The Air Force merely had to promise to build a decade later the West Coast base for the Shuttle and use it.

Furthermore, NASA was persistent, and very good at the long-term politics, as generals and defense secretaries came and went. It also helped to have good friends in the adversary's camp.

NASA's top technical leader of the early Apollo era, the distinguished Dr. Robert Seamans had in 1972 become Nixon's secretary of the Air Force. Despite Seamans's reported personal reservations, crucial support for the Shuttle from the DoD carried the day with a mercurial Richard Nixon.

But starting in 1977 and continuing through 1980, as scheduling problems worsened (see figure 12.5), the Shuttle found its greatest friend in the Pentagon. This was Dr. Hans Mark, the key Defense Department official for space and for the Shuttle. Mark became deputy secretary of the Air Force in 1977, with special space responsibilities, and then secretary in

1979. Mark, a Shuttle and Space Station messiah, had earlier directed NASA's Ames Research Center.

I had known Mark for many years. When visiting him in 1977 in his Pentagon office, I could not ignore the detailed model of the Shuttle that dominated his desk. The only difference between this model and those in the NASA offices was that Mark's was painted Air Force regulation *blue*. It was no secret that he advocated an Air Force–controlled Shuttle system. He later recounted with pride how he intervened with President Carter to save NASA's Shuttle program from the ravagings of its detractors. In his memoirs, [1] he describes how on November 14, 1979, he represented the DoD and USAF at a critical Shuttle meeting in the Cabinet Room of the White House. Carter, faced with strong and continuing support for the Shuttle by Defense Secretary Harold Brown, as well as by Mark, agreed to continue without change the delay-plagued Shuttle development. Carter's science adviser, Frank Press, was also present and approved, although he opposed additional new ventures like the Space Station.

However, Mark's memoirs indicate that a much more significant Shuttle meeting had taken place two years earlier, on December 16, 1977, in the office of the OMB's director, James McIntyre. McIntyre and the OMB were pushing to downgrade the Shuttle to an experimental program, with only three orbiters, operating exclusively from the Kennedy launch site. Such an approach would necessarily have kept expendable launch vehicles.

Events have demonstrated that this option would have been an inspired solution to the Shuttle problem. Development of a mixed fleet—and acknowledgment that the Shuttle was properly an R & D program—would have prevented the complete disaster that finally grounded America's efforts. Yet, according to Mark, Harold Brown, supported by Stansfield Turner, director of the CIA, deemed the OMB proposal unacceptable from the viewpoint of national security. Presumably, Brown and Turner were trying to protect the Shuttle's promised capability to launch extra-large reconnaissance payloads, from Vandenberg Air Force Base, at Lompoc, California.*

Brown's support (nourished tirelessly by Mark) for a Shuttle-based national launch program was decisive, because Carter aired significant

---

*Brown did include in his Shuttle position (personal communication, 1987) the procurement of long-lead-time parts for continuation of the Titan 3 line as backup. He thus provided launch vehicle insurance for the DoD for some years in the future. The initial success of the Shuttle, after Brown left office, made it more difficult for subsequent DoD officials similarly to maintain a backup capability. Hence Secretary of the Air Force Aldridge's intense battle for the Titan 4 in 1983–85 to provide full backup to the Shuttle in payload size and mass.

personal reservations regarding manned flight and because Vice-President Mondale had been one of the Shuttle's most vociferous critics in Congress.

Mark's successor in the Reagan administration, Edward ("Pete") Aldridge, had a far less cluttered agenda concerning national security. In 1985, after a bitter battle with NASA, Aldridge gained presidential permission to ignore the decade-old "Shuttle-only" national launch policy. He began to rebuild an independent launch capability for the Air Force. As a consequence of his foresight, industrial contractors were busy at work on an improved Titan-class launcher for the Air Force on January 28, 1986, when the *Challenger* accident blasted into oblivion the Shuttle-only launch policy of the United States.

But in 1978, when I buttonholed Yardley on the flight to Washington, NASA was in the saddle unchallenged. Yardley and Fletcher had woven a strong web of subsidy and political self-interest. Successive NASA administrators—Robert Frosch (under Carter) and James Beggs (under Reagan)—never considered it politically feasible to examine alternatives or modifications to that Shuttle-only policy, even with programmatic disaster looming.

As director of JPL, I was a member of NASA's senior management. In our quarterly meetings, and on many other occasions, we considered every facet of NASA's situation—unionization, congressional attitudes, major space projects, the works. And we spent a lot of time trying to perfect NASA's internal management processes. But we were *never* permitted by the administrator to discuss—much less challenge—the wisdom of the Shuttle-only policy. The only dialogue permitted about the Shuttle concerned how to make it work.

NASA's top leadership had become one-dimensional. By the late 1970s the agency leaders were wholly preoccupied with trying to deliver just the first flight, never mind an "operational" Space Shuttle. In fact, the Shuttle could never have flown as routinely or cheaply as they promised. But they failed to acknowledge these shortcomings even to the President. To them, NASA's credibility and availability were at stake, and protecting that credibility was the number one priority. NASA's purpose *became* the Shuttle—the triumph of means over ends. NASA's leadership doggedly waded deeper into the swamp until January 28, 1986, when credibility and purpose both went down in flames off the coast of Florida.

However, in 1978 I did not realize that NASA's fixation on the first successful flight of the Shuttle was such that it overrode an underlying sense of responsibility for future American space science and exploration. I still tried to work "through the system."

On March 5, 1979, *Voyager 1* brilliantly revealed to a ''planet-struck'' world the wondrous details of giant Jupiter and its extraordinary moons.

On July 5, 1979, *Voyager 2* repeated the spectacle and added intriguing new details.

On July 20, 1979, the third anniversary of Viking's landing on Mars was celebrated, along with the tenth anniversary of the *Apollo 11* landing on the Moon.

Also on July 20, 1979, just fifteen months after Yardley's patronizing response ''Hell, Bruce, we have eighteen months of pad,'' we at JPL learned by telephone from our friends at NASA in Washington that the Shuttle would not, in fact, be ready to launch our Jupiter Orbiter with Probe during the critical January 1982 window.

Our victory over Eddie Boland seemed hollow now, gutted by NASA itself.

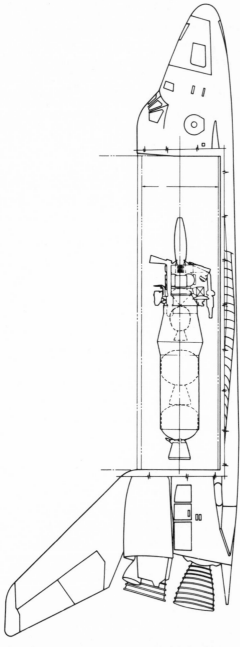

13.1 IN THE BELLY OF THE BEAST • The Galileo spacecraft in this diagram lies inside the cargo bay mounted atop the three-stage "planetary" inertial upper stage (IUS). This was the configuration planned when the mission was approved in 1977. Launch was set for early 1982 and arrival at Jupiter for mid-1985. Successive difficulties with both the Shuttle and its planned upper stages have led to four major redesigns of the spacecraft and to delays of seven years in the planned launch date and of ten and a half years in the scheduled arrival at Jupiter.

13. In the Belly of the Beast

NASA'S OBSESSION HAD CLAIMED its first victim. The Shuttle's performance and readiness were going to be lower than advertised. In July 1979 NASA officially scratched the 1982 Galileo launch to Jupiter from the Shuttle schedule. NASA termed its official Shuttle schedules "manifests," in yet another subliminal attempt to connect the unproven Shuttle with the notion of a routine shipping company. (See figure 12.5 for details.) However, no commercial shipping company could survive such a wildly optimistic schedule. The July 1979 "manifest" that postponed Galileo's departure still projected the first flight of the Shuttle within six months. The Shuttle *actually* flew for the first time twenty-one months later—in April 1981.

Worse yet, a less publicized technical catastrophe was building. The actual Shuttle was going to have a lower payload capacity than originally projected, and it would fly far less frequently. Furthermore, the Shuttle had always been designed to reach only a narrow range of low-altitude orbits. Therefore, a new system of additional rockets would have to be developed. These upper-stage rockets would need be carried along in the Shuttle's belly in order to propel a payload from that low Earth orbit to a high Earth orbit for communications satellite relay, or to escape Earth's gravity altogether on the way to other planets. (See figure 13.1).

This upper-stage rocket development was the less visible crisis. The U.S. Air Force had taken responsibility in the early 1970s for developing this interim upper stage (IUS). The IUS was deemed "interim" because the Shuttle was initially to carry only conventional, low-powered solid-fuel rockets to low Earth orbit. Lodged in the Shuttle's belly, they would safely be nudged clear of the spacecraft. Then, when they were automatically oriented in the proper direction, the rockets would be fired upon command to propel payloads to more-distant locations.

Any junior rocket engineer knows that it is not efficient to waste the expensive high-energy rocket propulsion of the Shuttle's liquid oxygen–liquid hydrogen main engines simply to lift low-efficiency solid-fuel stages for later use. High-energy rockets should be carried up in the belly. However, NASA was extremely conscious of safety. Any human-piloted machine is inevitably complex. Therefore, the Air Force opted (logically) not to introduce the supercold, liquefied gaseous fuels required for high-performance upper stages into the belly of the Shuttle—at least not so early in the Shuttle's development. Hence the term *interim upper stage* and the acronym *IUS*.*

Congress was watching too. Chairman Boland continued to seek soft targets at the periphery of the sacrosanct Shuttle program. By late 1979 he had zeroed in on this interim rocket inefficiency. Through language in NASA's appropriation bill, Boland attempted to force NASA to drop the IUS solid rockets and incorporate a higher-energy Centaur stage. NASA rejected Boland's arguments on rocket engineering, as well as his vociferous assault on the growing cost and schedule delays of the Air Force–managed IUS.

Why was the IUS slipping so badly? Proven, if inefficient, solid-fuel rocket propulsion technology had been chosen specifically because it was simpler and more reliable. But in the mid-1970s the Air Force seemingly caught the NASA management disease that subordinated hard-won rocket experience to untried, hopeful doctrines intended to achieve efficiency by decree. Development of the IUS guidance and control systems got out of hand. Component weights increased, and overall performance decreased. Unfortunately, like their NASA counterparts, the Air Force contract managers consistently swept these technical problems under the rug.

At every curve along this trail toward calamity, a solution was obvious to us at JPL. Galileo (JOP) clearly ought to have been moved off the Shuttle and back onto the Titan / Centaur rocket that had launched the Voyagers to Jupiter and beyond. Such a change would have meant a reduction in the spacecraft's size and a loss of some of its functions. But the JPL view was that the "downscaling" of Galileo was a small price to pay for freedom from this chaotic Shuttle / IUS development and from the growing shortfall in the basic Shuttle performance and availability.

*NASA and the Air Force quickly became so deeply mired in the job at hand that they could not stomach giving this out-of-control billion-dollar undertaking the simple label *interim*. Thus, on December 16, 1977, the Air Force quietly substituted the word *inertial* for *interim*, maintaining the same acronym, *IUS*.

In the late summer of 1979, I seized an opportunity to talk privately about the possibility of launching Galileo on a Titan / Centaur with Robert Frosch, Carter's NASA administrator (figures 2.2 and 8.3). Frosch, who has a Ph.D. in physics from Columbia, has the air of a detached intellectual. He delights in abstract discourse.*

Frosch merely accepted the political commitments and the doomed Shuttle policy of his NASA predecessor, Jim Fletcher.

"It simply is not politically feasible, Bruce," Frosch maintained.

The Shuttle had been sold and resold to Congress as a way to eliminate expendable launch vehicles. To request $125 million now for a Titan / Centaur vehicle to launch Galileo would undermine NASA's credibility. Furthermore, the Shuttle itself was running high in cost and well behind schedule. Better hold tight and hope it would work out.

So Galileo was delayed until 1984. Now that the most favorable 1982 opportunity had slipped away, Galileo would have to be redesigned and launched on two separate Shuttle IUS flights in 1984. One launch would propel the atmospheric probe remounted on a separate, new (and expensive) probe carrier, and the other would transport the orbiter spacecraft.† (See figure 13.1.)

Everyone knew that delaying Galileo from 1982 to 1984, using two Shuttle launches, and redesigning the spacecraft into two separately launched parts was going to cost NASA a lot more than resurrecting the Titan / Centaur. Still, NASA held tight, arguing publicly that it could not afford Titan / Centaur *and* the Shuttle. Total dollars were of less moment than NASA's political prestige.

John Casani and his disheartened design team set about dismembering their elegantly balanced dual-spin spacecraft. After two years of hard work, they were no closer to a Galileo launch than when they had started, in 1977. And, more salt in their wounds, Galileo was now referred to as a "problem" within NASA, a source of embarrassment. As Galileo's financial needs grew because of redesign and delays, it was seen to be freezing out other planetary missions.

The victim had become the villain.

By early 1980 American interest rates and inflation had skyrocketed.

---

*At NASA, Frosch was most enthusiastic about the very distant prospect of self-replicating robots (machines that could build identical copies of themselves from basic materials). While Shuttle problems mounted and the planetary program collapsed, he threw his energy into a summer study considering how such science fiction devices might one day help colonize space. Perhaps this retreat into abstraction was his way of surviving the crushing problems he faced at NASA and did not expect to solve.

†Redesigned to incorporate an auxiliary propulsion module needed to augment the less favorable gravity assist at Mars.

President Jimmy Carter had lost control of the nation's economy. Common sense demanded a cutback in government expenditures. Amid calls for independent action by Congress, Carter hurried to propose his own "$16 billion" cut: all civilian budgets for the fiscal year 1981 (the spending bill to be debated in Congress starting in February 1980) were to be reduced. Once again NASA declared Shuttle funds immune, citing national security.

If the Shuttle could not be touched, how was NASA to contribute to Carter's $16 billion cut? The answer was clear—hack away at space science. NASA Administrator Bob Frosch began to talk informally about a "reduced" Galileo. Yet the only way really to reduce its overall costs would be to launch it on an expendable rocket, a course of action Frosch had ruled out. So a reduced Galileo, in the near term, really meant further delay, immediate budget cuts, and perhaps eventual cancellation.

By January 26, 1980, Galileo hung by a thread. The critical cost-cutting session was scheduled in Frosch's conference room at NASA headquarters. I arrived in Washington on the red-eye flight at six-thirty in the morning, completely out of sorts and groggy, but too early for the meeting. After some breakfast, I wandered across Constitution Avenue from NASA headquarters to the spectacular National Air and Space Museum, filled with trophies from NASA's past. I stared through the expansive glass entrance. Up over my right shoulder loomed a towering mural depicting U.S. astronauts on the Moon. A diminutive lunar rover stood waiting in the gray-and-white background of the mural. In my sleepy state I realized that the central space-suited figure might be my friend Colonel David Scott, who commanded the *Apollo 15* mission and the first lunar rover. In that atmosphereless terrain the astronaut's face was necessarily covered by his space helmet and protective visor. Dramatically reflected in his visor was the cloudy blue image of distant Earth. The artist had caught the enduring drama and significance very well.

"That was only a decade ago," I said to myself. "What happened to the nation that did so extraordinary a thing as Apollo? Will the National Air and Space Museum live on as an anachronism, telling generations of Americans about a past they can greatly admire but not identify with?"

Then I went back across the street to NASA and sat down at 9:30 A.M. with the men who ran NASA, and there I helped once again to deflect the budgetary ax away from Galileo. No joy or pride rose in me this time. Galileo and the efforts of JPL's best and brightest were doomed to ride in the belly of a programmatic disaster—out of control. A technically competent President (Carter), his brilliant secretary of the Air Force (Hans Mark), a secretary of defense with a deep knowledge of space (Harold

Brown), an Air Force chief of staff with a Ph.D. in physics and a pioneer in space (General Lew Allen), and an honored science adviser (Frank Press)—all stood by and watched.

Perhaps there would be hope with a new President.

National elections and congressional sessions stimulate a sympathetic rhythm in American affairs of state—on Earth and in space. Good news abounds in even-numbered years. Richard Nixon started up the Shuttle in 1972. Reagan touted the Space Station in his January 1984 State of the Union Message and the futuristic aerospace plane in January 1986, when *Challenger* was barely cold in the Atlantic.*

If, however, a President truly wants a new U.S. space effort with major international implications, he will move forward without regard to the calendar. Witness Kennedy's May 1961 Apollo commitment and Reagan's March 1983 Strategic Defense Initiative.

In Washington bad news waits until Congress leaves town and the President's proposed new budget takes final form. Sad, not glad, tidings run with the Christmas holidays. NASA's Saturn 5 search for life on Mars (see chapter 2) succumbed in December 1965. And it was Christmastime when the OMB killed the full-blown Grand Tour concept in 1971 (see chapter 8).

So it fits this American political rhythm that new, very damaging revelations about the development of the Shuttle's IUS rocket system remained hidden from public view until just after the presidential election of 1980. Then the news leaked out—IUS was over budget and behind schedule.

Congressman Boland's 1979 attack on the IUS program as ill managed proved to be on target. Indeed, John Yardley soon had to cancel the special planetary version of the IUS that NASA had been counting on, because of growth in cost, delays in schedule, and loss of performance. Galileo would *not* be launched toward Jupiter by a Shuttle / IUS combination in two separate launches in April 1984. What now?

Would NASA shift to a new more capable, upper stage—follow Boland's lead and advocate the Centaur for the Shuttle? In that concept a modified version of the liquid oxygen–liquid hydrogen Centaur rocket

---

*President Ronald Reagan exhibited the surest political instincts about man in space of any President since John Kennedy. He (and his advisers) also exhibited the least technical understanding of any modern presidency. For eight years Reagan had been governor of California, the nation's leader in technology. When he was elected President in 1980, we at JPL carefully analyzed his reported circle of close friends and advisers. Among them, we could find not a *single* individual with any personal competence in science, engineering, or technology.

would be carried up to low orbit in the belly of the Shuttle (see figure 13.1). Centaur, the highly efficient upper stage of the rocket that powered the Voyagers toward Jupiter and the Vikings toward Mars, promised to augment the Shuttle's overall performance.

Or would NASA stick with the mediocre and poorly managed IUS project? Might switching horses lead to new troubles?

Carter's NASA administrator, Robert Frosch, having announced his imminent departure from NASA, returned to Washington in November 1980 from an extended trip to India and Japan. He was faced with two immediate challenges:

1) How to put the best face on continued slippage in the Shuttle schedule. (An unfounded worry, it turned out. The Shuttle was declared sacrosanct by Reagan's men, crucial to national security, and therefore off-limits to Budget Director David Stockman.)

2) How to deal with the growing sentiment within NASA, led by Andrew Stofan, the acting space science and applications administrator, to abandon the IUS development and start afresh with the Centaur upper stage adapted to be carried in the belly of the Shuttle. Centaur-in-Shuttle would propel Galileo and the twin American and European solar probes that constituted the International Solar Polar Mission (ISPM) toward Jupiter.* Discussions with the Air Force also revealed enthusiasm for defense use of the Centaur-in-Shuttle capability. Thus, a sister version of NASA's new Centaur could transport heavy military satellites from the Shuttle's low Earth orbit to geosynchronous orbit or to other demanding orbital locations.

Shuttle / Centaur was projected to be even more powerful than the now decommissioned Titan / Centaur. JPL could dream again of new Viking- and Voyager-class missions.

On January 15, 1981, in practically his last official act as NASA administrator, Bob Frosch committed NASA to the Shuttle / Centaur development.

The good news for JPL was that the projected Centaur-in-Shuttle combination would be so powerful that an all-up Galileo spacecraft could be launched directly to Jupiter with a single Shuttle launch. Thus, the expensive split-spacecraft and dual Shuttle launches required since late 1979 could be abandoned. Casani's design team could return to its original configuration.

The bad news for JPL was that the change would add another year's delay to Galileo. Centaur-in-Shuttle was a big new challenge for NASA.

---

*At Jupiter both probes would make a very close pass, then be wrenched around by Jupiter's powerful gravity and flung back toward the Sun.

TRYING TO FLY GALILEO • John Casani, the Galileo project manager, points out some features of the spacecraft in mid-1981 to Andrew Stofan, head of NASA's scientific programs (*center*), and James Beggs, NASA administrator. *NASA / JPL photo*

Liquefied hydrogen and liquefied oxygen (Centaur's superefficient propellants) would have to be handled ever so gingerly inside the belly of the human-piloted Shuttle. Cold liquefied gases are always boiling off vapor. For the Shuttle / Centaur configuration, those explosive hydrogen and oxygen vapors would have to be safely piped out of the Shuttle's belly through a complicated and removable plumbing system. Furthermore, liquefied hydrogen and liquefied oxygen are among the coldest substances available in any scientific laboratory. They constitute a dangerous cargo in the belly of an aerospace plane that must undergo complicated ground assembly and then survive the violent environment of its launch. Special (and weighty) insulation and supporting structures would have to be added to the Shuttle's innards. Substantial modification to its launch facility would also be required.

JOP, approved in July 1977 for launching to Jupiter in January 1982, now as Galileo, was targeted for a 1985 launch. Casani's disciplined engineering team, three and a half years in place, took up the task again, still as far from launch as when its project had begun.

Back at JPL from Washington after Frosch's climactic decision to switch to Centaur from IUS, I was spending most of my time trying to stabilize JPL's morale around the new NASA plan to return to a single Galileo spacecraft for launch with the Shuttle / Centaur, which meant a year's further delay. Barely a week after my return, Frank Colella, JPL's veteran press man, wanted to see me. "Bruce," he said, "have you heard anything about David Stockman's secret 'Black Book'? Bill Hines just called from Washington to tip us that he personally saw a copy which called for Galileo to be canceled."

Frank and I had known Bill Hines, the *Chicago Sun-Times*'s science and medical correspondent, for fifteen years as he covered the long string of planetary encounters at JPL. He was reliable, accurate, and a good friend. And he was right. Galileo was on Budget Director Stockman's secret hit list, part of an immediate 9 percent ($629 million) cut targeted in the NASA budget inherited from Carter's OMB.

Overall, a 9 percent cut for NASA was gentle compared with the traumatic ones prescribed for most other civilian agencies and programs. The school lunch program, the Small Business Administration, the Import-Export Bank, and the Department of Education were all slated for elimination.

But NASA's Shuttle program swallowed up well over half the total NASA budget, and once again the Shuttle was shielded from critical financial review on grounds of national security. Even Stockman could not challenge it. So NASA's overall 9 percent cut had to bite much deeper into NASA's space-science and other non-Shuttle expenditures.

How to fight off this newest threat? Perhaps we should plead our case with the new White House, still wrapped in patriotic bunting and oratory. I had already tested that a bit, trying to interest Reagan's men in a Halley's comet mission. No Luck. I had not turned up a single space advocate among the President's top men.

Our only hope was to provoke hawkish Republicans, now in control of the Senate, to intercede on our behalf. Some new political geography had to be mastered quickly. Stockman aimed to implement his cuts, and to kill Galileo, by amending the spending plan just going to Congress in February 1981.

Rapid contacts were made with senior Republican senators like Barry Goldwater, a consistent space supporter. John Tower, of Texas, chairman of the Senate Armed Services Committee, lent a sympathetic ear. Most important, the new president pro tem of the Senate, Strom Thurmond, swung into action. Our appeal was simple and technological: Galileo represents American prestige, to be sure, but also the world's leadership in intelligent spacecraft technology. The Jupiter orbiter will be on its own very far away. It must be able to make critical decisions concerning its state of being to an unprecedented degree. Such Galileo technology is also important to future defense space systems that might have to operate during nuclear "hostilities." To cancel Galileo will be a major setback for America's future prestige in space and for our military space capability.

Our "silver bullet" in this contest proved to be a letter we drafted for Senator Thurmond, who dispatched it personally to Stockman.

But there was a flip side to all this, with troubling personal questions. Thurmond had always symbolized hate-ridden southern racism for me. Our strongest supporter in the House of Representatives was the Republican congressman John Rousselot, and he had been an *officer* of the John Birch Society. In 1964 I had worked for the Democrats in southern California to help defeat Goldwater and his cowboy-style approach to the presidency. These were uncomfortable political bedfellows indeed.

In fact, the Republican Right in Congress really did care about American leadership in space, much more so than the President's men or the Democrats. In just a few weeks we created enough heat to force Stockman to back off from killing Galileo, for the time being.

Back on Capitol Hill, NASA's Centaur plan encountered serious constituency problems: it did not sit well with NASA's Marshall Space Flight Center, in Huntsville, Alabama. Marshall is only one of many government laboratories scattered throughout the country, but a big one. And in rural northern Alabama, Marshall is a big deal. It became a nationally prominent rocket laboratory after Wernher von Braun and his German rocket pioneers, spirited out of Germany by the American Army in 1945, were settled there by the American government. The Army facility, long situated in Huntsville, rapidly gained stature and capability under von Braun's influence.

By 1957 von Braun's team had advanced steadily from their original V-2 missile technology, which terrorized London in 1944, to the nuclear-tipped Jupiter rocket, American's first entry into the ballistic missile phase of the arms race. When Russia flung Sputnik into low Earth orbit on October 4, 1957, America was caught unprepared. But less than four months later, von Braun's team within General Medaris's Army Ballistic Missile Agency, and Bill Pickering's elite space rocket and electronic group at JPL, put America into space with *Explorer 1*. Jim Van Allen's Geiger counters encountered unexpectedly intense radiation—the now famous Van Allen radiation belts—and American space exploration was born.

Both Army labs were soon drafted to be charter members of the National Aeronautics and Space Administration. By the end of 1961, follow-

AMERICA ENTERS THE SPACE AGE • William Pickering, James Van Allen, and Wernher von Braun triumphantly raise a scale model of *Explorer 1,* the first U.S. satellite, at a news conference in Washington, D.C. The satellite was launched on January 31, 1958. *Explorer 1* was built by Pickering's JPL and launched by von Braun's Army rocket development group in Huntsvile, Alabama, using a Juno rocket, a model of which is in front of him (it is diagrammed in figure 2.2, p. 52). Van Allen supplied the key scientific instrument that detected the unexpectedly intense radiation surrounding the Earth, now known as the Van Allen radiation belt. *NASA / JPL photo*

ing Kennedy's Apollo commitment, von Braun's team started building the world's largest rocket, the Saturn 5 Moon rocket.

This was heady stuff for Alabama, and big business too. Every major aerospace contractor set up shop in Huntsville to supply the thousands of contract workers needed to supplement the government work force at the now renamed Marshall Space Flight Center and to compete for contracts on the propulsion stages and other parts of the Saturn series. Schools, suburbs, shopping centers, airports, and cosmopolitan ideas grew steadily.

But the completion of the Apollo program in the early 1970s signaled a difficult period for the Marshall Center. The shrewdly entrepreneurial von Braun was struck down by cancer. The German crowd grew old and faded away. The new engineers were far less experienced in the affairs of the nation and the world than von Braun's survivors of the Third Reich had been.

Then, in 1972, NASA's Johnson Space Center, in Houston, Texas, seized overall leadership of the new Space Shuttle program. Houston and Huntsville had become fierce rivals during the Apollo program in the 1960s. Now Marshall's proud work force had to take direction from its archrival Houston as well as from Washington. Even so, Marshall's responsibility included the Shuttle main engine and development of the giant solid-fuel rocket boosters ultimately to be manufactured in segments and sealed together with O-rings.

Marshall's mood became more and more defensive. Relentless efforts to maintain employment levels replaced von Braun's dream of the stars. Von Braun's splashy national charm and appeal to the executive branch gave way to traditional southern politics, focusing on congressional committee seniority and the assuring of local federal payrolls.

By spring 1981 Representative Ronnie Flippo had captured chairmanship of the House subcommittee that authorizes all NASA funds. Support from that subcommittee is indispensable for any NASA administrator. The most notable feature of Flippo's central Alabama district is the Marshall Space Flight Center and its thousands of government employees, contractors, and their families. The NASA subcommittee under Flippo looked at NASA through an Alabama prism.

And that view was troubled by a minor by-product of Bob Frosch's January 15, 1981, decision to drop the Air Force IUS upper stage and opt instead for a NASA-managed Centaur-in-Shuttle development. A shift in NASA center responsibilities was required—from the IUS at Marshall to the Centaur at NASA's Lewis Research Center, in Ohio.*

It was clear that the Lewis Research Center should manage the development of the Centaur, for it had done so since the rocket's beginning, in the early 1960s, when it was first mated with the Atlas intercontinental ballistic missile carrier. By the early 1970s Lewis had mated the delicate Centaur to the more powerful Titan 3 military satellite launcher, enabling Vikings to be propelled toward Mars and Voyagers toward the outer planets.

---

*By 1988 Flippo had long since disappeared from the space scene. Representative Bill Nelson, whose Florida district includes Cape Kennedy and environs and to whom NASA gave a ride on the Shuttle before the *Challenger* disaster, had in the interim won that subcommittee chairmanship.

In contrast, Marshall Space Flight Center's role regarding IUS had been relatively minor. It looked over the Air Force's shoulder, as the Air Force looked over the shoulder of the Boeing Company, its prime contractor. In practice, a few engineers from Marshall dutifully attended periodic reviews and coordination meetings, where they tried to represent NASA's interests. Mostly, they just filed voluminous documents.* Additionally, contract and procurement specialists at Marshall struggled with cumbersome bureaucratic mechanisms required in order to transfer NASA funds to the Air Force and, ultimately, to Boeing. A handful of job assignments and an undistinguished technical role for the once proud developers of the Saturn 5 were all that Marshall had at stake in the switch from IUS to Centaur.

Nevertheless, Flippo launched an all-out fight to reverse NASA's Shuttle / Centaur decision as soon as Congress reconvened in February 1981. In Congress the men from Alabama chipped away at that decision, while David Stockman tried to lop off NASA's planetary operation altogether.

Galileo, Centaur, and the American part of the joint mission to explore the polar regions of the Sun (christened by Bob Frosch the International Solar Polar Mission, ISPM) were all targets for Stockman's ax. This was no time for wisdom or the long view. Frosch had left NASA on January 20, 1981, and Reagan's own head of NASA had not even been selected.† Something had to be sacrificed to the political passions of the moment. The remaining NASA leaders made the best deal they could and reached a compromise in February over the Carter–cum–Black Book budget being debated in Congress. Galileo would stay; JPL's centerpiece project dodged a bullet once again. NASA could start a contract with the General Dynamics Corporation for the Centaur as well.

But the American part of ISPM would be canceled. Never mind that

---

*Imagine the difficulties faced by John Casani and his Galileo team at JPL in obtaining necessary information about the IUS through this long, contorted chain of management. Schedule slippages, cost increases, and, especially, performance declines of the IUS were critically important to JPL's spacecraft development. To keep up, a JPL representative had to meet with the Marshall Center personnel to try to understand what was happening under the Air Force contract with Boeing. Worst of all, Marshall was renowned throughout NASA for its extraordinary reluctance to share information—especially bad news. It was quite a change from when von Braun's Huntsville and Pickering's JPL got together to put up America's first satellite, *Explorer 1*, in ninety days.

By the late-1970s JPL usually had better success in discovering what was happening inside the OMB and in the White House than in learning what was going on inside the Marshall Center.

†Reagan's White House delayed announcing the new appointments to head NASA until after *Columbia*'s first Shuttle flight, on April 12, 1981, presumably to ensure that if it failed it would be Carter's failure.

much money had already been spent, or that half the instruments aboard the scrapped American ISPM spacecraft were paid for and developed by European countries, or that the United States had been working since 1975 with the European Space Agency on this project. Never mind that the United States had already backed out of a planned comet mission with the Europeans (see chapter 16). Never mind that delays in our Shuttle development had increased the costs to our European partners for developing and using a dedicated Shuttle instrument platform—the Space Lab.

NASA would still launch the remaining single European ISPM spacecraft* on the projected new Shuttle / Centaur combination. †

Stockman got his symbolic pound of flesh—$50 million or so saved by killing the American ISPM. Galileo could still push ahead on its reconversion to the single spacecraft and single launch version. NASA got its new Centaur rocket development managed by the Lewis Research Center in Ohio despite Flippo's opposition. And Western Europe got another indication of how little the United States cared about our North Atlantic partnership in space.

During the mid-1981 respite following the Galileo-Centaur-ISPM compromise, the incoming NASA administrator, James Beggs, and his deputy, the former Air Force secretary Hans Mark, were confirmed together in a gentlemanly way by the Senate. Both men used their confirmation hearings to commit themselves, and NASA, to the next big step in space as they saw it—the Space Station. The Senators applauded.

The fifty-two-year-old Mark made a very strong impact. He no longer had a blue Shuttle model on his desk, but his crew cut still symbolized his priorities. His words and actions were chilling to the very spirit of a civilian NASA.

On his visits to all the NASA centers, Mark lectured engineer and manager alike on NASA's future as a servant of high-priority defense

---

*Later renamed Ulysses.

†Robert Allnutt, then a key player at NASA, states (in a personal communication) that Stockman's OMB told NASA it could not cut the Shuttle budget as an alternative way to meet the budget reduction. But something had to go. NASA felt there was some hope Congress might restore the American ISPM mission. If Galileo were offered up to Stockman instead, however, Chairman Boland might well block any congressional effort to restore Galileo (once JOP), remembering our well-publicized victory over him in getting JOP started in 1977.

It should also be noted that Galileo justified the Centaur development much better than did ISPM. Hence, propulsion politics also favored sacrificing ISPM rather than Galileo.

In any event, Chairman Boland was far more concerned with Stockman's assaults on public housing and other social programs under his jurisdiction. Congress did not restore the American ISPM.

needs. The Shuttle as the single national launch system, he argued, removed the basis for separate civilian and military space programs. (This conclusion blithely ignored the separate civilian program mandated by the 1958 Space Act that chartered NASA.) Mark was in fact as confusing to NASA in 1981 as was David Stockman.

Mark carried these defense-oriented views to JPL—a joint NASA-Caltech enterprise. I had to reassure JPL's key personnel after he flew on to his next stop. All the while, he repeatedly impressed upon me personally that my responsibility as director of JPL was to protect its capabilities by a vigorous shift to defense work. He argued, and surely believed, that there was no longer an independent constituency for a civilian space program.

In any case, Mark had written off the American planetary program long before Stockman drew his dagger. As director of NASA's Ames Research Center in the early 1970s, Mark had mandated that the center reduce its emphasis on planetary exploration, declaring that the country was no longer interested in such endeavors.

Thus, I would not have been surprised (had I known) that on October 8, 1981, he discreetly circulated within NASA headquarters a memo calling for "a de-emphasis" on planetary science until NASA could complete its *next* goal *after* the Shuttle—the Space Station.

Mark, a brilliant and dedicated disciple of Edward Teller, a leader in the development of the hydrogen bomb, seems to have shared Teller's views about the inevitability of confrontation with the USSR. Much of that thinking (and much of Mark's charm) is displayed in his letter to me of September 1976:

Dear Bruce:

Many thanks for sending me your paper on "Science and the New Theology." It is really a brilliant job—so brilliant that I am even going to buy your book to see what more you have to say about the subject!...

While I am in general agreement with your objectives, I have to confess that I was left with an uncomfortable feeling after I finished reading the paper. This is probably due to our different backgrounds; yours being Christian, New World, and optimistic, and mine Jewish, old World, and pessimistic. I don't look upon the situation today, for example, as being as unique as you say. Sections of the human race have been there before, and because these portions of the human race did not *know* of the existence of other people, the psychological conditions we see today have existed before. The Polynesians who left Bora Bora fifteen hundred years ago to go to Hawaii knew all about overpopulation and pollution. The Jews at Masada at about the same time understood what it meant to be powerless.

I thoroughly agree with you that we must work toward a society of the kind where people have more choices and the freedom to make the choices than they have now about what to do with themselves, and yet at the same time maintain high productivity. *I believe that today the biggest obstacle in the way of doing this on a worldwide basis is the Soviet Union and the other nations with aggressive collectivist totalitarian governments. Unfortunately, I believe we will have a confrontation* with these people before we can really do what you want to do* [italics added]. . . .

Now, as deputy administrator of NASA, he made no secret of his view that American international prestige must be bolstered by powerful technological demonstrations of manned space flight with the Shuttle and the Space Station, and definitely not by new missions of planetary exploration.

Dr. George Keyworth, another Teller disciple, had just risen from the obscurity of middle management at the Los Alamos National Laboratory to prominence as Reagan's science adviser. (Although the Reagan administration had downgraded the science adviser's role, Keyworth still was the highest-ranking person in the White House concerning matters of science and technology.) Keyworth bitterly opposed Beggs and Mark on the Space Station initiative but was ultimately outmaneuvered by them—witness Reagan's commitment to that concept in his 1984 State of the Union Message.

Keyworth and Mark, however, agreed in other areas: both were early and dedicated "Star Warriors." And Keyworth independently expressed his distaste for major American planetary exploration, favoring instead a shift to Shuttle-based space astronomy as the primary NASA science activity.

On Sunday, April 12, 1981, the Space Shuttle finally flew. When back on Earth, the crew was summoned for a well-publicized presidential "photo opportunity" at the White House. High national priority for manned space flight could no longer be in doubt.

Shuttle politics or not, David Stockman still wanted to get NASA out of planetary exploration. On November 24, 1981, he directed NASA to cancel Galileo and Centaur and to stop preparing for future planetary missions like that to map Venus with the novel "synthetic aperture" radar pioneered by JPL with its 1977 SEASAT mission (discussed at the end of chapter 7). Furthermore, NASA was to cut to the bone its world renowned Deep Space Net and other activities linked to the 1986 Uranus encounter by *Voyager 2*. Such was the Reagan administration's response to Voyager's brilliant views of Saturn, just concluded in August 1981.

Only NASA's administrator, Jim Beggs, along with a small, divided

*In reviewing this letter on October 20, 1988, Mark noted, "It is just possible that we may have had the confrontation with the Russians that I foresaw then during the past four years."

scientific constituency and our JPL "underground" political advocacy,*
could keep life in America's journey into space for this century.

So we mounted yet another political campaign to keep a little hope and
excitement in our collective future. Again members of the Republican
Right in Congress were our strongest supporters, even though they now
faced an intransigent White House, prepared to let Republicans in Congress running for reelection in 1982 bear the brunt of the political reaction
to Stockman's cuts.

Stockman did prevent any new planetary start-ups in 1982. But in late
December we managed anyhow to assure good preparation for *Voyager
2*'s eventual encounter with Uranus. And we beat him off once again
regarding Galileo. It survived.

Gone, alas, was Centaur.

Imagine the perspective then from JPL. Early 1982 was the original
launch time for JOP / Galileo. John Casani and his engineers had expected
to be celebrating that launch at the Kennedy Space Center and to be flying
toward the Jupiter encounter in 1985. Instead, early 1982 found them still
grounded at JPL, directed once again to modify their unfinished Galileo
spacecraft so that it could be designed for launch toward Jupiter on a very
lengthy journey involving a flyby of Earth before finally setting out to
Jupiter. Without Centaur the low-performance, two-stage IUS was back in
play.

Weary of trying to blunt Stockman, NASA now officially opposed the
Centaur stage for the Shuttle. Nevertheless, many observers noted Reagan's consistent interest in the Shuttle and its "photo opportunities,"
which had replaced news conferences as his primary means of communi-
cating with the American people. The President was even flown to the
fourth Shuttle landing, on July 4, 1982, at Edwards Air Force Base, in the
California desert, where he smiled proudly as the cameras captured him in
the bright sunlight flanked by blue-suited astronauts.

Since 1982 was an election year, congressional incumbents thirsted for
good political news. On July 15, 1982, Chairman Boland of the House
appropriations subcommittee† demonstrated his budget clout to everyone
when the House unexpectedly overturned Stockman's cancellation of Centaur. NASA headquarters was pleasantly surprised and NASA's Lewis

---

*Plus the newly formed Planetary Society, discussed in chapter 20 and in the appendix.
† Aided by the subcommittee member William Lowery, whose California district included
the General Dynamics plant where Centaur is manufactured. Some of those close to these
events feel that the actual politics was even more Byzantine than the rendition provided
here.

Research Center overjoyed. The Marshall Space Flight Center and its representative Ronnie Flippo had been outwitted.

JPL was still on a yo-yo. The good news now was that the Galileo team could once again forget the last six months of spacecraft redesign and dust off the set of drawings for a machine that could fly directly to Jupiter on the Shuttle / Centaur.

But there was a catch. Funding battles take time. The latest schedule disruption had pushed Centaur readiness past April 1985. Galileo and the European ISPM spacecraft, renamed Ulysses, were now rescheduled for May 1986, four and a quarter years later than the original launch date. (See figure 12.2.)

What else could befall this snake-bit project?

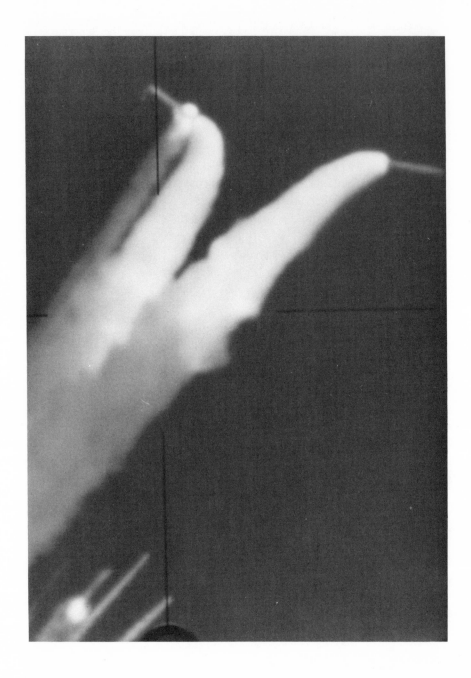

14.1 THE CHALLENGER EXPLOSION • In this photograph from a high-resolution tracking camera, the two solid-fuel rocket motors, moving freely after an explosion has blown the Shuttle orbiter *Challenger* apart, continue to propel themselves in gradually curving paths. Debris from the orbiter is evident in the lower left-hand corner. *NASA/Johnson Space Center photo*

14. Collision with Reality

REAGAN'S NEW HEAD OF NASA, James Beggs, possessed great experience in technical management and a determination to make NASA succeed. A Naval Academy graduate and a Harvard M.B.A., Beggs also knew Washington from many sides. During Nixon's years he had been under secretary of transportation. Earlier, he had been head of NASA's Office of Advanced Technology. He had also sharpened his management tools at the giant defense contractor General Dynamics, where he had risen to near the top. Beggs's informal and supportive style was well received by a dispirited NASA, whose self-image had steadily deteriorated in the decade following Apollo.

The Shuttle finally flew just before Beggs took over. Now, in 1981, he strove to make the Shuttle system operational, practical, and affordable. He bravely sought to deliver on NASA's old promises.

It is axiomatic in aerospace that a good manager finds good people and delegates to them. Right off the bat, Beggs drew upon his close relationship while at General Dynamics with one of the Air Force's most highly regarded young generals, James ("Abe") Abrahamson.* They had worked together on the Air Force's F-16 fighter plane, a General Dynamics product. Abrahamson was now persuaded to take over the Shuttle program after John Yardley's departure following the Shuttle's first flights in 1981.

Rounding out the top management trio was Deputy Administrator Hans Mark, with long experience at NASA and in the Air force, where he also worked with Abrahamson. Together, the three began the painful process of weaning the Shuttle program away from its developers at the Johnson

---

*Abrahamson was pulled off the Shuttle in 1983 to run Reagan's highest-priority new program, the Strategic Defense Initiative (SDI). He announced his retirement from the Air Force in the fall of 1988.

Space Flight Center, in Houston, Texas. The bulk of the operational effort was now to move to the Kennedy Space Center, in Florida, where the Shuttle orbiter, payloads, solid-fuel rocket boosters, and giant external tanks were to be integrated on a routine basis.

Abrahamson had special personal credentials for the ticklish task of restructuring the emotion-ridden man-in-space endeavor. He had been an Air Force test pilot in the mid-1960s and then been tabbed to be a military astronaut for the Air Force's own Manned Orbiting Laboratory, until its cancellation in 1969. He was clearly a true believer with the "right stuff."

Beggs himself focused on the key problem—how to turn the Shuttle from an overly long R & D program into a routine operational space trucking system. But he quickly confronted the intrinsic NASA dilemma. He would have to transform NASA itself from an R & D agency into an operational one if NASA's pledges on the Shuttle were ever to be re-deemed.

Jim Beggs is a likable guy who wears a well-tailored Stetson cowboy hat and quotes Shakespeare and Mark Twain in his speeches. However, underneath all the charm and goodwill, Beggs is a "company man." He may never have questioned whether a cost-effective Shuttle operation was in fact possible. And I doubt that he thought through the consequences to NASA, to himself, and to the country, if the Shuttle's "promise" proved an expensive illusion. Like technical managers generally, he accepted the job he was given—and the baggage that came with it.

Fate would not deal kindly with this man. Not only did the Shuttle's crash take place at the end of his tenure, but Beggs was simultaneously ensnared (and victimized) in a flawed criminal prosecution stemming from government contracting issues at General Dynamics.

All of that was years in the future when he came aboard in 1981. During his first four years at NASA, he kept a steady hand on the tiller. Galileo and Centaur progressed without major dislocation or rescheduling, after Centaur's restoration in July 1982. Casani's team at JPL took advantage of unaccustomed stability to ensure that Galileo would fulfill a vast array of scientific objectives at Jupiter.

In fact, the Galileo team expanded its aspirations to include a look at a glamorous new kid in the sky—an asteroid. A computer search through hundreds of asteroid orbits revealed that in 1984 Galileo's path outward from Earth could intersect the path of the 125-mile-thick asteroid Amphitrite. From as close as 6,200 miles, Galileo's powerful camera and spectrometers could give a dandy first look at that asteroid on December 6, 1986, six months after the craft's scheduled blast-off from the Kennedy Space Center on the new Shuttle / Centaur combination.

This prospect appealed strongly to planetary scientists and engineers who had been waiting a decade for Shuttle to finally start launching planetary spacecraft.

In December 1985 the Galileo spacecraft was nestled into an enormous box and shipped by truck convoy across the country. Payload Preparation Facility Number 1 at the Kennedy Space Center was intended to be Galileo's final earthly residence, before joining an unfueled Centaur rocket waiting in the Vertical Processing Facility.

In a building close by rested the Ulysses spacecraft, still waiting to be flung out toward Jupiter on its journey over the poles of the Sun. Soon, it too would rest upon another new Centaur. The spacecraft and Centaur teams kept an eye on the calendar. In early spring 1986 each spacecraft and Centaur combination had to be carefully moved—still in a vertical configuration—to Launch Pads 39A and 39B. Then the Centaur / Galileo combination would be edged into the abdomen of the Shuttle orbiter Atlantis. It's sister ship *Challenger* would swallow up Ulysses resting atop its Centaur.

In early May, Shuttle Commander Rick Hauck would be at the controls of *Challenger* as it thundered aloft with its cargo of Centaur and Ulysses. Ulysses would be fired off toward Jupiter as soon as possible so that Commander David Walker could get the go-ahead to launch Atlantis—and Galileo. Jupiter and Earth would be in a favorable alignment for barely a month. Everything had to go without delay if both missions were really going to get off successfully.

Why were there *duplicate* Shuttle / Centaur facilities at the Kennedy Space Center? The reasons date back to 1982. In his enthusiasm for the new Centaur project, the congressional kingpin Eddie Boland mandated *parallel* conversions of *two* Shuttle launch facilities and *two* Shuttle orbiters. This expensive and challenging approach was necessary if both Galileo and Ulysses were to be launched within the same three-week period in May 1986, when Jupiter and Earth would be favorably positioned for interplanetary travel. (Launch opportunities to Jupiter occur every thirteen months and last for only a few weeks.) The performance margins were so close that each Shuttle orbiter would after lift-off have to be quickly powered up to maximum thrust level, referred to in NASA jargon as ''109 percent of nominal.'' Dual facilities would also support a vigorous future program of Shuttle / Centaurs.

A technically more conservative, and affordable, approach would have been to convert only one Shuttle orbiter and only one Shuttle launch facility to handle the demanding Centaur. That way Galileo would have blasted off in May 1986, but Ulysses would have waited for the next Jupiter opportunity, in June 1987.* Evidently, the heavy planned Shuttle schedule of twenty-four flights per year, as well as the proposed Air Force Centaur payloads, justified dual facilities—especially if Chairman Boland wanted them.

By the end of 1985 nearly $700 million had been invested in this aggressive Centaur-for-Shuttle development.

More excitement lay ahead in 1986. In March, well before the Galileo and Ulysses blast-offs would end the eight-year hiatus in new American space exploration, a specially timed Shuttle flight would carry into low Earth orbit selected instruments with which to observe Halley's comet. It was unfortunate that America was not planning a space flight to probe the cometary nucleus, as Europe, Russia, and Japan were doing. But Earth orbit was at least a step in the right direction.

Later, in September 1986, NASA's greatest space-science endeavor ever—the twelve-ton Hubble Space Telescope—was finally scheduled to be lifted by Shuttle into Earth orbit and, once there, to be deployed by astronauts to reveal the cosmos in greater detail than ever before.

No wonder Jim Beggs buoyantly proclaimed 1986 the year of space science.

But Jim's positive facade concealed a personal crisis. It began with the need to find a new deputy administrator after Hans Mark left NASA in September 1984 to become chancellor of the University of Texas. Traditionally, NASA's deputy administrator has been a general manager—usually a technocrat—and normally an appointment in which the administrator has had a strong say. But Reagan's men in the White House were not long on protocol. Mark, a friend of congressional conservatives, had in fact been chosen deputy in 1981 before Beggs was chosen administrator. But Mark was at least properly experienced for the job. Now, after Mark's departure, ideologues in the White House sensed an opportunity to advance a *real* conservative to a governmental position of greater responsibility.

Their choice for deputy administrator was William Graham, a Reagan appointee to the Arms Control and Disarmament Agency and a director of the politically active Committee on the Present Danger. Graham had started at the Rand Corporation, gone on to a Rand spin-off called Research and Development Associates, and then joined the Heritage Foundation. In all these jobs he had focused on U.S. preparation for nuclear war. Bright, doctrinaire, and knowledgeable about policy analysis, Graham had no

*The European Space Agency would have vigorously opposed such a change. But Ulysses was originally scheduled to be launched in 1983, a year after Galileo.

experience in running a technical organization of NASA's size and complexity. Furthermore, he had no known interest in civilian space activities and objectives.

To Beggs, and to NASA's constituency, Graham just did not seem qualified to be general manager of NASA or to promote its civilian role in space.

So Jim Beggs stonewalled the White House. If it could not agree on someone he felt was acceptable, he would be his own deputy for a while.

The battle dragged on behind the scenes from 1984 into 1985. Then a new threat arose. The Justice Department investigated Beggs and three other former General Dynamics executives concerning criminal charges of federal fraud. The government did not accuse him of having sought personal gain. Rather, it charged that he and the others had misallocated costs on a federal contract in the late 1970s. Normally, this would be a civil matter, to be argued by lawyers and accountants. But the Reagan administration was conspicuously prosecuting fraud and overcharging by defense contractors—presumably to make their whopping peacetime defense budgets more palatable. Perhaps Beggs had been caught in this political vortex. But he was also in bad odor with White House conservatives over Bill Graham. Could this be retribution, a way to get Graham into the top NASA job eventually? Beggs's situation was not improved by his bitter opposition to then Under Secretary of the Air Force Aldridge's farsighted attempts to procure new military Titan launchers as backup for the uncertain Shuttle.

By fall 1985 the noose around Beggs's neck was tightening. President Reagan officially nominated Graham to be deputy NASA administrator. The Senate confirmed Graham, and he moved into Hans Mark's old office on the top floor of NASA headquarters. Beggs and Graham did not speak; across the seventy-five feet that separated their offices, a silent chasm yawned.

On December 2, 1985, the roof caved in on Jim Beggs. A federal grand jury in Los Angeles indicted him and three ex-colleagues from General Dynamics on criminal charges of fraud.*

But Beggs refused to resign. Instead, he took a leave of absence and came to his office daily. After only nine days at NASA, and a virtual stranger to its ways and people, Graham was named acting administrator.

Casani and the Galileo team were busy over Christmas moving their prized spacecraft to Florida. Few took much notice of goings-on back in

---

*The government dropped the case in 1987, virtually acknowledging that it should never have been prosecuted in the first place. One of Attorney General Ed Meese's last acts was to write Beggs a public letter of apology.

the NASA "palace." More real to them were the myriad details of technical achievement.

Good engineering requires frequent consultation and broad review. Ever since 1982 NASA's Lewis Research Center had called together the Shuttle / Centaur review board once a quarter. Senior propulsion experts from the Lewis Center in Cleveland, key Shuttle managers from NASA's Manned Space Flight Center in Houston, rocket veterans from Kennedy, and Casani and other payload experts from JPL sat down with the General Dynamics Corporation, which had built previous Centaurs and was building this one. At their quarterly review in October 1985, the members decided to meet next at the Kennedy Space Center. The day chosen was January 28, 1986. That way they could personally view the final Centaur preparations, squeezed in amid the usual frenzy of Shuttle activity.

Only nine flights were accomplished in 1985—up from five in 1984, but far short of the eventual goal of twenty-four flights per year around which the Shuttle program was designed. Beggs and Mark had kept relentless pressure on NASA to build up the flight rate and minimize the turnaround time. But now things were getting tight. *Columbia,* on flight 61-C, had been delayed from its planned December 18, 1985, launch by two potentially catastrophic errors made by the ground crews. Then bad weather had hit, further delaying lift-off until January 12. The local congressman Bill Nelson was one of its seven-man crew. More bad weather at Kennedy forced additional delays in its planned return from orbit. *Columbia* finally landed at California's Edwards Air Force Base (not back at the Florida shuttle base) on January 18, whence it would have to be ferried back to Florida on NASA's specially modified Boeing 747, adding yet more delay to preparations for *Columbia*'s next flight.

Meanwhile, flight 51-L of *Challenger* was undergoing equally frustrating final preparations. *Columbia*'s tardy return pushed back *Challenger*'s launch date four days further, to Superbowl Sunday, January 26. Then an (incorrect) forecast of bad weather caused another rescheduling, to Monday, January 27, also forcing Vice-President Bush to cancel a planned visit. On Monday morning another set of ground-crew errors caused more critical hours to be lost, while the flight crew waited impatiently.

Commander Dick Scobee's previous *Challenger* flight, in April 1984, had started more smoothly. On that highly successful mission a solar observing satellite (intentionally launched into a low Shuttle-compatible orbit years earlier) was repaired and returned to service by *Challenger*'s crew. For *Challenger*'s upcoming flight, however, no such heroics were planned. A $100 million NASA TDRSS communications satellite mated to its IUS rocket would be released. Both items, it was hoped, would perform better than their Shuttle-launched predecessors. The IUS rocket as well as the TDRSS spacecraft suffered problems. Now NASA desperately needed this TDRSS satellite to be successfully operating in its geostationary orbit.

Michael Smith, a hotshot Navy pilot, would be sitting on Scobee's right as pilot. This was his first Shuttle flight. The mission specialists Ron McNair, Judy Resnik, and Ellison Onizuka had all flown previously and, like Scobee, experienced launch delays. The payload specialist Gregory Jarvis, a rookie astronaut from Hughes Aircraft, was especially keen to fly. Reportedly, he had been bumped twice before, once to make room for Senator Jake Garn and later to accommodate Representative Bill Nelson.

Christa McAuliffe had won out over thousands of other teachers to be America's first teacher in space. She hoped to use the drama and excitement of her space voyage, including a live video hookup, to motivate young people to work toward higher personal achievement. She had trained hard and now awaited her greatest moment as a teacher.

But Monday's planned launch was finally scrubbed just after noon. They would try again on Tuesday morning, January 28.

It was an unusually cold Florida morning, even for January, as Casani and other board members entered an old, familiar engineering building at the Kennedy Space Center and converged on the second-floor conference room. As they filled their Styrofoam coffee cups and exchanged friendly greetings, a closed-circuit-TV monitor in the corner showed a Space Shuttle on Pad 39B. It gleamed in the morning light, vapor trails streaming from its gantry tower, and stood nearly ready to blast off from the launch pad adjacent to where Galileo would fly three and a half months later.

The Centaur quarterly review was briefly adjourned so that everyone could crowd into the adjacent corner office, where full-length windows provided a spectacular northeast view. *Challenger,* which would also take off in May with Centaur and Ulysses, was clearly in view, but neither Galileo nor Ulysses could afford a two weeks' delay like this flight. Nevertheless, their spirits were lifted by the dramatic activities unfolding there, beautifully set off by brilliant white coral sand beaches and the green tropical Florida ocean. The day was clear, bright, and cold.

The last minutes of the countdown proceeded smoothly. These veterans of space flight had witnessed dozens and dozens of launches, successful and otherwise. Nevertheless, each tensed a little as the male voice over the NASA net evenly called out, "Three, two, one ... we have ignition."

Challenger lifted slowly off the pad, accelerated, and rolled over on its back, as planned. The characteristic white trail of smoke curved gently seaward. At sixty seconds into the launch, the clear voice of Commander Dick Scobee noted the passage through maximum acceleration. *Challenger,* in fact, was scheduled to reach only 104 percent of full power on this flight, well short of the 109 percent still required for Galileo.

Suddenly, a dead silence pervaded the room. A white-and-orange cloud bloomed where *Challenger* had been an instant before. Everyone in that room had seen rockets explode in the early days, but none of those had carried men and women.

The Centaur review board was not alone in its shocked reaction to *Challenger*'s cataclysm. Powerful tracking cameras routinely recorded Shuttle launches in vivid color, and that video was fed live or replayed to whatever TV networks would air it. This time the demand skyrocketed. *Challenger*'s spectacular demise was played and replayed on every television screen in America. *Challenger*'s death mask instantly became a pop-art image, rivaling even astronaut scenes on the Moon or beautifully ringed Saturn as the symbol of space.

The national outpouring of grief was unprecedented and genuine. The *Challenger* seven could hardly have been a more representative group of American martyrs—three races, four religions, male and female, military and civilian, married and single, fathers, and a mother. Virtually everyone could empathize with victims and survivors alike.

Stunned and leaderless, NASA reacted to the *Challenger* disaster like a punch-drunk fighter. "Keep on schedule" was the initial policy communicated to the Galileo project from NASA headquarters. Perhaps *Challenger*'s explosion would be traced to a trivial, easily corrected problem. It was not to be. A presidential commission to investigate *Challenger*'s demise was immediately formed. Chairman William Rogers soon concluded that "the decision to launch was flawed." By then, a sobered NASA had already acknowledged that Galileo could not be launched in May 1986. Galileo and Ulysses would have to slip at least thirteen months more, to June 1987, when Earth and Jupiter next would be in favorable locations.

This was a blow, but the unraveling of the Shuttle program, and of its handmaiden NASA, was just beginning. Televised excerpts of the Rogers Commission revealed a weak and weary NASA management. The Marshall Space Flight Center seemed to have chosen bureaucratic process over common sense in blindly moving Shuttle's solid-fuel rocket motors along. The proximate cause of the *Challenger* disaster was mediocrity at NASA.

In the emotion-charged arena of television, a new perspective emerged. The nation uncritically accepted its President's theatrical view of the Shuttle as a craft whose prime purpose was to transport brave Americans safely to and from that beckoning promise of space.

News reports bared the long-suppressed fears of astronauts about the Shuttle's fragile landing gear, its brakes, the lack of an emergency-escape system, and weaknesses in the powerful main engines. In an atmosphere encompassing grief, outrage, even betrayal, a new national Shuttle policy developed. The safety of the astronauts became the only consideration. The words *cheap, easy, routine,* and *operational* disappeared from NASA's lexicon. The important thing was "to get the Shuttle flying again"— *safely*. The cost and effectiveness of its missions were no longer paramount. Form triumphed over function, theater over reality. The Shuttle had always catered to the need for fantasy in us all. Now it served no other purpose.

The Rogers Commission, for example, was never asked by the President whether the tasks of *Challenger*'s crew justified the risking of their lives.* Nor did they investigate whether the work could have been done with two or three crew members, as in a high-performance military aircraft,† or with none at all.

By January 1986 Ronald Reagan had led the most technically advanced nation in the world for five years. In all that time he remained isolated from independent advice on scientific and technology matters. A brief flirtation in early 1983 with Edward Teller,‡ "the father of the H-bomb," encouraged Reagan's enduring embrace of "Star Wars" technology. When Reagan announced SDI in a televised speech in March 1983, he took nearly all of his White House advisers by complete surprise.

At the time when *Challenger*'s fireball ignited a national inquiry into

---

No one in the federal government publicly raised the obvious policy issue: somehow safety must be balanced against the *purposes* of manned flight. Even the best-managed program of manned space flight will carry substantial risk. Thus, the key policy need is to determine what space results really warrant the risking of human life. Had this criterion been applied at any point in the Shuttle's history, the number of planned Shuttle flights and the number of crew members would have been drastically reduced. That such a policy is not being applied formally *after* the trauma of *Challenger* is difficult to understand.

† A small flight crew would have permitted a capsule escape system like the one that saved test pilots in the development of the B-70. But "Airliners don't carry parachutes" was the slogan I got for an answer when I raised the issue earlier in offhand conversation at NASA.

‡ The governor of California is automatically a member of the Board of Regents of the University of California. As a regent, Reagan had a stormy relationship with the UC system. However, the University of California is also responsible for the Livermore and Los Alamos nuclear weapons research laboratories. Teller, a principal Livermore personality, is the leading scientific spokesman for the strategic view that U.S. nuclear supremacy must be attained to prepare for the inevitable conflict with the USSR. Teller was probably one of the few UC figures whom Reagan knew on sight.

the Shuttle, even the downgraded position of presidential science adviser was vacant. The nuclear bomb and laser specialist George Keyworth had resigned months earlier to become a Washington consultant. White House Chief of Staff Donald Regan had not found a satisfactory successor. A technically ignorant White House fell back upon an existing National Security Council committee composed of senior bureaucrats from the Defense Department, the National Security Council, NASA, the Commerce Department, the CIA, and the Transportation Department.

This uninspired approach to a space-policy crisis excluded the kind of expert participation that the formation of a special presidential task force would have permitted. Thus, there were no timely inputs from the National Academies of Sciences and Engineering, from the presidents of major scientific and technical universities, or from leaders of prestigious corporations like IBM, Polaroid, TRW, Kodak, GE, GM, or AT&T. Also shut out were respected technical figures from previous administrations—the former secretary of defense Harold Brown; the former NASA administrator Thomas Paine (also chairman of the President's National Commission on Space); the former science adviser Frank Press; the former energy, defense, and CIA head James Schlesinger; the former Air Force chief of staff Lew Allen; or the former CIA deputy director Admiral Bobby Inman. All these men had relevant experience and expertise, but all seemed to be viewed with suspicion by the White House, as men not to be trusted near the inner sanctum of power and policy.

Reagan's faceless bureaucrats composed the Senior Interagency Group—Space, SIG(Space). They worked under the newly appointed national security adviser, Admiral John Poindexter, Reagan's fifth national security adviser in as many years. Poindexter would also last only a year, when his pivotal role in the Iran-contra scandal became known. Subsequent televised hearings and other revelations leave little doubt that the fate of NASA was not high on Poindexter's list of priorities in the spring of 1986. Actual responsibility for SIG(Space) fell more to Donald Regan, the presidential chief of staff, a former Wall Street financier. SIG(Space) fashioned America's response to programmatic and policy disaster in space. Its dominant criteria rapidly became clear: adjudicate bureaucratic turf, and rationalize short-term political factors (1986 was an even-numbered year, and Republican control of the Senate was in jeopardy).

SIG(Space) first confirmed military primacy in space. The Defense Department got the green light to develop a new fleet of expendable rockets—medium, large, and extra large. This was actually an inevitable and well-founded governmental decision because, even after *Challenger,* NASA still opposed expendable rockets. The DoD would be top dog in U.S. space with priority access to both expendables and the Shuttle. For the first time since its birth, NASA would neither develop nor supply new rockets.

Next, the White House ended NASA's launch subsidies of commercial communications satellites. As was described earlier, NASA had set its prices for commercial launches artificially low even as its own costs soared. NASA was in fact subsidizing the launch of foreign and domestic commercial satellites with U.S. taxpayers' money. Five years earlier David Stockman had attacked such subsidies by the once sacred Import-Export Bank and a dozen other government institutions. Now, after *Challenger,* the Shuttle mystique was finally pierced. Conservative fiscal doctrine could apply even to NASA. The White House directed that no new commercial payloads be carried by the Space Shuttle. American communications satellite developers would have to compete for launches as best they could against foreign government-subsidized competitors.

In just a few months SIG(Space) and Donald Regan dismembered more than a decade and a half of NASA's carefully crafted monopoly of the launching of American spacecraft. So far so good. The Reagan White House was excellent at striking down pretentious civilian governmental ambitions.

But what about the future? Following the draconian measures that repudiated the failed Shuttle-only national space policy of the 1970s NASA administrator James Fletcher, the White House was at last positioned to set a new direction for NASA. Reagan had a chance now to put his own mark on America's journey into space. But his staff had to reconcile post-*Challenger* technical and economic reality with Reagan's instinctive support for Americans in space.

Answers to three major questions were overdue. Together they would point a new direction for NASA and for American hopes and visions.

• Who would be appointed at NASA to lead the agency out of the post-*Challenger* chaos?
• Should *Challenger* be replaced with a new Shuttle orbiter?
• Should the nearly completed second Shuttle base at Vandenberg Air Force Base, in California, be finished?

On March 6, 1986, the White House named James Fletcher* the new NASA administrator. Why this extraordinary step backward? First, Fletcher

*Reagan personally pressured Fletcher to accept this responsibility.

SECOND TIME AROUND • President Ronald Reagan looks on as Vice-President George Bush swears in James Fletcher as NASA administrator on May 12, 1986, in the Roosevelt Room of the White House. Members of Fletcher's family witness the event. Fletcher also was NASA administrator from 1971 to 1977 and successfully promulgated the U.S. "Shuttle only" space policy. *Official White House photo*

was one of only a few technical administrators with knowledge of space who could pass the White House "litmus test."* Second, Fletcher could be confirmed easily by Congress (the White House favorite, Bill Graham, could not be). Key congressmen and their space committees had consistently benefited politically by supporting the Shuttle fantasy. Congress had little interest in repudiating Shuttle policy; its motto was "Let's get the Shuttle flying again." Jim Fletcher was a predictable, known quantity.

Two White House decisions regarding space quickly followed, both political. Vandenberg Air Force Base and the Shuttle's prime contractor, Rockwell, are located in California, where Republicans wanted badly to defeat the Democratic Senate leader Alan Cranston. Cranston had been a

---

*Fletcher not only passed the "litmus test," thanks to his conservative political leanings, but had also chaired the technical review panel that assessed the feasibility of SDI. Its classified report was said to include considerable skepticism even on the part of that carefully selected team. Fletcher's public statements, however, provided a bland assurance that, technically, the President's goal of confident protection of American cities through use of new laser and space technology was feasible.

consistent Shuttle supporter, and the Reagan administration did not want to give him a live local issue by cutting back on the Shuttle fantasy in California.

A new Shuttle orbiter would indeed be purchased from Rockwell at a cost of $2 billion, newscasters in Los Angeles cheerfully announced. Two thousand southern California jobs went along with that decision. The White House did not emphasize that *Challenger*'s ''replacement'' would not go into service before late 1991. By then, a fleet of new *expendable* launch vehicles would be carrying most of America's military, commercial, and scientific payloads.

And what of the nearly completed military Shuttle base at Vandenberg Air Force Base? Political compromises in the early 1970s had committed a mostly uninterested Air Force to expend $3 billion later in the 1970s for a manned space facility that almost no one in the military wanted. It would cost $300 million to $400 million a year, or more, simply to maintain operational status.

Shuttles to be launched from Vandenberg were now years away. The same logic that led Air Force Under Secretary Aldridge in the early 1980s to develop for the Air Force an independent unmanned launch capability now made Vandenberg's fate as a Shuttle base crystal clear—shut it down and take a write-off. If ''Star Wars'' or other missions needed the Shuttle's big belly and astronauts for early tests and demonstrations, the Air Force had all it needed at the Kennedy Space Center or with Aldridge's big new Titan rockets.

To shut down the Vandenberg Shuttle base in 1986 would highlight the weakness of Reagan's decision to build another Shuttle orbiter. Three Shuttle orbiters remained. Kennedy could service them efficiently but not really more. Yet Cranston would surely blast the Republicans if Vandenberg were closed. So the Air Force announced that the Vandenberg Shuttle base would be placed on ''caretaker'' status and that it would suffer only a limited reduction in employment. Blue-suited generals reassured the local businesses (and voters) that the Air Force's commitment to the Shuttle was as strong as ever.

But on November 4, 1986, Cranston beat Silicon Valley's Congressman Ed Zschau in a close race. And, sure enough, in December, Washington's month of ''bad tidings,'' new reductions were announced in the Vandenberg Shuttle work force. A de facto policy of phasing the Shuttle base out seems to be under way. Fortunately for local business, Vandenberg will boom again in the 1990s, as the center of a renaissance in Air Force expendable launch vehicles.

Vandenberg's future in space, and the Air Force's, is much brighter than NASA's.

The planned Shuttle / Centaur launch in June 1987 carried many old liabilities. What if the Shuttle were forced to abort a mission before it could release its explosive Centaur rocket cargo? Even empty, the Shuttle is the world's worst glider. It loses a mile of altitude for every three miles of flight toward its touchdown point. Its landing gear and brakes have never been considered robust by engineers or astronauts. With an additional twelve to fifteen tons of unreleased (and highly explosive) cargo in its belly, the speed on approach and the stresses on the landing gear at touchdown would substantially increase. Damage to the fragile landing gear—and a catastrophic accident—became a greater possibility.

Such a scenario was always part of Shuttle reality, even if NASA officials had blithely expanded the term *routine* to include such derring-do as night landings by the Shuttle on the single, long runway at the Kennedy Space Center, with its highly variable weather. After the *Challenger* explosion, however, America wanted its astronauts safe.

To make matters worse for Galileo, *Challenger* had exploded seconds after reaching ''104 percent'' of nominal thrust. It will be a long time before NASA again flies the Shuttle above 104 percent of nominal thrust. But in order to send Galileo to Jupiter in May 1986, Shuttle had to reach 109 percent and to do so with maximum lift-off weight.

The maximum lift-off weight permitted for the Shuttle orbiter after *Challenger* plunged from its original full-up weight of 65,000 pounds to 50,000 pounds. Suddenly, JPL realized that the new Shuttle / Centaur system carried on the now down-rated Shuttle would significantly exceed these new lift-off weight limits. The only solution would be to remove a part of the liquid hydrogen–liquid oxygen components of Centaur's fuel tanks. A half-empty tank in Centaur meant a Galileo propelled too slowly to overcome Earth's gravity, catch Jupiter, and begin its exquisite gravitational tour of Jupiter's moons. The once heralded Shuttle / Centaur would now be unable to launch Galileo directly to Jupiter at all.

This was where we had been between 1979 to 1982, when the Shuttle / IUS lacked the power to reach Jupiter. At that time JPL had had to break Galileo apart so that NASA could plan to launch it on two separate Shuttle / IUS combinations. After *Challenger,* however, NASA's enthusiasm for dual Galileo launches cooled. So JPL once again proposed a long windup before the long, fast fall to Jupiter. First, the Shuttle / Centaur would shoot Galileo out into empty space, whence it would fall slowly back toward a carefully contrived close encounter with Earth. The resultant gravity assist from Earth (just like that used by *Mariner 10* at Venus and by Voyager at Jupiter) would impart enough extra velocity to Galileo to enable it to coast outward all the way to Jupiter. This complex trajectory added an unproductive three years to the total flight, but it would get Galileo there. JPL called the trajectory Δ VEGA, for Delta Velocity and Earth Gravity Assist. It was better than nothing.

Meanwhile, Centaur-in-Shuttle found new detractors—the astronauts. Shaken by *Challenger*'s fiery fate in a cloud of liquid hydrogen–liquid oxygen fuel, Shuttle's crew began to shy away from riding ahead of those same propellants in the cargo compartment.

That was the last straw. On June 19, 1986, following the advice of his staff, NASA Administrator James Fletcher canceled Centaur-in-Shuttle. NASA took another write-off—this one for $700 million—on its Shuttle fantasy.

Even then, Fletcher would not acknowledge what everyone else knew— that planetary missions were intrinsically incompatible with the Shuttle. American planetary exploration remained hostage to the dying embers of NASA's Shuttle fantasy.

For John Casani and the Galileo team at JPL, the situation during mid-1986 became desperate. The projected lift-off performance and the permissible landing weights of the Shuttle declined further following the cancellation of Centaur. Suddenly, even a gravity swing-by of Earth (Δ VEGA) could not provide enough extra acceleration for the spacecraft to reach Jupiter. What trick was left? Clever trajectory designers found another way for Galileo to be launched from an IUS carried in the belly of the Shuttle and still end up at Jupiter. But this new technique required velocity increments from two passages by Earth, plus another one by Venus (Venus, Earth, Earth Gravity Assist—VEEGA). The alternative? Outright cancellation of Galileo. That action was too painful to JPL, and too embarrassing to NASA, to be seriously considered.

In June 1986 Galileo was rescheduled for launching by the Shuttle in October 1989 on a path inward toward the Sun and thence to an intersection with Venus in February 1990 (see figure 14.2). There, a carefully aimed close passage by Venus would accelerate Galileo back toward Earth for a close flyby of its home planet in December 1990. Then, like an ice show performer racing toward a row of barrels, Galileo would swing around the Sun, whipping toward a second close encounter with Earth's gravitational field in December 1992. If these complicated and lengthy maneuvers succeeded, Galileo would at last be aimed toward Jupiter, there to drop off its aging atmospheric probe and position its oft-redesigned orbiter in September 1995, just eleven years later than originally planned, and eighteen years after the project had triumphed in the full House of Representatives.

Galileo returned to JPL in February 1987, like a fallen warrior brought home for final rites. Instead of burial, though, a miracle of rejuvenation is being attempted. Extensive work is under way to recondition it to survive the intense solar heat at Venus's orbit. Furthermore, Galileo's electronic components are already beginning to exceed their planned total shelf life.

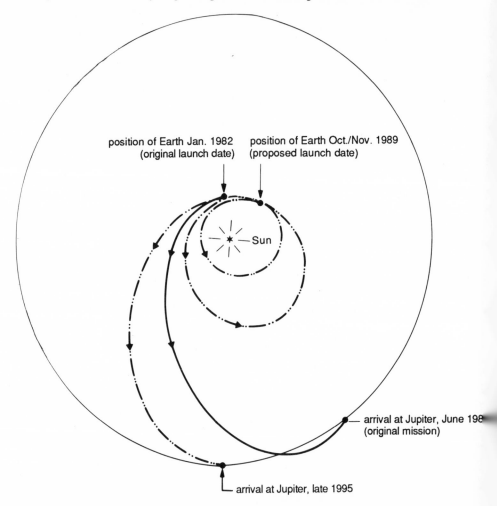

14.2 THE SLOW BOAT TO JUPITER • The original trajectory leaving Earth in January 1982 and arriving at Jupiter in June 1985 is shown as a solid line. The dashed line depicts the current plan, according to which an October–November 1989 launch projects Galileo first inward toward Venus, in order to gain energy by a close passage by that planet. Then its path carries it by Earth twice more to gather additional energy through carefully designed close passages. Finally, in December 1992, it will begin its journey outward to Jupiter, arriving there in late 1995.

Thus, major electronic reworking is also required. Most serious of all, the critical plutonium-fueled nuclear batteries (RTGs),* assembled in the late 1970s, are approaching their original design lifetimes.

Galileo is hemmed in on every side by dark clouds, so towering that even the most innovative mission analysis and engineering may not be able to overcome them. Even as Casani and his aging team struggled through Galileo's most challenging transformation, their nemesis—the Shuttle development—readied another blow. In July 1987 NASA delayed the first flight of the Shuttle subsequent to *Challenger*'s explosion from February 18, 1988, to June 2, 1988. Then, in March 1988, the flight was delayed first to August 1988 and later to the end of September. Subsequent flights are also delayed (see figure 12.2). Galileo's "slot" on the manifest remains October 1989, but it is now only the seventh scheduled flight in the post-*Challenger* series, rather than the fifteenth.

Perhaps Galileo will reach Jupiter in 1996 and start radioing back new secrets about the Sun's greatest companion. I hope so. But if it does not, perhaps that much modifed Galileo spacecraft should finally be housed in the National Air and Space Museum. There it could serve as a billion-dollar monument to a time in America when political mediocrity triumphed over technical competence and dedication.

* Additional complications ensued when *Challenger*'s explosion forced a re-analysis of the possibility that a similar explosion on Galileo would contaminate Earth with plutonium. Nuclear batteries power Galileo (as they did Voyager and *Pioneers 10* and *11*) because there is very little sunlight at Jupiter. Plutonium is their principal constituent.

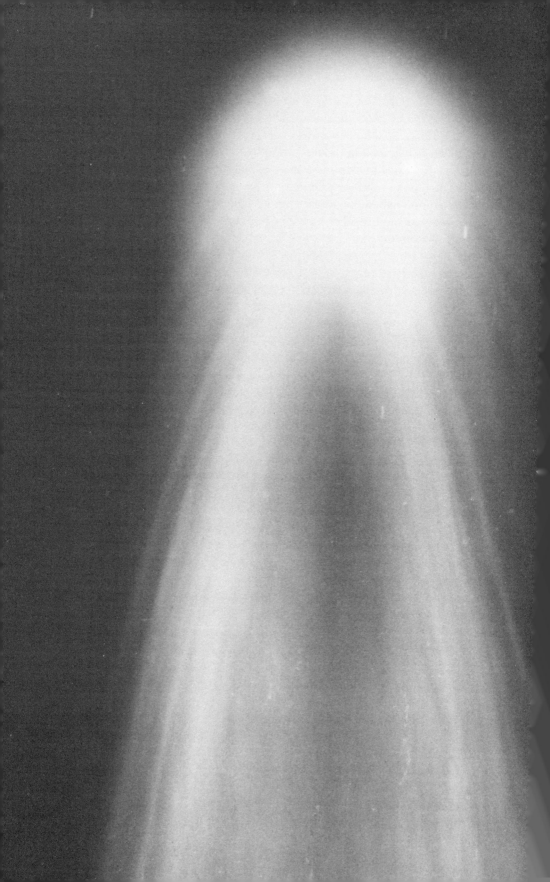

PART FIVE

Comet Tales

15.1 HALLEY'S COMET IN 1910 • The bright coma of Halley's comet is seen at the head of a divided tail in this famous photograph taken at the Mt. Wilson Observatory in May 1910. The giant cloud of dust and gas at the comet's head, termed its coma, is over 100,000 miles (160,000 kilometers) across. Dust and gas spew out from a solid nucleus, which is far too small to be resolved from Earth. Photo courtesy of the Mt. Wilson and Las Campanas Observatories, of the Carnegie Institution of Washington.

15.2

Chasing Halley's Comet

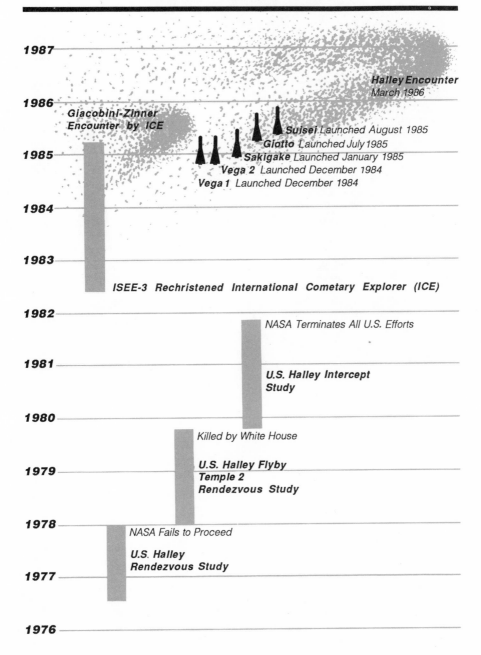

1987

Halley Encounter
March 1986

1986

Giacobini-Zinner
Encounter by ICE

Suisei Launched August 1985

Giotto Launched July 1985

1985
Sakigake Launched January 1985

Vega 2 Launched December 1984

Vega 1 Launched December 1984

1984

1983

ISEE-3 Rechristened International Cometary Explorer (ICE)

1982
NASA Terminates All U.S. Efforts

1981
**U.S. Halley Intercept
Study**

1980
Killed by White House

1979
**U.S. Halley Flyby
Temple 2
Rendezvous Study**

1978
NASA Fails to Proceed

**U.S. Halley
Rendezvous Study**

1977

1976

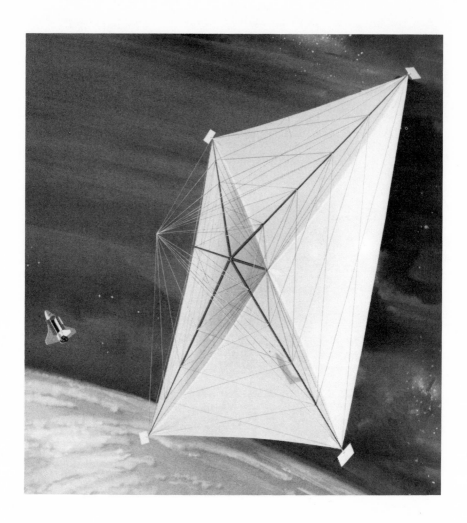

15.3 SOLAR SAILING TO HALLEY • An artist's conception in late 1976 of a giant solar sail that JPL proposed to build. The sail would have been intricately folded up and then carried into orbit by the Shuttle. There it would be unfurled and properly oriented. The pressure of the Sun's radiation reflecting off its highly reflecting surface would gradually push the sail onto a path to join with Halley's comet in early 1986. Once flying "formation" with the comet had been achieved, a scientific spacecraft would have been released and the large sail jettisoned.

15. To Catch a Comet

"WHEN WAS THE LAST TIME anyone told you you were not being imaginative enough?" I asked Dr. Johnny Gates, veteran designer of planetary missions, who answered with a broad grin.

We sat in my office in JPL's administration building. It was June 1976, and I was trying to tackle JPL's most serious problem now that the Shuttle had replaced the Moon as NASA's primary goal—how to plan future planetary missions that appealed to both scientists and the public.

Dr. Thomas Paine, NASA's boss through the halcyon days of the Apollo lunar voyages, had fought hard to keep lunar exploration going. He had also proposed rocketing Americans to Mars as the next "giant step."* But his successor, James Fletcher, wanted NASA in the human-piloted space trucking business.

Fortunately, an extraordinary family of robotic explorers was spawned in the final three years before Apollo's fire dimmed. As a consequence, Fletcher's 1971–77 reign at NASA was notable chiefly for the dramatic Jupiter missions of *Pioneers 10* and *11, Mariner 10*'s first look at Mercury by way of Venus, and, for a grand finale, Viking's search for life on Mars. Even the Grand Tour concept, which proved too big for post-Apollo realities, carried enough Apollo-era momentum to midwife the live birth of the Mariner / Jupiter / Saturn (MJS) project in 1972. MJS, renamed Voyager, would dazzle the world through the end of the 1980s.

As Apollo faded into NASA's luminous past, new planetary endeavors suffered from reduced priority for space *exploration* of any kind. Planetary *science* became just another area of scientific self-interest, struggling for a seat at NASA's overcrowded space-science table. The dominant figure in

*The grandeur of Paine's vision for America in space emerged fully when he became chairman of the National Commission on Space, in 1985–86.

that crowd was space astronomy. Its twin advantages—a large scientific constituency and undemanding launch vehicle needs—fit neatly into Fletcher's Shuttle focus for NASA.*

By the time I took the reins at JPL, in April 1976, the advocacy of bold, new planetary adventures had largely withered away at both JPL and NASA headquarters in Washington. To be sure, some of the bright and committed space scientists drawn to JPL (and to all parts of a younger NASA) in the 1960s still had stars in their eyes in 1976. But at NASA few listened now, and JPL's own institutional energies were largely devoted to making the upcoming Viking and Voyager missions successful.

"We've got to move fast," I told Johnny Gates. "The greatest collection of science and space journalists since Apollo is moving in on the lab. We've got to take advantage of that opportunity."

Those journalists were restlessly waiting for Viking's first look at the surface of Mars—and the first chance to encounter extraterrestrial life. Until that first landing, though, the reporters in our press room needed some news—*any* news—to feed to their impatient editors. Here was a perfect sounding board.

Gates and I quickly assembled JPL's most imaginative scientists and engineers. Their assignment was to come up with leads on exciting new missions.

Thus, the Purple Pigeons were born. I coined this flamboyant term to symbolize the imaginative thinking we hoped to get from this group. Purple Pigeons would be missions of high scientific content, but they would also have excitement and drama to garner public support. By contrast, Gray Mice, as we called them, were scientific missions of interest mainly to a few scientific disciplines. One such valuable but narrow scientific mission—which I tried, unsuccessfully, to sell in 1976—involved the placing of a polar-orbiting satellite about the Moon in order to get better measurements of its geochemical and other properties. (These fanciful categorizations were ultimately acknowledged by my fellow scientists. In March 1977, at the Lunar and Planetary Science Conference in Houston, Texas, where most of the scientists especially concerned with the Moon gather annually, I was given a special award—a T-shirt on which a gray mouse was clearly victorious over a prostrate purple pigeon.)

JPL's Purple Pigeon factory warmed to its task. In three months the group stamped out enough challenging projects to keep Earth's planetary explorers busy into the next century:

- Wheeled robots on Mars, roving about, gathering samples of the red

*Substantial opportunities for astronomical research can be created simply by placing dedicated instrument packages above Earth's atmosphere, even in the low-altitude, low-inclination orbits dictated by the Shuttle's limited range of operations.

planet's soil and rocks, to be automatically returned to Earth. This is now the subject of intense study as a possible joint U.S.-USSR endeavor in the late 1990s (see chapter 22).

• A long-lived scientific observatory, set down on one of Jupiter's large moons (a logical follow-up to Galileo).

• A Galileo-like orbiter of Saturn, delivering an entry probe into its pearl-colored atmosphere and a longer-lived probe to be parachuted into the hydrocarbon-rich atmosphere of Titan. This concept evolved into a mission concept termed Casini, which is under consideration by NASA and the European Space Agency as a possible joint mission after the end of this century.

• A high-resolution radar-mapping orbiter of Venus—a concept now embodied in the Magellan spacecraft, which awaits its delayed visit to Venus, perhaps in 1990. The Soviet spacecraft *Veneras 15* and *16* carried out an earlier version of this mission in 1983.

• And the biggest hit of all—solar sailing to a rendezvous with Halley's comet in late 1985 and early 1986, as it made its seventy-six-year swoop through the inner Solar System.

Mark Twain had been born and had died during the two preceding passages of this most active of all the predictable comets. Surely America—the leader in space and proud of its frontier origins, so well captured in Twain's writings—would not let pass this opportunity to touch and taste Halley's dust.

A sublime insight of modern physics is that moonbeams, sunlight, and all other light rays exert a tiny pressure on reflecting surfaces. In the 1920s the Russian space visionaries Tsiolkovsky and Tsander recognized that, in principle, sunlight could exert enough pressure on very thin, very large reflectors in space to propel them, and attached payloads, on journeys through the Solar System. Since then, flight by sunlight, a kind of space yachting, has fascinated space scientists. (See figure 15.3.) Freeman Dyson and Robert Forward speculated, in a 1980 Caltech workshop on interstellar propulsion, that massive arrays of giant lasers located on the Moon could propel a sail more than six hundred miles (one thousand kilometers) in diameter and have it carry a robot emissary all the way to another star.

Jerome Wright, an intense, red-haired space engineer, was working quietly at Battelle Memorial Institute in 1975. His job was to analyze deep-space transportation technologies, even those with little practical significance, like solar sailing. NASA's space technology planners wanted such analyses to improve their long-range view—and to cope with brilliant outsiders like Richard Garwin, of IBM, who had challenged Administrator Fletcher on solar sailing.

Jerry had an insight. Solar sailing worked better as the sail approached the Sun. At Mercury, for instance, solar sailing could be ten times more efficient than at Earth. Perhaps a giant solar sail—ten football fields across—could be orbited from Earth inward, nearly to Mercury's orbit, there to build up velocity and hurl itself up out of the plane of the planets and even into the opposite direction of their orbital motion about the Sun.

He hit on such a bizarre maneuver because that is the path Halley's comet takes every seventy-six years (see figure 15.4).

Halley spews out a great cloud of gas and dust (its coma) and hence is usually conspicuous during its periodic passage by the Sun. Sky gazers, from the time of the ancient Chinese and Babylonians down to the present day, have described this awesome spectacle. (In 1986 Halley's passage was at an unfavorable orientation for earthly viewing, especially in the Northern Hemisphere. In contrast, the spectacular 1910 apparition was the most exciting public event of the year.) Such unpredicted astronomical phenomena were once widely believed to presage momentous changes on Earth. In 1692 Isaac Newton demystified the comet. He recognized from centuries of historical reports that a comet had appeared in the same part of the sky and then moved in a similar way through the constellations at intervals of about seventy-six years. His friend Edmund Halley in 1705 concluded correctly that it was the same comet returning and, on that basis, predicted the year of its reappearance.

The great comet returned as Halley had predicted, assuring lasting fame for Halley and giving all the advance notice that mission planners could ever desire. There was time to design the solar sails that would permit the scientific payload to "reach out and touch" Halley's mysterious nucleus. Then, after a blazing passage by the Sun, the sail and payload would accompany the comet out beyond Neptune. In the year 2061 the (by now) silent spacecraft would return with Halley's comet to the inner Solar System for the next apparition.

Now, *there* was an idea worthy of Mark Twain's frontier energy, curiosity, and optimism. A rendezvous with Halley's nucleus meant that the spacecraft could make a relatively leisurely approach to its target, with ample time to record the unusual physical processes taking place there. In contrast, an ordinary flyby spacecraft on an intersecting trajectory would flash through the diffuse coma surrounding Halley's solid nucleus at an astonishing closing velocity of about forty-four miles (seventy kilometers) per second.* These extreme encounter circumstances for a Halley flyby

*The very high closing velocity is a consequence of Halley's retrograde and very eccentric orbit. This closing speed is approximately ten times higher than that of any previous spacecraft encounter. (See figure 15.4).

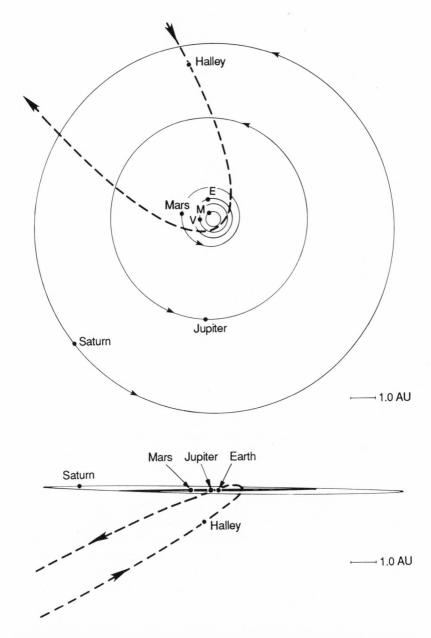

15.4 HALLEY'S PATH THROUGH THE SOLAR SYSTEM • Halley's comet travels on an elongated elliptical path that is highly inclined to the plane of the planetary motions about the Sun (*lower diagram*). In addition, it travels in the opposite sense of the planet's motion, in a retrograde orbit (*upper diagram*). In 1948, when at its most distant point from the Sun (its aphelion), Halley's comet was slightly more distant than is the planet Neptune. The distance from the sun to the earth, one *astronomical unit* (AU), is indicated for scale.

made a rendezvous with the nucleus of the comet—not just a flyby—crucial to the winning of a scientific endorsement.*

I thought solar sailing was a perfect Purple Pigeon. When used to send a spacecraft to rendezvous with Halley's comet, it became the Purplest Pigeon of all. The sail would be very large and, of course, reflective—so bright as to be visible to the eye in *daylight* at some locations. Indeed, *Time* magazine picked up on our studies and publicized the concept on March 14, 1977. I thought we could easily sell the public on this exciting project. So, in September 1976, those in the Purple Pigeon factory at JPL were put to work on an intense feasibility study of solar sailing to Halley's comet.

Johnny Gates always kept a sharp eye out for creative zealots. His organization already had enticed Jerry Wright to JPL six months before the color purple became popular there. In September 1976 Wright was named the chief mission analyst of our Halley's solar sail.

But a concept is one thing and a real sail another. What of space structures, thermal control, attitude control, and all the other critical features of the totally new sail concept? And who should head the feasibility study? For a recommendation, I relied on one of JPL's most experienced and colorful figures, a tall, cigar-smoking Texan, Jack James.

Jack had been project manager of *Mariner 4,* and he well understood that JPL's glory days would be numbered unless our space engineering laboratory could adapt to national changes and learn how to sell its proposed projects.†

But I had a problem with Jack's choice as head of the solar sailing project, Louis Friedman. Friedman, a six-year veteran of mission design at JPL, was of the same mold as Gates's other protégés—exceptionally bright and dedicated to the dream of space. However, Friedman was also a rebellious child of the 1960s, a frizzy-headed, antiauthority figure who delighted in scoffing at institutional procedures, ranging from parking regulations on up. This had brought Friedman and me into major conflict, three months earlier. Some JPL employees, led by Friedman, had bypassed NASA and gone directly to the Office of Management and Budget to promote "Project Columbus." Columbus was a hopeful solicitation to Congress and the OMB for a ten-year commitment—one new spaceflight

---

*From here through chapter 17, I draw heavily on the detailed narrative prepared by John Logsdon, of George Washington University. Logsdon also had access to most of my papers in his preparation.

† In the passionate 1960s, Jack, as JPL assistant laboratory director (that is, vice-president), had such a deep belief in space and in America that he wrote a personal letter to President Lyndon Johnson expressing his concerns about both. This was not standard procedure in a tightly run organization like NASA!

per year leading up to 1992, the five hundredth anniversary of Christopher Columbus's voyage. The idea was not practical.* Worse, in my "honeymoon" with NASA, I wanted to demonstrate organizational loyalty and faithfulness to bureaucratic channels. I needed to counterbalance my own earlier reputation for independence.

Knowing Friedman only slightly, I moved to fire him on the spot, as a signal to NASA of JPL's tighter ship. Fortunately, Jack James's cooler head prevailed. So Friedman survived and now reappeared as the proposed leader of my new pet project.† Because of my respect for James's judgment, I yielded.

Friedman did an outstanding job of gearing up the solar-sailing initiative. Giant strides were taken in identifying ultrathin plastic sheeting (one-tenth the thickness of Saran Wrap) that could withstand the intense solar irradiation and the harsh vacuum of space. Great sailing ships, ten football fields (one kilometer) across, took shape on computer drawing boards.

Somehow, though, money had to be found quickly if solar sailing was going to have a chance. No new money was forthcoming. We would have to scoop up unspent funds from other NASA space-technology tasks, a practice that created new internal enemies for the sail project.

In fact, to my surprise, sailing to Halley's comet was drawing more detractors than supporters within NASA. Few at the other NASA centers showed enthusiasm for a new technical challenge, especially with an un-manned system. Moreover, this was JPL's show. Others saw it mainly as additional pressure on NASA's resources. The most determined opposition came from NASA's own Office of Advanced Technology, which had invested about $75 million over fifteen years trying to develop a totally different approach to low-thrust‡ propulsion. In this approach, panels of solar cells as long as a football field would be unfurled in space (instead of the even larger sail). The large electric current created as the Sun's light fell upon the solar panels would power a battery of miniature particle beam accelerators. The electrically charged particles—in this case, ionized mercury like that in mercury vapor street lamps—would be accelerated elec-

*Both Congress and the OMB jealously guard their role in approving each new space endeavor (unless included in an Apollo-like, long-term goal). Each planetary mission is a "new start." Long-term, open-ended, Columbus-like concepts conflict with the political reality for Congress to maintain fiscal and political control.

†Within three years I fully institutionalized free-lance political activities far exceeding those of Friedman's. And in late 1979 Friedman and I joined Carl Sagan in founding the Planetary Society, to carry the fight for planetary exploration directly to the public. Since then Lou has become a close day-to-day collaborator.

‡The low thrust in such schemes is applied continuously over months or years, thereby achieving an accumulated orbital effect larger than that available from conventional solid- or liquid-fuel rockets, which burn fiercely, but only for minutes.

trostatically and expelled from a nozzle to create continuous thrust. The solar electric propulsion system (or SEPS) concept is complex, and it lacked the aesthetic and popular appeal of the idea of solar sailing. But large, sustained funding by NASA over fifteen years had built a broad constituency—three NASA centers (Lewis, Marshall, and JPL) as well as Boeing, Lockheed, TRW, and other aerospace companies.

NASA's Office of Advanced Technology sought persistently to find a niche for SEPS in Earth orbit and always failed. The intense radiation trapped in the globe-girdling Van Allen belts is very bad for solar panels. Ordinary solid- or liquid-fuel rockets can handle things in Earth orbit just fine, so SEPS had become an expensive prototype technology in search of a deep-space mission, for instance, a comet rendezvous. There its virtues would really count—a low-thrust capability to slowly modify the shape and orientation of the path of a spacecraft about the sun.

If upstart solar sailing were selected to power a trip to the heart of Halley's comet, fifteen years of SEPS effort would probably be abandoned.

On September 30, 1976, I stood to present the case for solar sailing in James Fletcher's blue-carpeted conference room at NASA headquarters. Three built-in back-lit screens stood ready behind me. We had carefully prepared and rehearsed the story of JPL's solar sail to Halley's comet. Now we needed NASA's endorsement. Fletcher sat in the middle of the large semicircular table facing me. His key headquarters aides and the directors of some of the other NASA centers flanked him. Three rows of theater-style seats in the rear were packed with lower-level NASA officials who had a stake in the outcome.

The presentation went perfectly. I finished up enthusiastically, and Fletcher evinced a strong positive reaction—in front of his top aides. That should swing the balance, I thought. "We've got it."

Then as Fletcher stood to leave, he looked over to his staff and said, "Of course, we'll also have to consider SEPS."

Solar sailing never had a chance from that minute on. The final choice dragged out until the following July, two months after Fletcher had left NASA. But when Fletcher tossed our solar-sailing concept into a shootout with NASA's own perennial favorite, solar electric propulsion, the die was cast. Given $75 million over fifteen years, NASA technologists were more comfortable contemplating SEPS's known shortcomings than the unknown problems surely waiting to be encountered in the much newer solar-sail concept. Fletcher, in his one-sentence directions to NASA headquarters, gave no weight to solar sailing's broad imaginative appeal, even though he personally reacted as enthusiastically as anyone.

I did not see Fletcher again until August 1978, at a space promotion in Lyons, France, that featured a covey of astronauts and cosmonauts as well as scientists like Carl Sagan. Albert Ducrocq, France's premier exponent of space and exploration, had organized a lively affair to promote space. In the midst of it was Fletcher, who, after the usual pleasantries, asked, "What ever happened to solar sailing, Bruce? That was a great idea." He did not even realize that he had doomed it.

Very well. If solar electric propulsion is the only way NASA will catch Halley's comet, so be it. A rendezvous with Halley's comet by any means is still a Purple Pigeon. Mark Twain's legacy could still be fulfilled.

Once Fletcher declined to give the project priority, however, NASA politics took over. The Shuttle czar, John Yardley, decided that SEPS was a "propulsion stage" and assigned its development to the Marshall Space Flight Center. Marshall was already struggling to deliver the Shuttle's solid-fuel rocket boosters, and the IUS upper stages loomed as the next challenge. Few there cared about Halley's comet, and even fewer cared to risk any new technical challenges. The predictable took place. Bureaucracy and complexity increased on the SEPS project; enthusiasm and imagination decreased.

Marshall's estimate of the cost of catching Halley's comet skyrocketed. SEPS alone would cost $200 million to $300 million. Worse, its projected performance by Marshall slid and was barely adequate to overtaking Halley at all. The total cost of a rendezvous with the comet by means of SEPS had grown larger than that of either of the other well-established new-start candidates for NASA—the Gamma Ray Observatory (GRO) or JPL's own Venus Orbiting Imaging Radar (VOIR).

The Halley rendezvous finally faded completely from NASA's bureaucratic process in September 1977, just fifteen months after my hopeful conversation with Johnny Gates. America was not going to stretch humankind's reach to Halley's comet at all.

The Purplest Pigeon was dead.

16.1 PROTON LAUNCHES VEGA • The powerful Proton launcher (shown diagrammatically in figure 2.2, p. 53) was photographed blasting off from the Baikonur Cosmodrome in December 1984. It placed one of two identical VEGA spacecraft on a path that intersected in March 1986 with the nucleus of Halley's comet. *Official Soviet photo*

16. Falling Behind

THE RENDEZVOUS WITH Halley's comet was dead for seventy-six years. But NASA's SEPS stage survived the death of that rendezvous in incubation at Marshall Space Flight Center. Propulsion politics can confer immortality on a well-entrenched technology. What should NASA do next? Find a more easily encountered comet for a later rendezvous. However, a computer search of the orbits of known comets revealed only a few, drab possibilities, all burned out from countless close passages of the Sun.

Unfortunately, volatile-rich cometary bodies do not behave as scientists would wish them to. Pristine comets orbit quite leisurely in a great spherical shell thousands of times farther from the Sun than is Earth.[2] On occasion two icy bodies will, by chance, collide within this cloud of debris from the Sun's own birth. Or a faint gravitational tug from a distant passing star will send one of these fossils falling slowly toward the Sun. Gradually, over millennia, the cold, dark object accelerates. Finally, a brilliant display will light the skies of Earth, Venus, and Mars for a few days as the new comet makes its close passage of the Sun.

For many of these comets this first pass by the Sun will be their last. The gravitational pull of Jupiter (and sometimes of other planets) often adds a little extra speed to the icy projectiles' paths—just as Jupiter slung *Pioneers 10* and *11* and *Voyagers 1* and *2* outward to explore more-distant realms. That extra push casts them out of the Solar System altogether, to become wanderers in the cold, silent depths of interstellar space.

Other pristine comets, however, are imperceptibly slowed by Jupiter's gravitational tug, and this assures their return to pass by the Sun for at least one more time. Such a comet may return again and again—to match wits with Jupiter, the ruler, to see whether it will stay, or leave forever. Each repeated encounter with Jupiter increases the prospects for the comet's permanent tenure in the Solar System. And the frequency of fiery encounters with the Sun speeds up as Jupiter repeatedly sucks energy from the comet's orbit. Over millions of years Jupiter's great gravity field domesticates once-pristine comets, reshaping the wild orbits initiated by the original chance fall toward the Sun. Little by little, the mature domesticated comets mimic the orbits of asteroids. Indeed, some asteroids are quite likely dead comets, stripped of their primordial organic-rich ices and other volatiles by repeated close encounters with the Sun.

Here is the dilemma that confronts a comet explorer. The most attractive cometary nucleus for robotic rendezvous, and for in situ chemical analysis, is one least altered from its pristine state. Halley is the best example, the most luminous and volatile-rich comet whose orbit path and time of return can be predicted for a spacecraft encounter. But undomesticated comets like Halley are mostly still in "wild" orbits. To make a spacecraft from Earth catch such a comet on its wild ride is so demanding technically that it requires something as bizarre as a giant solar sail.

Other comets are easier to reach and more predictable because Jupiter has shrunken and reshaped their orbits during many passages about the Sun. But most of their original volatiles have been lost as a result of prolonged solar heating. Sampling such cosmic slag by means of robots would be scientifically interesting, of course, but that slag is so heated and devolatilized that it is only a residuum of the original cometary material. The chemical signatures of the birth of the Solar System have been altered or even lost.

Furthermore, pieces of such old, devolatilized comets may already have been analyzed, right here on Earth. Objects in the crowded asteroid belt collide frequently. Their debris is splattered throughout the inner Solar system, and some strike Earth. These are meteorites.* A few contain as much as 10 percent dark organic-rich matrix and are called carbonaceous chondrites. Present opinion holds that some carbonaceous chondrites may be crusty residues of long-dead comets, now masquerading as dark asteroids. A Halley rendezvous could have confirmed that.

But now that rendezvous was a dead issue. Those who had argued for a rendezvous rather than for a flyby faced a huge obstacle. Only the old, nearly dead comet Encke could be chased, in 1987 or 1989, or the slightly more luminous comet Tempel 2, in 1988 and every three years thereafter. No others were brought into range by the projected capability of the expen-

---

*In addition, some microscopic sphericles recovered in stratospheric sampling and from polar ice sheets are believed to be tiny bits of old, dead comets swept up by Earth's atmosphere.[3]

sive Marshall SEPS stage. Rendezvous with just any comet is not a Purple Pigeon.

Then inspiration struck. Perhaps a SEPS-powered spacecraft traveling toward a rendezvous with the comet Tempel 2 in 1988 could be aimed near enough to the comet Halley that in 1986 it could release a flyby probe. That probe would then speed daringly close to the Halley nucleus. Meanwhile, a full retinue of remote-sensing instruments would monitor Halley from the distant, SEPS-powered mothership.

This idea of a comet mission attracted both broad public appeal and strong scientific support, the Halley Flyby Tempel 2 Rendezvous (HFB / T2R.) It combined long-duration direct investigation of the older Tempel 2 nucleus with a close flyby as well as distant remote-sensing observations of the large, volatile-rich comet Halley. Comparative cometology could take place. The public would get a Halley flyby; the specialists would have the opportunity to study a comet during a rendezvous. The Marshall Space Flight Center would get a customer for the SEPS stage and the aerospace industry a fat contract.

Then someone else—perhaps stimulated by independent German studies—saw how to make the big package even more salable to NASA and the White House.[4] The Europeans could build the Halley flyby probe to be carried by the American SEPS-powered mothership. That way, U.S. costs would be reduced, and there would be some political benefit to the United States for bringing in its Western European allies as junior partners, while the crucial rendezvous science would remain in American hands. So here was a bright new concept for budget time in 1978.

Halley / Tempel 2 had a lot going for it, but it was not at the head of the line. Advocacies of new science missions were piling up as unprogrammed Shuttle costs mounted and launch opportunities diminished. In late 1976 the massive Space Telescope project had won White House approval, along with JPL's Jupiter Orbiter with Probe (soon to be renamed Galileo). A year later the International Solar Polar Mission (ISPM) slipped through an increasingly restrained financial environment as President Jimmy Carter, who had promised "no new dreams," was learning the high cost of the Shuttle dream he had inherited from Richard Nixon.

A setback for all space science followed in the fall of 1978. The Carter White House approved neither the Venus radar endeavor (VOIR) nor the proposed Gamma Ray Observatory (GRO). The space-science line did not move forward at all that year, and the possibility of gently touching a comet once again dimmed.

So, by early 1979, we at JPL began thinking seriously about breaking Halley / Tempel 2 into two separate parts. First, a Halley flyby would be developed at relatively low cost. Then the more expensive SEPS-powered rendezvous mission for Comet Tempel 2 would be brought along for a launch a year or two later. That seemed the only way to spread the cost and gain new mission approval. Otherwise, the expensive Halley / Tempel 2 rendezvous mission would have to leapfrog VOIR and push the strongly supported GRO out of the way. After no new starts in 1978, surely only one new science endeavor *at most* would gain support by the White House at budget time in the fall of 1979.

When NASA tried out the idea of splitting the Halley / Tempel 2 mission on its Comet Science Working Group, the answer was swift and uncompromising. The working group "was unanimous for the HFB / T2R mission as a first choice" and would "recommend an in-depth study of additional options (such as Halley flyby) ... only if the HFB / T2R failed to achieve a new start."[5]

So NASA was stuck, obtaining no flexibility from its scientific advisers in the face of a budget shortage. With JPL's concurrence, it decided to ask the White House only for the SEPS stage in the budget process in the fall of 1979. GRO and VOIR would, of course, be requested again. The Tempel 2 rendezvous spacecraft, and its European-developed Halley probe, would be proposed by NASA to the White House the following year. Perhaps the queue would be shorter then. It was reasonable to hope that if the White House committed itself in 1979 to the development of SEPS, a go-ahead for the Tempel 2 mission in the next year or two would logically follow. There was no other planned use of the SEPS.

This sophisticated strategy of NASA's kept the near-term budget obligations low and still capitalized on the strong special-interest support for SEPS business for Alabama's Marshall Space Flight Center. Representative Ronnie Flippo, the Marshall area's congressman, sat on the House authorizing committee, while in the Senate authorizing committee Senator Howell Heflin also kept a sharp eye out for new business for his home state.

In September 1979 Carter's OMB required each agency to rank its funding requests by priority. NASA submitted a budget containing eighty items, from continuation of the Shuttle on down. The science favorite, GRO, ranked twenty-ninth, SEPS a surprising thirty-sixth, and VOIR a distant fifty-sixth. SEPS (leading, it was hoped, to the start of a comet rendezvous mission the following year) was now ranked ahead of VOIR for the first time.

But Carter's OMB had its own surprises; it mandated no new starts, for science or other realms—for the second year in a row. The Shuttle's insatiable appetite for dollars was the culprit.

On November 29, 1979, NASA Administrator Robert Frosch and Deputy Administrator Alan Lovelace pitched directly to Jimmy Carter the notion that GRO and SEPS were essential to maintain U.S. space-science leadership. They had by then dropped the VOIR solicitation for 1979 to try to strengthen their overall position. Carter asked Frosch to assess the gamma ray mission and the SEPS / comet mission. Frosch, a nuclear physicist by training, replied, "Halley is very spectacular and in many ways important. ... If you asked me where I think there is a Nobel Prize, it's probably more likely from the data from GRO."

Carter, a naval nuclear engineer by training, had been reading a popular book on black holes by Walter Sullivan, of the *New York Times*. He agreed that both were good projects but noted that he was "partial to the gamma ray thing because of the connection with the black hole problem."[6]

Thus did the dream of the Halley Flyby / Tempel 2 Rendezvous abruptly end—in the Oval Office itself, wrapped in musings by the most scientifically literate President since Thomas Jefferson. Ironically, GRO still awaits its Shuttle launch, now scheduled for 1990, whereas HFB / T2R would have gotten off before *Challenger*. Carter's personal reservations about the Shuttle were well founded. But he was trapped between the open-ended Shuttle commitments of his predecessors and the short-term focus of his own top aides. Through it all, he never learned to deal with the symbolism of space. Jimmy Carter was too honest to mislead but too narrow to lead.

We at JPL were disappointed, though not surprised. In fact, starting a year earlier, when NASA bowed to the comet scientists' stubborn insistence on a comet rendezvous—or nothing—JPL quietly conceived of a Halley flyby spacecraft to be built up from spare Galileo hardware. This approach was cost-effective, and it ensured exceptional imaging and other remote sensing. Galileo's hardware represented the state of the art in space exploration. Furthermore, we put adversity in one mission to good use in another by showing how the Halley Intercept Mission, christened HIM, could be handled by the Galileo mission operations team. HIM would be launched in mid-1985 to encounter Halley in March 1986. The Galileo spacecraft would at that time be quietly cruising to Jupiter. The same team could operate both.

Although no other country could match America's capability for a Halley mission, many were making plans for real projects. The European Space Agency (ESA) started designing a short-duration probe of Halley's nucleus to be carried on the Halley Flyby / Tempel 2 Rendezvous mission. Then Japan had quietly announced its first deep-space endeavor—Suisei—a diminutive probe of the large coma of Halley. Suisei (whose name means "comet" in Japanese) would pass over 100,000 miles (160,000 kilometers) from the Halley nucleus but would be well positioned to monitor the comet's influence on the electrical and magnetic properties of space. It would also record the appearance of the coma every few hours over an interval of weeks.

Most alarming to me was news about the Russians. My French friend Jacques Blamont reported that the Soviets sought to extend their next Venus mission (in which the French were involved), slated for launch in December 1984 and for arrival in June 1985, to go on to encounter Halley in March 1986. I was skeptical. This would be a uniquely ambitious endeavor for the Soviet Union. But Blamont said a new force in Soviet space science, Roald Sagdeev, director of Moscow's Institute for Space Research, was rising.

The Japanese and the Soviet-French efforts to explore Halley's comet were joined by an independent European one. The European Space Agency (ESA) was left in the lurch at the end of 1979 by the demise of the Halley Flyby/Tempel 2 Rendezvous, for which ESA had been developing a short-lived Halley probe. By then, many European scientists had had enough of their junior partnership with NASA. Key NASA decisions affecting Europe seemed to be totally dominated by domestic—even intramural—political factors.

On January 31, 1980, ESA concluded in a memorandum to its scientists that "it is extremely unlikely that a new collaborative mission between NASA and ESA can be set up" and added that "very preliminary ideas about a possible low cost European backup mission" had been put forth.[7] In fact, nineteen European scientists proposed in a January 29 telex that a simple spinning spacecraft (of a type previously flown to study Earth's magnetosphere) be adapted to make a close pass of Halley's nucleus. The spinning spacecraft would be called Giotto, after the Italian artist who included Halley's comet in a medieval religious painting.

The primary objective of the spinning ESA probe Giotto would be to sample the ionized particles and dust spewing out from the solid heart of Halley. American scientific prejudice notwithstanding, the European scientists felt that such sampling, even at very high velocities, could reveal the composition of the nucleus itself. Some kind of camera would also be carried. But photographing a comet at a closing velocity of forty-four miles (seventy kilometers) per second and doing so from a spinning spacecraft, was a tenuous proposition. ESA had never flown a deep-space mission, much less one as difficult for picture taking as Giotto.

Giotto quickly gained support from the European space-science and industrial communities. Nevertheless, in March 1980 ESA still offered NASA the possibility of collaboration on its mission if NASA would launch Giotto on the reliable Delta expendable rocket. The European Ariane rocket could also launch Giotto, but this would greatly increase European cost. Furthermore, Ariane had flown only once at that time.

According to NASA's notoriously optimistic schedule (see figure 12.5), the Shuttle would be flying twenty-four flights a year by 1986. The Delta rocket was to be phased out before the 1985 Halley launch opportunity. So NASA insisted that Giotto ride "free" on the Shuttle instead. ESA gave a firm no to that offer. For it to fly on the Shuttle, Giotto would have to undergo expensive modifications. In addition, the Europeans had reservations about the Shuttle, which had not yet flown.

On May 21, 1980, the Giotto issue came to a head in the same NASA conference room where, two and a half years earlier, Jim Fletcher had doomed solar sailing to Halley's comet. This time I listened to the NASA space-science boss, Tim Mutch, plead with *his* bosses for a much more modest NASA step toward Halley. "Just plan to keep the Delta line going long enough to launch Giotto in 1985," he argued. That way, he explained, American scientists could carry out first-class comet science, and for a small fraction of the cost for a dedicated U.S. mission. America, though not the leader, would at least be a participant in the Halley "Armada."*

The meeting did not go well. Alan Lovelace, NASA's hard-driving deputy administrator who carried the day-to-day burden of getting the Shuttle flying, put it bluntly: "If we let *them* have one, then *everyone* else will want one too." That killed any realistic chance of U.S.-ESA cooperation on Giotto—sacrificed on a minor altar in the temple of the Shuttle.

On July 9, 1980, ESA slated Giotto as its next scientific mission. It would be an all-European endeavor now launched on the Ariane rocket.

Three months later, on October 22, 1980, NASA Administrator Frosch had an eleventh-hour change of heart concerning Giotto. (He had meanwhile failed to include any NASA Halley mission in his September 1980 budget submission to the OMB.) Frosch told the director general of ESA, Erick Quistgard, that NASA was now "prepared to provide a Delta launch vehicle to launch the Giotto spacecraft." Quistgard firmly declined, noting that the Delta offer was "controversial in Europe" because its acceptance now would "imply the dismissal of the Ariane as a Giotto carrier." To make the new political circumstances perfectly clear, he allowed that the

---

*A moniker coined by the historically minded Jacques Blamont.

ESA Science Advisory Committee "wished to underline Europe's capability of carrying out this mission in its entirety."[8] Another defeat for the spirit of the Atlantic alliance.

The demise of the Halley Flyby/Tempel 2 Rendezvous exacerbated JPL's tenuous role within NASA. The NASA brass were increasingly concerned about their ability to "support" JPL, the only one of the (then) eleven NASA centers not staffed by government employees. Despite our close day-to-day partnership, JPL was still an expendable NASA contractor compared with NASA's own civil-service cadre.

And things were not going well with the Department of Energy, the second federal sponsor for JPL. Jimmy Carter had earnestly declared the struggle for self-sufficiency in energy to be "the moral equivalent of war." Bud Schurmeier had led a thoughtful and vigorous attempt to use JPL to help fight that "war." But energy research was not one of America's political priorities in the late 1970s. High inflation, rising interest rates, and unemployment were far more important to the average American. Worst of all, in Congress, energy funding replaced the Rivers and Harbors Act as the focus of pork barrel politics, making it especially difficult for JPL to secure challenging work if it relied only on its reputation and expertise. Was it time to look to the Defense Department for challenging work?

JPL had, in fact, been spawned out of a wartime partnership between Caltech and the U.S. Army. Von Karman's students secured their first contract, for $1,000, in 1940 from the Army Signal Corps. Right after World War II ended, JPL jumped into the guided missile business in a big way. Most of JPL's top managers had cut their teeth on America's first surface-to-surface missiles, Corporal and Sergeant. Parks, Schurmeier, James, Giberson, even Heacock and Casani had all started with crew cuts, photo ID badges, and a uniformed sponsor. Indeed, that in-house competence built up in service to the Army enabled JPL to push America into space with *Explorer 1* in 1958.

By 1980 those looked like "the good old days." Many questioned whether really challenging technical work for JPL might still be available in defense.

But Voyager's extraordinary journey past Jupiter in 1979 opened a gateway. Those finicky Voyager robots were the most autonomous satellites in the world—high-tech defense satellites included. Decades of continued escalation of Soviet and U.S. strategic forces had made space communications and reconnaissance critical to American security. Voyager's autonomy pointed the way to new satellite robustness. Voyager had, after all, been designed to survive Jupiter's lethal radiation belts. Defense planners believed a similar hardening was necessary to make military satellites resistant to enemy nuclear blasts that might be set off in Earth orbit.

So I began negotiations to transfer Voyager's hard-won lore to Air Force satellite designers. Lieutenant General Richard Henry, commander of the Air Force's Space and Missiles Systems Organization (SAMSO), was an old hand in space technology. He relished the idea of getting JPL involved in Air Force work. SAMSO is headquartered in El Segundo, forty miles southwest of Pasadena. Barely a decade earlier I had sat in SAMSO's heavily guarded vault, along with other members of the space panel of the President's Science Advisory Council. There we had watched a sobering slide show of a massive Soviet buildup in practically every area of strategic weaponry.

Now, in June 1980, I was back at SAMSO headquarters, in General Henry's blue-carpeted conference room, banked with scale models of some of the exquisite satellites SAMSO had created over the years. With me was Dr. Marvin ("Murph") Goldberger, Harold Brown's successor as president of Caltech. His presence was important to convey to General Henry the seriousness of our intent to work with the Air Force. That part went very well. But the next step was a bigger challenge. Murph and I would in a few days face a Caltech faculty, highly skeptical about JPL's searching for wartime work in times of peace.

The Hall of Associates in Caltech's hallowed faculty club (Einstein once slept upstairs) was packed. Over a third of Caltech's 250 faculty members had gathered on a warm June afternoon two days before the 1980 graduation exercises. A year earlier I had been the graduation speaker, exhorting my comfortable professorial colleagues to embrace the future rather than resist it, to welcome institutional change and experimentation while tiny, elitist Caltech was at the top of the heap of research and teaching institutions. But "The Challenge of Success"[9] did not strike a responsive chord. Caltech faculty members are often told they are great, they think they are great, and they usually can get all the money they need to carry out research. Why change?

This June, I had to be more persuasive in winning approval for a specific change at Caltech. I had to get the Caltech faculty to accept the idea of JPL's helping the Air Force develop better autonomous spacecraft. In order to do the Air Force job, JPL specialists would have to learn about the purposes and the designs of some Air Force satellites. They had to have access to top-secret information; JPL would have to take one small step back toward its classified, DoD origins.

In 1980 Vietnam was still a dissonant memory, especially in the university community. Fortunately for my purposes, Caltech is a largely apolitical university. Its only god is Science and its only fashion individual research excellence. There are no campus political activists, student or faculty, such as those that threw Harvard, Berkeley, and Stanford into turmoil in the 1960s and 1970s. Most Caltech faculty are technology oriented. Furthermore, they do not see institutions in stark black-and-white the way politically oriented people sometimes do.

In fact, the most telling faculty objection to my plan came from my friend John Benton, a tough-minded medieval historian who showed a keen sense of the present. He argued that individuals grow to emulate the ideas and attitudes of their sponsors. What then would be the long-term impact of defense sponsorship (in peacetime) at JPL? This indeed was my own deep reservation. How long would that special JPL culture flourish— nourished by an open exploring of the planets—if technological excellence was achieved by one guy's becoming better than the other at killing? How long could optimism survive, surrounded by pessimistic assumptions?

In Pasadena, that June of 1980, we debated in microcosm a central issue that America would face through the eighties. Should America lead by example or by force?

As it did in America in the eighties, force triumphed. My expedient— and truthful—arguments were roundly supported. JPL won faculty acceptance of its need to take on a small amount of classified Air Force research to help tide over spacecraft expertise until NASA once again voyaged into deep space.

Unfortunately, I was less effective in lobbying NASA on behalf of the Halley Intercept Mission. NASA seemed content to watch other nations lead the way to Halley. By August, HIM was generally considered dead. To make matters worse, Tim Mutch, less than a year after coming to NASA headquarters (on leave from Brown University), and an explorer of Earth as well as of Mars, died tragically in early October while leading a group of students on a climbing trip in the Himalayan Mountains. His engineering stand-in, Andy Stofan, deputy associate administrator for space science, like so many at NASA, was primarily interested in new launch vehicles— especially the Shuttle/Centaur. Although an effective advocate and a highly capable engineer, Stofan understood means better than ends.*

---

*After selling the Shuttle/Centaur concept, Stofan was appointed by James Beggs to be director of the Lewis Research Center, the center designated to develop that new rocket stage. When Jim Fletcher became NASA administrator again in mid-1986, he persuaded

In September 1980 Bob Frosch submitted NASA's proposed budget to the OMB. VOIR was there, as expected, but not HIM. Three weeks later he announced his own imminent departure.

NASA, like Jimmy Carter's presidency, was down and out. Technically, there was still time to develop an American mission to Halley's comet, but not through NASA's initiative. It would have to be taken directly to the next occupant of the White House.

Andy to come back to Washington and lead the Space Station effort. Stofan shepherded the project through an official, if muted, start in Congress in the fall of 1987 then left NASA for the aerospace industry.

17.1 ARIANE LAUNCHES GIOTTO • The midsize Ariane launcher (shown diagrammatically in figure 2.2, p. 52) blasted off from the Kourou, French Guiana, launch site on July 2, 1985, and placed the Giotto spacecraft of the European Space Agency on a path to intersect the nucleus of Halley's comet in March 1986. *Photo courtesy of ESA*

17. Halley and the White House

CALTECH ATTRACTS prominent and well-connected figures to its board of trustees. The biggest trustees' meeting of the year is held over a late October weekend in Palm Springs, where diamond-studded dining is mixed with tennis and customized education is mixed with Caltech business. At the 1980 gathering the mood was especially festive. The former California governor Ronald Reagan's presidential victory seemed close at hand. A number of the trustees were his close friends and major contributors. A few were members of the kitchen cabinet.

On Saturday morning, October 25, my Caltech assignment was to provide an educational and entertaining slide show with commentary concerning *Voyager 1*'s imminent encounter with Saturn. But JPL is business, too, Caltech's business. Caltech receives an annual fee for running JPL. So I also soberly laid out the Shuttle-caused decline of NASA, the developing threats to JPL's viability as a NASA center, and the importance of the Halley's comet mission for the United States and for JPL.

The trustee Earle Jorgensen quizzed me further as we sipped coffee in the sunshine outside the meeting room. Jorgensen's tanned appearance, still blondish hair, and crisp tennis game belied his being nearly eighty years of age. Having immigrated to the United States from his native Denmark as a young cabin boy, he had built a fortune in the specialty-metals business. I knew Jorgensen better than some of the other trustees because he had recruited me two years earlier to serve with him as a director of the Kerr McGee Corporation, headed by his fellow Caltech trustee Dean McGee. Conversations that October morning revealed that the Jorgensens were frequent companions of the Reagans.

I decided to make a crash effort to enlist Jorgensen's help. At my home, on Sunday, November 2, some key JPL staff and I put the final touches on a two-page epistle entitled "Halley's Comet: A Symbol of America's Future." It would have to be our best shot at piquing the interest of the next President and his closest advisers—perhaps even an item for his first State of the Union Message.

It was unabashedly chauvinistic, concluding with a strident appeal for presidential action:

> NASA needs to be asked by the President to add the Halley's Comet mission to its objectives in order to renew the Agency's sense of its national importance. Our country needs a commitment to the future symbolized by the Halley mission in order to believe again in that future. Friends and adversaries alike must know that we are not abandoning our leadership in space exploration, nor by implication that we are withdrawing from other areas of international competition.

> If the new Administration doesn't act swiftly, history will record that our Western friends and Soviet adversaries were first to explore Halley's Comet. And America will have to wait for another chance in the year 2061.

On Tuesday, November 4, as tens of millions of Americans voted all over America, I met with Jorgensen at his steelyard office in South Central Los Angeles. Yes, he liked our paper and would see what he could do. It was good that I got there early, because he would be leaving soon with his wife to go to Reagan's private victory party.

Jorgensen was true to his word. That very night he passed our letter on to Michael Deaver, Reagan's closest personal aide.

The following Tuesday, November 11, 1980, Voyager skimmed by Saturn's cloud-shrouded moon Titan. On the twelfth it swooped by Saturn itself. Increasingly detailed views of Saturn's rings and moons flowed from JPL's computers. Press attendance shot up (especially to fill in the post-election news gap). Here was a press agent's dream.

In little more than a week, I gave interviews to journalists from the *New York Times*, the *Wall Street Journal*, the *Washington Post*, the *Los Angeles Times*, the *Chicago Sun-Times*, the *Boston Globe*, the *Miami Herald*, the *San Francisco Examiner*, the *San Francisco Chronicle*, and the *Philadelphia Inquirer*. Both *Time* and *Newsweek* made Saturn their cover story, and there were features in the *New Yorker, Newsday, Discover, Science 80,* and *Science.* ABC provided extensive television coverage, as did CBS radio and the BBC.

Interest in the future direction of the country had been enhanced by the presidential election and now combined with the glorious visual encounter with Saturn to stimulate feature stories asking, "Where next is the United States headed in space?" Halley was my answer.

The extensive news coverage and my personal interviews had an unexpected positive side effect at NASA headquarters. Andy Stofan and other key officials struggling to keep things together in Washington felt more vulnerable than ever. Administrator Bob Frosch had given notice and was traveling overseas. A new face would soon be in the White House, alien both to Washington and to space. Down at Kennedy double shifts of laborers were madly glueing tiles on a Shuttle orbiter that might or might not fly.

On Thursday, November 13, Stofan, showing a good systems engineer's interest in elucidating all the options, responded faster than most of his peers at NASA headquarters to the new reality looming on Pennsylvania Avenue. He acknowledged to me that the concept of the Halley Intercept Mission should be kept alive a while longer.

On that same day, the Caltech trustee Stan Rawn, Jr., a member of my JPL advisory council, sent off a "Dear George" letter to Vice-President Bush, urging his attention to the Halley possibility. He attached a copy of our "Halley's Comet" piece.

Key JPL staff* scratched hard to figure out how to reach every member of the new White House staff just as soon as he arrived in Washington. Another Caltech trustee, Mary Scranton, was very helpful. Mary, the wife of the former Pennsylvania governor and one-time Republican presidential hopeful William Scranton, could open many important doors. She contacted the incoming transportation secretary, Drew Lewis, an old Pennsylvania political ally. Lewis promptly assigned a key aide to advise us on approaching the new faces appearing at 1600 Pennsylvania Avenue. A meeting with the new chief of policy development, Marty Anderson, was set up (actually materializing in March).

Then Stan Rawn learned that his good friend James Baker III was to be a member of the troika running the Reagan inner sanctum—with Reagan's cronies Michael Deaver and Edwin Meese. A "Dear Jim" letter went from Rawn on February 3, reiterating the earlier Halley arguments and noting that rumors of impending budget cuts threatened the entire U.S. planetary enterprise.

In the three months since my election day visit to Jorgensen's office, we had touched practically all the bases in the White House. The two conspicuous omissions were Ed Meese and the Office of Management and

---

*Led by John Beckman, a founding member of the Purple Pigeon factory. I am indebted to John for keeping safely the notebooks that documented our activities in 1980–81 and that even included quotes from key personages. We tape-recorded the main telephone reports that those of us in Washington and elsewhere called in each day. John also recorded key interviews at JPL. His job was to extract the main points from the tape-recorded phone calls, meetings, and interviews and to create a daily agenda. Hence the notebooks, which were not available to Logsdon for his Halley's comet narrative.

Budget (OMB), now under the direction of David Stockman, a Harvard-educated congressman known for his devotion to something called supply-side economics.

On February 5, 1981, we got a stunning glimpse at the new OMB through Jim Van Allen. Van Allen was then taking a year off from his duties at the University of Iowa to write a memoir describing the beginnings of the space age.[10] For this purpose he was housed at the National Air and Space Museum, in Washington. He naturally ran into Washington friends from time to time. On one such occasion he received a casual invitation to ''come in and chat'' with Hugh Loweth, a senior OMB official who handled the NASA, NSF, and other science budgets.

When Van Allen arrived at Loweth's office, a block from the White House, he immediately sensed tension. His friend Loweth (a man highly regarded by a whole generation of American scientists) and two other familiar persons greeted him—but more like poker players than old acquaintances. What was the game? Why was the conversation so restrained?

''Van,'' said Loweth, ''we have problems in OMB with space priorities.''

Van Allen responded by emphasizing the importance of planetary missions, especially Galileo, and expressing his disappointment in the treatment of HIM. Loweth, after reminding Van Allen that he had been a friend of science for a long time, admitted that he had a terrible problem: Stockman was demanding draconian cuts in space science. It would not be possible to maintain the pattern of one new-project start a year.

''What would happen if there were no new starts for five years?'' Loweth asked. Then he added the clincher: ''What if JPL were to disappear?''

Any hopes we might have had that Reagan's election would improve the environment for new planetary missions vanished when we learned of Van Allen's conversation at the OMB.

Stockman had a long hit list, it turned out, and turmoil soon reigned at NASA. Gone was the American part of the International Solar Polar Mission; gone was the Venus radar mission, VOIR; and gone was the hard-won Centaur stage for the Shuttle.

Ironically, the Halley Intercept Mission was the *only* new NASA endeavor that the OMB staff was even considering. A friendly congressional staffer, on very good terms with the OMB, brought us that news—along with the comment ''OMB doesn't trust NASA, especially concerning Centaur.''

Now, the trick was somehow to keep HIM alive as an option through 1981. That would still leave three and a half years to design and build the spacecraft before its launch in mid-1985. Our politically oriented appeal to the White House should get a better hearing after the first round of Stockman's bloodletting was over. The Russians really were coming. That was a message hard to ignore, even by those who saw America's international leadership solely in terms of force and power.

Into this chaotic NASA world then stepped Hans Mark. In March 1981 he reemerged from Reagan's landslide as the designated new deputy administrator of NASA. Mark's true-blue Teller pedigree and his cultivation of congressional conservatives overshadowed his key role in the "liberal" Carter Defense Department.*

Hans would not be confirmed until June, along with the new NASA administrator, Jim Beggs. But, with or without portfolio, Hans tirelessly worked his own launch vehicle agenda, trying to overturn NASA's recent Shuttle/Centaur decision. He favored a move back to the Air Force IUS stage mounted within the Shuttle (see chapter 13). Mark was interested in JPL's Halley flyby perhaps in part because it might have helped rationalize splitting up the big Galileo mission to fit once more onto two Shuttle IUS launches. In any case, Mark promised to raise the HIM issue with Reagan's national security adviser, Richard Allen.

On March 26 I finally met one of the new people at the Reagan White House. Martin Anderson had worked at the Heritage Foundation, the conservative Washington, D.C., think tank, before being asked to come to the White House. From his office in the White House basement, Anderson did the work on staff policy issues for his boss, Ed Meese, who determined which issues Reagan saw.

Stockman's key aide from the OMB, Fred Khedouri, was at our meeting to keep things under control. JPL had prepared a special rendition of "The Russians Are Coming," not just emphasizing the Halley encounter but presenting new data on the upcoming Soviet Venus radar mission as well. This was fortunate because I rapidly discovered that competition with the Russians struck the single responsive chord. Otherwise, Anderson seemed interested only in "privatizing" NASA. In a government-wide search for functions to turn over to the private sector, he was trying to build an analogy between NASA, primarily an R & D organization, and civilian operational functions like the postal or prison systems. He seemed unaware that no business sector was being excluded by NASA's in-house activities. In fact, the profit for the private sector was in selling goods and services to NASA, for which the aerospace industry was well organized and moti-

---

*Quite a remarkable feat. Hans was probably the only ranking Carter appointee to survive into Reagan's first administration, where even middle-of-the-road Republicans were usually blackballed.

vated. Except in minor cases, such as civilian remote-sensing operations, privatization simply did not make sense for NASA.

Later that week I returned to California from my eighth East Coast "marketing" sojourn in four months. I was still a bit stunned from my encounter with the personable ideologue Marty Anderson. Was *this* kind of mentality going to rule the country now?

Most of all I wondered whether we had slipped our message, that the Russians were coming, to all the key places in the White House? Vice-President Bush had responded to Stan Rawn's "Dear George" letter of November 13 with a pleasant but noncommittal "Dear Stan" response. His caution was to be expected. As Reagan's recent rival in the primary election, Bush was still viewed suspiciously by conservatives for his characterization of supply-side economics as "voodoo economics."

Reagan's personal aide, Michael Deaver, had been reached through Earle Jorgensen on election day. Presumably, he had forwarded our missive appropriately, but we did not know where. James Baker III, the newly appointed White House chief of staff, had responded on March 5 to Stan Rawn's February 3 letter apologizing for the delay, but commenting only vaguely on Halley. Further inquiry at his office made it clear that Baker deferred to Ed Meese on NASA. Baker focused on economic matters. National Security Adviser Richard Allen probably had been, or shortly would be, exposed to "Halley and the Russians Are Coming." We hoped he would be supportive if the issue was ever raised by Meese.

OMB Director Stockman would oppose any new NASA endeavor. In fact, he had already recommended "waiting another seventy-six years" to do the Halley mission. However, we knew of support within Stockman's staff for HIM as the single high-visibility but affordable new NASA endeavor. Thus, in the long run he might be persuaded to accept this one, small program even while cutting many others.

That left Ed Meese as the key. We had to get our message to him directly and soon. But how? Our underground drive to the White House was bogging down. NASA's ax hung over the HIM effort and soon would fall.

On May 20 a direct connection to Ed Meese turned up. A top JPL manager knew a California political insider, Robert Finch. Finch had been Nixon's first secretary of health, education, and welfare and had also served Reagan while he was governor of California. Finch really did know Meese, as well as Reagan's close political adviser Lyn Nofziger. He would get the "Halley and the Russians Are Coming" story to Meese. He felt there was a good chance Meese would take it to Reagan.

On June 26, 1981, Finch reviewed our material in the White House with Meese and Nofziger. They liked it. Yes, it would go to the President within a week or ten days.

But the days dragged by. No news. Finally, on July 8, Finch called us to say that Meese was upset by the overruns NASA was showing and that he "wanted to understand the NASA situation before going to the President."

Welcome to the world of the Shuttle, Ed.

Once the euphoria had worn off after the Shuttle's first flight, on April 12, 1981, the Reagan White House received its first Shuttle budget jolt. NASA would need an extra billion or so in order to launch the defense payloads on schedule.

Stockman was adamant. Nothing else for NASA.

At almost the same time, I had independently learned that the OMB was indeed "working a big new Shuttle problem." Now I was sickened to realize that our Halley proposal had been within days of reaching Reagan's desk—with Meese's endorsement. Reagan would surely have snapped it up, I felt. Then the same old Shuttle problem reached out to poison the deliberations.

Finch reported that the Halley briefing books had been taken to the President, although it was not clear to him whether Reagan actually looked at the material. It *was* clear to us that HIM still dangled, alive if unsure, at the very entrance of the Oval Office.

A week later, on July 28, 1981, I had my first private meeting with Reagan's new NASA administrator, Jim Beggs. We sat together in the same pleasant office where I had spun out comet tales for previous NASA administrators—Jim Fletcher and Bob Frosch. Now I pushed hard with Beggs for the Halley flyby mission as a unique, affordable opportunity for the United States and for NASA. And I had some new luster to add to the HIM story: it had just been recognized that a Halley flyby spacecraft could follow an orbit that eventually brought it back to Earth. (Shades of *Mariner 10* and the repeated passages by Mercury, I thought to myself.)

This new orbital possibility was immediately termed HER, Halley Earth Return. The special excitement about HER was that it could be used to bring some of the actual atoms and molecules from Halley's dust cloud back to Earth.* The same kind of dust shield we planned for HIM—a two-layer affair that harmlessly exploded the rapidly impinging dust particles on the first shield and then collected the vaporized debris on a second, stronger layer behind it—afforded an elegant way to sample the comet

---

*But only after a lengthy Sun-circling trip; the return to Earth would have been in August 1990, over four and a half years after the Halley encounter.

itself. The second layer of the spacecraft's dust shield could be designed to *preserve* Halley's dust debris free from contamination. That shield would later be rolled up like a window shade and squeezed into a small reentry package, like those used to parachute exposed film from spy satellites back to Earth. Waiting to analyze the contents would be in this case not generals but an eager delegation of the world's leading geochemists.

Later that busy day (July 28), I met Reagan's new science adviser, George Keyworth. He had come to the second-floor corner office of the Old Executive Office building (vacated five months earlier by Carter's science adviser, Frank Press) from the Los Alamos National Laboratory, in New Mexico. Keyworth had been a midlevel figure there, relatively unknown outside the nuclear weapons scientific establishment. Earnest, loyal, and hardworking, Keyworth nevertheless did not enjoy the prestige of Carter's Frank Press or of Nixon's science advisers, Ed David and Lee DuBridge. Clearly, Keyworth was not a potential threat to the policies or ambitions of any major figure in the Reagan White House.

His job, I presumed, was to help Ed Meese's staff in technical areas. What I did not realize, as I chatted with him, was that he already was the White House contact with Beggs concerning the Halley's comet mission. Our efforts earlier in the year to enlist Marty Anderson had yielded some results after all. Anderson reportedly had raised the Halley question with Keyworth, stating, "I know exactly what the President wants with the space program, and I am worried about Halley's comet. We do not seem to be in the position to make a strong American effort when Halley arrives."[11]

Our Anderson thrust, reinforced by the dealings with Meese and Nofziger, produced some stirring in the White House. On August 5 Keyworth wrote to Beggs asking him officially to "develop several options for carrying out a U.S. mission to Halley's Comet."[12]

On August 17 Beggs responded officially, noting that "the idea of taking a close look at Halley's comet has caught the public imagination" and that "the administration should make a positive response."[13]

But there was a problem. Both sides were trying to push the cost and responsibility of the Halley mission onto the other's territory. Keyworth in his letter expressed the hope that "most, if not all of these options would fall within existing budgetary guidelines." Beggs was now free to back the Halley mission, but to do so he would have to swallow the added Halley costs.

Beggs's agenda, like Fletcher's and Frosch's, was the Shuttle, and now also the Space Station, objectives to which Halley did not directly contribute. So he passed the buck. NASA officially asked the Space Science Board to evaluate whether a U.S. Halley mission would provide enough "science increment" over Giotto and other planned foreign efforts to justify the $300 million or more involved. The official answer, in a letter dated August 7, reiterated the board's long-standing position that good comet science required a rendezvous mission. It did admit, "The proposed Halley Intercept Mission (HIM), while scientifically excellent, does not provide a major capability to advance our understanding of comets beyond that expected of Giotto."[14] Beggs got what he needed—another scientific put-down for HIM.

However, the science advisory committee did respond enthusiastically to the new HER ideas about obtaining actual samples of volatilized dust particles from Halley. In fact, we at JPL were busy planning how to combine HIM and HER (without getting far enough to develop an appropriate acronym). "Altogether we see the possible sample-return mission as an opportunity with a high potential scientific return and with some—but yet unknown—degree of risk of partial or complete failure. . . . Addition of the Halley Earth Return mission would constitute a commendable United States response to the Halley apparition," stated the committee.

As if according to a Hollywood script, the moment our Halley advocacy climaxed in Washington, *Voyager 2*'s encounter with Saturn lured nearly all the Washington players to Pasadena. The only ones missing were Reagan and Bush and OMB Director Stockman (who stayed home, we later learned, to prepare a devastating new assault on all planetary exploration).

As Meese stepped off his helicopter to a red-carpet greeting at JPL, he was followed by Deputy White House Chief of Staff Richard Darman, Cabinet Secretary Craig Fuller, and Presidential Assistant Joseph Canzeri. They were joined shortly by Science Adviser Keyworth and NASA Administrator Beggs in JPL's impressive Space Flight Operations Center. There, surrounded by color TV monitors showing beautiful, never-before-seen views of Saturn, Ed Stone, of Caltech, spewed out lucid scientific commentary to this powerful audience. And I spewed out "The Last Picture Show" and "The Russians Are Coming," punctuated with exhortations for a Halley mission.

Just one year after NASA headquarters had consigned the Halley mission once and for all to a silent bureaucratic death, it was playing center stage before a blue-ribbon White House audience.

But we still had to produce a specific Halley package that both Beggs and Keyworth could endorse. On September 1, I wrote Beggs proposing that high-quality imaging be combined with an inexpensive and moderate-risk sample-return package. That pairing would combine the maximum scientific priority of sample return with the popular and politically appealing imaging. Thus, America's mission would complement, not compete with, Giotto's in situ physical analyses. But the Soviets would face an unbeatable scientific and popular combination.

Here was a package for everybody. NASA's fractious scientific advisers would be mollified by the inclusion of sample return. The White House could not help noticing NASA's responsiveness. JPL, approaching institutional instability as the hammer blows of Stockman and the Shuttle continually battered the work force's hopes, would finally receive tangible evidence of a renewed national interest in its primary product. And the in-house space mission expertise at JPL would be saved for at least another four years.

Most important, Halley's comet would become a symbol for all America, signifying its return to leadership in space exploration rather than its decay.

On September 17, I wrote to Stofan once again to emphasize the growing institutional crisis at JPL. However, it made no difference. NASA had already given the White House its answer the day before. Beggs's September 16 letter to Keyworth laid it out in nonnegotiable terms. It was no for any Halley mission without a big NASA budget increase. Taking refuge in the narrowness of his scientific advisers, Beggs tossed the ball directly back to Keyworth:

Following completion of our budget review at NASA, we have not included a Halley Comet mission in the FY1983 recommended budget. This is primarily because in the prevailing climate of fiscal restraint, the cost is not justified considering the probable science return ranked against competing science priorities.

Then he transformed that "ball" into a live hand grenade, laying down a two-week ultimatum to the White House:

As you know, the development schedule for either Halley option requires a near-term decision in order to get underway by January of 1982 and requires a substantial increment of FY82 funding. If I do not receive a positive Administration commitment to a Halley Comet mission by October 1, 1981, I will have to terminate all efforts in support of such a mission.

I was stunned as I read a copy of this hostile NASA response to the White House, pirated out of NASA headquarters by a high-ranking NASA friend (who himself soon left that rapidly declining institution). Neither American space leadership, it seemed, nor continuation of planetary exploration, nor JPL, nor good relations with the White House warranted a conciliatory effort by NASA on Halley's behalf. NASA's answer was flatly no.

On September 30 Stofan, "with deepest regrets," officially directed JPL "to discontinue all ... activities associated with the Halley Intercept Mission."

It was all over.

18. Trajectory Change

THERE WAS NO TIME to lament the death of HIM or HER or the end of the golden age of American planetary exploration.

JPL itself was in danger. Stockman's second budget assault, beginning in September 1981, was far more threatening than his first had been. Ronald Reagan insisted on increasing defense spending and on decreasing taxes. Stockman had to make unprecedented cuts in civilian government expenditures. Now he strove to lop off *all* planetary exploration. He reasoned, probably correctly, that the only way to shrink NASA substantially was to cut entire categories of work—along with the NASA centers involved. NASA's own actions (and Deputy Administrator Hans Mark's words) clearly demonstrated that planetary efforts were NASA's most expendable category.

JPL had to negotiate a long-term defense role, not just an occasional exotic task like transplanting Voyager's smarts into future Air Force space robots. Otherwise, JPL staff would begin to make their own decisions. JPL employees live within commuting range of the largest concentration of aerospace activity in the world. Nothing tangible holds JPLers in Pasadena. Only belief in their work.

18.1 JPL • The NASA/Caltech Jet Propulsion Laboratory occupies 145 crowded acres adjacent to a sand- and boulder-strewn dry riverbed at the foot of the rugged San Gabriel Mountains. In 1938 the Caltech graduate students of Theodor von Kármán selected this site, at the time uninhabited, to test pioneering concepts of rocket propulsion. They secured their first federal funding in 1940, a $1,000 contract from the U.S. Army Signal Corps. However, because "Rocket Research Laboratory" sounded too fanciful then, the technically equivalent term "Jet Propulsion" was chosen instead. JPL shifted from military-rocket research to scientific-spacecraft development in 1958, after designing, manufacturing, and flying the United States's first successful satellite, *Explorer 1*. *NASA/JPL photo*

But what defense role should we seek?

We knew that the Air Force's space efforts were already well supported by two dedicated nonprofit laboratories—MIT's Lincoln Laboratories, outside Boston, and the Aerospace Corporation, located right beside General Henry's Space Division Headquarters, in El Segundo. Beyond them a covey of Air Force civil service laboratories were busy throughout the country. The Air Force was well covered. (In fall of 1981, Reagan's Strategic Defense Initiative was two years away.)

The Navy was even less accessible to JPL. It had long been in partnership with the Applied Physics Laboratory (APL) of Johns Hopkins University, in Maryland. And the famous Naval Research Laboratory (NRL), outside Washington, D.C., was a vital in-house resource, among many others.

Only the U.S. Army needed a long-term relationship with nonprofit JPL. The Army lacked the high-tech traditions of its sister services, especially after losing Pickering's JPL and von Braun's missile group to NASA in 1958. By 1981 it was awash with expensive, complex, and independently conceived electronic and computer developments. Each of them was touted to revolutionize tomorrow's battlefield or training ground. What the Army lacked was battlefield *systems*. It needed an experienced and trusted nonprofit institution to help it develop sophisticated information systems for unsophisticated battlefield users—soldiers.

Another attraction for JPL is that Army developments generally warrant a security classification lower than that of the other services. Battlefield technology is more quickly revealed to adversaries than is, say, sophisticated spy-satellite technology. Hundreds of thousands of soldiers and civilians must build and use a new battlefield communications system. By contrast, only a few thousand Americans at most will be exposed to the inner workings of a new spy satellite.

But there *were* complaints from Caltech. Working for the Army just wasn't "fine enough." JPL had designed the best interplanetary spacecraft in the world. Why couldn't JPL now contribute to the best spy satellites, instead of mucking around with some noisy, dusty, and perhaps even bloody battlefield?

At 4 P.M. on October 20, 1981, I met with my Caltech colleagues. The sole topic was our proposal to move JPL into as much as 30 percent defense work on a long-term basis. A significant fraction of that 30 percent might eventually be classified, leading to the prospect of special access to buildings at JPL, of conspicuous photo ID badges, and of governmental pressure for uniform employee "clearability."*

---

*Meaning U.S. citizenship and, in some cases, no close relatives in Communist countries.

There was more faculty reluctance this time than in June 1980. But the necessity was higher. NASA had just struck back at Stockman by vowing to eliminate entire categories of non-Shuttle activities if forced to meet proposed OMB cuts over the ensuing years. Planetary exploration was at the top of NASA's cut list, and if it was eliminated, "JPL would become surplus to NASA's needs."[15] NASA was playing hardball with Stockman, and JPL was the ball.

Once again, in Caltech's dark, wood-paneled faculty club, pragmatism triumphed over idealism. A reluctant faculty accepted the less lustrous and more defense-oriented future for JPL.

At 10 A.M. on Friday, October 23, in a small meeting room at a fancy hotel in Palm Springs, I sat with the newly formed Trustees Committee on JPL. Earlier in the year I had persuaded Caltech's President Goldberger to form this ad hoc trustee group in anticipation of the crisis we now faced. Although JPL is larger and more populous than the entire rest of the Caltech campus and its budget much larger, administratively it was still treated as one of the many divisions of the campus. The JPL director was not even a vice-president of Caltech, nor was there any direct oversight by the trustees. In 1976, on becoming director, I had created the JPL Advisory Council, comprising faculty, trustees, and distinguished public members, to provide an informal board of directors. But by 1981 the Caltech trustees simply had to take more direct responsibility.

The trustees on the committee grasped the mounting crisis. Now they were ready to act, recognizing that a significant change in direction for JPL was essential.

Stanton Avery, chairman of Caltech's board of trustees (and founder of Avery International), opened up by wryly noting, "Optimists say this is the best of all possible worlds ... and pessimists agree." It was his gentle way of emphasizing the uncertainty in all best case–worst case analyses and of urging flexibility. Robert McNamara, Kennedy's and Johnson's secretary of defense and later president of the World Bank, endorsed the move to defense, as did Lew Wasserman, the long-time boss of MCA (Music Corporation of America) and a highly regarded observer of the national scene, with legendary skills in conflict resolution.

Successive endorsements continued as each trustee spoke in turn. Si Ramo (the R in TRW), approved, as did the former Caltech president (and Nixon science adviser) Lee DuBridge, under whom JPL had first metamorphosed from an Army caterpillar into a NASA butterfly. The independent investor and stalwart JPL supporter Stanley Rawn, Jr., added his support. Rube Mettler, a Caltech Ph.D. who had risen to become chairman and CEO of TRW (and would later succeed Avery as chairman of the Caltech board), likewise endorsed JPL's new direction. The former federal

THREE PRESIDENTS OF CALTECH • Lee DuBridge (1946–68) is flanked (*left*) by Harold Brown (1969–76) and by Marvin ("Murph") Goldberger (1978–87). All were in attendance at the Jupiter encounter of *Voyager 1* on March 1979. Thomas Everhart succeeded Goldberger and is the current president of Caltech. *NASA / JPL photo*

judge Shirley Hufstedler, a Pasadena resident once again after having served as Carter's secretary of education, supported it as "a reality check" in the face of near-term difficulties, as did Mary Scranton, chairman of this ad hoc committee.

Given these key trustees' unanimous support, there would clearly be no problem obtaining the full board's formal approval later that day.

Only one committee member demurred—Caltech's President Goldberger, the man who would have to help implement the change. Murph knew a lot about defense, having worked on early atomic bombs and having later served as a scientific consultant for the Pentagon. But his experience at Princeton University in the turbulent sixties had left him profoundly ambivalent about mixing academic and defense work. Now his personal professional focus was arms control, not arms development. He freely acknowledged the pragmatism of a JPL move toward defense, but his heart was not in it.

That afternoon, when the proposal was presented to the full board, reluctance showed through once again. Caltech's provost, Jack Roberts, a renowned chemist, in responding to a question from the floor, correctly characterized the faculty's support as "reluctant." That, however, was the wrong term to use with that singular group of strong-minded business leaders. Like a rhinoceros sensing an intruder, Fred Hartley, the outspoken head of the Union Oil Company, whirled to attack what he discerned to be flagging patriotism at Caltech. Not to be outdone on matters of country, a powerful railroad CEO sitting one row back took aim as well. Other gray heads in the audience began to stir. Things were getting out of hand.

I was still at the speaker's rostrum, in a position to deflect the stampede before anyone got gored. Yes, Caltech *is* committed to serving the national interest at JPL, I said. Now that the trustees are approving this practical new Caltech policy, proper JPL support of the DoD is assured.

But was it really? A hunting license was one thing. Bringing home a major defense sponsor for JPL was quite another. *Caltech* really had to be committed in order for *JPL* to pull that one off.

At least, NASA now openly supported JPL's new defense initiative—too openly on some occasions. On Tuesday, January 12, 1982, most of the top NASA brass gathered in JPL's largest meeting room for the annual performance review, or "report card," as it was called by the JPL staff. This novel annual exchange (invented in the late sixties by Dr. John Naugle, then NASA's top scientist) periodically discharged the tensions between JPL and NASA headquarters.

This time Deputy Administrator Hans Mark headed the NASA delegation. The Caltech administration was represented by Provost Jack Roberts. There was a gulf between those two.

No major problems emerged as each NASA headquarters office presented its report, and the appropriate JPL officials responded. Finally, it was time for Mark to sum up, presumably with some general remarks to set a constructive tone for the coming year.

But Mark was not in the mood for an inspiring "togetherness" speech. Instead, he upbraided me. I had placed JPL in grave danger by obsessively pursuing planetary missions, delaying too long the inevitable move into defense work.

I was stunned, as were many others in the room. The deputy head of NASA was chastising me publicly for pursuing too relentlessly a key part of NASA's charter! Slowly, I responded. "Hans," I said, "we took a risk in pursuing planetary missions because we *believe* in them.''

I felt something being passed to me under the table from the person sitting at my left, Bob Allnutt, then one of NASA's top officials. It was a piece of yellow writing paper folded into the shape of a medal. On it he had printed in bold letters ''HERO.''

I was not alone in believing that NASA's mission should be civilian achievement.

In fact, though, Mark's vision of the future was coming true. I had become a character in *his* vision, one already sketched out in a remarkable letter of April 9, 1974, seven years earlier. That letter was addressed to the head of the Caltech committee that was then searching for a successor to JPL's director, Bill Pickering. In it, Mark took himself out of consideration, because he foresaw (correctly) the approaching dilemma for Caltech over defense work at JPL, knowing that he would inevitably differ profoundly with the general Caltech view. ''Dear Bob,'' he wrote (to Robert P. Sharp, the committee chairman),

I must admit that I have had a very difficult time in sorting out my thoughts on the JPL situation. . . .

There is no question in my mind that JPL is a great laboratory. The recent *Mariner 10* results are another indication that this, indeed, is true. I must say that I am most flattered to be considered as a candidate for the Directorship of such an institution independent of whether I am finally offered the job.

The basic problem faced by the laboratory is that its purely NASA business will probably decline. This means that the laboratory must either be cut back or that it must find new business. . . .

The major opportunities for new business lie in the Department of Defense. This fact, in turn, leads me back to the dilemma that we explored during my visit to Pasadena last month. I am referring to the relationship between JPL and Caltech. My feeling is that the relationship that now exists between these two institutions should be thoroughly reexamined. As I see it there are two problems:

1. At present Caltech limits JPL's ability to seek new business by restricting classified work that can be done at the laboratory. This restriction must be lifted as soon as possible. I believe that the political situation in the country is changing so that such a move will be acceptable in the not too distant future. In any event, I am convinced that it is the right thing to do, not only for the nation, but also for the talented people at JPL who might otherwise have to leave and join institutions where classified work is permitted.

2. The connection that presently exists between the faculty at Caltech and the laboratory is too weak. As far as I can tell, the relationship is now restricted to only a few people. It may be true that many professors at Caltech, indeed, have some dealings with the Jet Propulsion Laboratory, but very few have the kind of relationship that is characteristic of someone like Bruce Murray. In my opinion, it

is relationships of this kind that really count in the making of a successful university-operated Federal laboratory. . . .

. . . Let me make just one final comment. If, as a matter of policy, you and your committee believe that JPL should remain part of Caltech, then I would suggest in the strongest possible terms that you select as the next Director, an active senior professor who is currently on the Caltech faculty. I believe only in this way will you be able to assure the proper relationships between the faculty and the laboratory are established. *I also believe that only a man who is highly respected by the faculty at Caltech could persuade the Institute to lift the restriction on classified work* [italics added].

In a handwritten note to me, dated October 14, 1981, transmitting a copy of his April 1974 letter, Hans summarized our different views of the possibility—or the impossibility—of America's continuing planetary exploration:

. . . Where you and I have differed over the years is in our judgment of whether the popular support enjoyed by the planetary exploration program can be translated into the necessary long-term political support to assure a stable level of funding large enough to carry out what the planetary community thinks of as an adequate program. I have never believed that this could be achieved and I still do not believe that it can be done. It is for this reason that I have urged—and that I continue to urge—that the leadership at JPL must take immediate and aggressive steps to get a strong and stable defense related program going at JPL. After having watched "big science" closely in the United States for almost three decades, there is no doubt in my mind at all that national defense is the only truly *stable* source of large research and development funds. . . .

Even now as I read over Hans's deterministic view of the inevitable decline of American planetary exploration, I think of key points where history might have turned differently and America's golden age of exploration have continued on:

• If in 1972 a more perceptive technocrat than Jim Fletcher had headed Nixon's NASA and had opted for a mixed-fleet approach to launch vehicles from the beginning instead of the flawed "Shuttle only" policy.

• If in 1977 almost anyone other than the Shuttle-phile Hans Mark had become undersecretary of the Air Force, he would very probably have moved toward a mixed launch fleet for the Air Force, as Pete Aldridge did later. This, in turn, would have maintained expendable-launch-vehicle alternatives and examples for the civilian program.

• If in 1977 or 1978, or even later, then Secretary of Defense Harold Brown had opted to help his boss President Jimmy Carter fashion practical space alternatives for the country that included a mixed fleet. With a more rational national space policy, successive NASA administrators might not have sacrificed American planetary exploration to the impossible Shuttle dream.

• If in 1977 I had blindly followed the "cause," rather than my own analyses of probable outcomes, and accepted Frank Press's request to become Carter's NASA administrator. I could not have prevented the Shuttle programmatic disaster—the vested interests were too strong—but I would have minimized the side effects. Certainly, I would have fought to put planetary missions on expendable vehicles.

• If in late 1979 (or earlier), when NASA was finally ready to back a Halley flyby, NASA's scientific adviser had endorsed that practical necessity rather than continuing to oppose it.

• If in July 1981 the next big Shuttle budget "surprise" had hit the White House just a week or two later, *after* Meese had taken the Halley proposal to Reagan.

• If in September 1981 Jim Beggs had been able to see far enough beyond his looming battle with the White House and David Stockman to support a U.S. Halley mission, instead of playing hardball with JPL and the whole planetary effort.

But back at the crisis-ridden JPL, in February 1982, any debating of determinism with Hans Mark's shadow was irrelevant. JPL needed a serious DoD sponsor, and quick.

We had a big one about ready to sign up. Reagan's undersecretary of the Army, James R. Ambrose, had worked with JPL decades before and formed a good impression. Now he wanted JPL to expedite the Army's application of the most advanced technology into a coherent set of operational systems to carry out its mission. Ambrose, sometimes referred to as *the* project manager of the Army, was moving toward a detailed memorandum of understanding with JPL. High-technology demonstration projects were planned. Better yet, the Army explicitly acknowledged JPL's need for substantial in-house support of high-technology efforts on a stable basis.

Here, finally, was JPL's savior.

Now, finally, I could plan to leave.

Leaving had been on my mind for some time. Ever since Shuttle trouble surfaced in mid-1979 to threaten the Galileo Jupiter mission, I had been working at the limit of my capacities, leading what proved to be a strategic retreat from JPL's golden age. I could take satisfaction that a modest American planetary exploration effort survived. It might have been worse. And JPL was still intact and would be stable with its new defense partnerships.

But watching up close the strangulation of America's space future was very distressing to me. I preferred to view it from afar—for instance, from China. In fact, I had quietly planned a trip to China for Suzanne and me in November 1982.

However, before such an alluring daydream could be seriously pursued, one big roadblock to JPL's advance remained. That Caltech "reluctance" concerning JPL's foray into defense work, which was conspicuous at the October 1981 trustee meeting in Palm Springs, continued to affect day-to-day decision making. Specific defense initiatives that JPL developed often had to wind their way through a diffuse and somewhat unpredictable approval process on the Caltech campus.

I outlined the problem in a terse confidential paper that awaited the JPL trustees as they gathered at JPL on March 18, 1982, for their semiannual meeting.

That essay began by formally announcing my decision to leave JPL later in 1982. It went on to suggest that it was really sensible for Caltech to bring in a new man at this point to run the new, multimission JPL of the 1980s. Changed external realities required changed internal arrangements at Caltech. The next JPL director must be a full vice-president of the institute, the current ad hoc trustee committee should be replaced by a formal JPL standing committee of the board with oversight responsibilities, and increased financial authority should be delegated to the JPL director.

The high-powered, busy people who composed the trustee committee on JPL had found time to make the journey to Pasadena for this March 18 meeting. They were not pleased to be forced to deal with new complexities as well. They felt they had done *their* job in Palm Springs in October when they approved the new JPL policy.

But they came through brilliantly. By October 1982 they had recruited an uncommonly well qualified new JPL director. Lew Allen had just completed a distinguished technical and military career, his last post being that of Air Force chief of staff. With a Ph.D. in physics, he pioneered early military space applications and later headed the high-technology National Security Agency. He was highly regarded by nearly everybody in Washington.

Allen was named a vice-president of Caltech *and* director of JPL. Trustee oversight was duly strengthened, and other procedural changes followed.*

* With a top-flight new director, and a clarified role in the Caltech family, JPL has prospered since 1982, becoming strong enough even to withstand the excruciating further disappointments and delays triggered by *Challenger*'s explosion in early 1986.

Most critical for my plans, on April 1, 1982, President Goldberger had signed the memorandum of understanding that committed JPL to long-term support of the Army.

On the next day, I gave my last "state of the lab" talk and announced the upcoming changes.

Saying good-bye over the next few months* to my many loyal, long-time friends and colleagues was difficult. Nonetheless, my step became lighter each week as I began to feel in charge of my own life once again. I had almost forgotten what an interesting world we live in and how privileged we are to experience it.

---

*A one-year Caltech sabbatical became effective July 1982. November of that year was spent in China and Japan. The following April, May, and June, the whole family resided in Paris. Three-year-old Jonathan even attended the United Nations Nursery School. Would that all kids, and parents, could share such an experience.

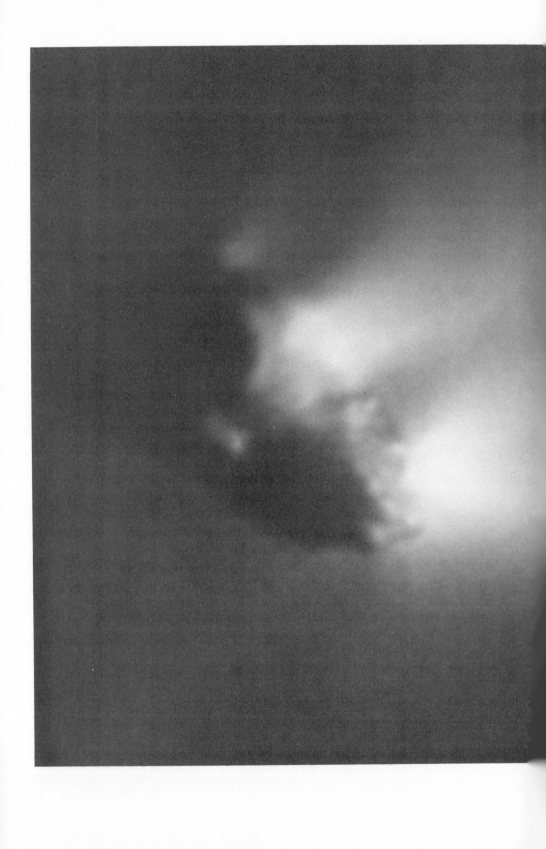

19. Encounter with Halley

THE GRAY AND COLD Kazakhstan steppe brightened momentarily at noon on Saturday, December 15, 1984, as the exhaust column pushed the Proton launcher skyward (see figure 16.1). Half as long as a football field and weighing nearly as much as a 747 airliner, the reliable four-stage Russian rocket smoothly powered the five-ton *VEGA 1* spacecraft precisely toward its initial target, the planet Venus.

Six days later, from an identical launch pad half a mile away, *VEGA 1*'s clone, *VEGA 2*, was flung safely into space, also on its way to Venus.

Access to space had become quite routine in Russia. The exact VEGA launch dates had been fixed into the weekday work schedule of the Baikonur Cosmodrome years earlier. *VEGA 1*'s Saturday launch marked an exceptional accommodation to the timing requirements for a flight to Venus and then on to Halley's comet; of the hundred or so Soviet launches each year, one-third are on Wednesday. Virtually none are on Sunday and very few on Saturday.

Even as the Proton launch team celebrated its eleventh and twelfth successful launches in 1984 and prepared for the New Year's holidays, a small Japanese team from the University of Tokyo worked feverishly through the midwinter darkness on the southeast coast of Japan's southern-

---

19.1 HALLEY'S NUCLEUS FROM GIOTTO • The European Space Agency spacecraft Giotto flashed through the dusty coma of Halley to pass within 370 miles (590 kilometers) of the dark, irregularly shaped nucleus. Here, the object, some 10 miles (16 kilometers) long, is illuminated by sunlight coming from the right. Bright jets of dust spewing out from the comet dominate the right-hand side of the picture. The solid nucleus itself is much darker than the dust jets and can be discerned in the middle of the picture. The camera experiment was under the direction of H. U. Keller of the Max Planck Institute for Aeronomy in West Germany. Image was processed by Alan Delamere and Harold Reitsema of the Ball Aerospace Corporation.

most island, Kyushu. There, at the Kagoshima launch site, an experimental spacecraft named Sakigake (Pioneer), weighing only 3 percent as much as the giant Soviet VEGA craft, was fixed atop the third stage of the small M-3SII rocket.

Two years earlier, on a visit to the Halley project leader, K. Hirao, at Tokyo University, I had watched a disconcertingly casual and very noisy test of a pyrotechnically powered mechanism. Its purpose was to impart a critical spinning motion to the Japanese Halley spacecraft at the end of the rocket launch phase. Now that "spin table" would have to work for real on Sakigake at 8:30 A.M. on January 5, 1985. The launch of a somewhat better-instrumented sister craft, Suisei (Comet), was planned for the following August.*

As if to highlight the contrast with a huge, rigid Soviet launch, Sakigake's blast-off slipped two days because of minor difficulties with the three-stage solid-fuel launcher. The performance margins for Sakigake's mission were so narrow that this two-day delay pushed the eventual miss distance with Halley's comet from 3.7 million to 4.4 million miles (6 to 7 million kilometers).

Suisei also fell two extra days later for launch because of heavy weather. It would, however, cut corners on its way to the Halley encounter by flying a "fast track" that passed just inside the orbit of Venus. (Venus itself was elsewhere on its orbital track, well out of Suisei's observation range.)

In mid-1985, though, Venus did become better known than ever—as a result of the VEGA's passage. On June 11 and 14 each VEGA ejected a large instrumented lander that descended to the scorching surface of Venus. Even more important, a novel balloon with special instrumentation was flung down into Venus's atmosphere from each of the passing VEGAs. Instruments and battery-powered radios in the balloon chronicled their windblown trajectories for over two days.

After ejecting the four scientific packages at Venus, the VEGAs, like *Mariner 10* a dozen years earlier, each passed precisely through a gravity "gateway" located near the massive planet Venus. But the VEGAs, unlike *Mariner 10,* were not thrown inward toward the Sun. Instead, their paths were gently reshaped so that they would ultimately intersect the path of Halley's comet the following March, well around on the other side of the Sun.

---

*The University of Tokyo group, by accepting limitations of scale, had managed up to then to maintain an independent launch, tracking, and spacecraft capability for space science, funded by the Ministry of Education. A larger, NASA-like agency runs a grander program pursuing new launch vehicles, weather satellites, satellites, and other practical uses of space communication.

Meanwhile, the two VEGA balloons were tracked by JPL's Deep Space Net (DSN) and by other stations over the globe, as they drifted and bobbed high in the Venus atmosphere. Some American specialists, like my Caltech colleague Andrew Ingersoll, were involved with the Soviet-French team in deciphering the balloons' radio signals and unveiling Venus's atmospheric secrets. As is usual with important scientific space events, a press conference at JPL was called to publicize the valuable American tracking and scientific contributions (promoted largely by the indefatigable Jacques Blamont). But, at the last minute, orders from Washington canceled the event. The notion of U.S. tracking and scientific support for a USSR mission was not pleasing to the Reagan White House then. Better that such cooperative space efforts go unrecognized by the American public.

Halley's comet itself first came back under earthly observation in October 1982, even as I was urging the faculty, administration, and trustees of Caltech to endorse JPL's own institutional "trajectory change" into a multimission laboratory. My longtime Mars, Venus, and Mercury associate Ed Danielson recorded Halley's return. In 1977 Ed had teamed up with Caltech's Jim Westphal, another old associate, to develop the world's most sensitive astronomical camera for the Space Telescope. Then, in October 1982, Ed and the Caltech graduate student David Jewitt attached some of that supersensitive camera technology to the focal plane of Earth's best telescope, the 200-inch at Mt. Palomar. With the aid of state-of-the-art computer processing, they picked out Halley's tiny reflection from out beyond Saturn's orbit.

Danielson's pictures became the first contribution to the archives of the International Halley Watch (IHW), a worldwide coordination and collection of ground-based Halley observations. The IHW was originally conceived by Louis Friedman after his solar-sailing program died at JPL. Now, as the five robotic explorers were targeted at Halley, the IHW provided key data regarding its orbit, as well as an unprecedentedly diverse and valuable set of telescope observations to help unravel the Halley scientific puzzle.

There was special personal meaning for Ed Danielson in capturing Halley's first portrait since 1910. Back in 1978, when the proposed Halley Flyby / Tempel 2 Rendezvous mission was under serious study, Ed and a former student of mine, Michael Malin, had invented a very clever way to acquire detailed pictures of Halley's speeding nucleus from the small spinning probe to be released from the SEPS-powered mother ship. As we have seen, the SEPS, the mother ship, and any American spacecraft went away. But Danielson's and Malin's novel camera idea survived as the basis for the camera in the spinning Giotto spacecraft developed by the European Space Agency. Using America's Ball Aerospace Company for detailed design and implementation of the Danielson-Malin concept and relying on state-of-the-art CCD sensors from the Texas Instruments Corporation, the camera became an essential ingredient of that ambitiously conceived and very high-technology spacecraft.

Hiding behind a specially designed dust shield, Giotto was to be aimed much deeper into the hazardous dust-filled coma of Halley than the Soviet VEGAs were. From its superior vantage point very near the ablating nucleus, it was to diagnose the composition of the dust and gases and to snatch, if it could, our best view of the mysterious nucleus.

Giotto, packed in the gleaming nose of the fourteenth Ariane launch vehicle, lifted off at 8:30 A.M., July 2, 1985, from a jungle clearing in French Guiana. The Ariane launch base is located in that remote environment because the launching of satellites from near the equator yields a benefit in their performance.

All of Earth's emissaries to Halley were now on their way.

A modest American observation program from Earth orbit, intended for launch in February 1986, also accompanied the developing Space Shuttle program. But it perished along with seven astronauts in the *Challenger* disaster and the subsequent grounding of the Shuttle fleet.

America did contribute to cometary investigations by spacecraft in another, ingenious way: an old U.S. spacecraft, known as *ISEE 3*, was retrofitted *on orbit* to become a comet probe. ISEE (the International Sun-Earth Explorer mission) originated as a late-1970s collaboration between NASA and the European Space Agency. Three spacecraft were deployed to discern the interaction of Earth's strong magnetic field with the solar wind, that brisk flux of electrically charged particles and magnetic fields that flows steadily out from the Sun. *ISEE 3* was launched in August 1978 and positioned as a scientific control point along Earth's orbital track about the Sun. There it remained for four years, steadily monitoring the solar wind prior to its interaction with Earth. Scientists at the Goddard Space Flight Center, in Maryland, and their ESA colleagues used these "control" data to sort out the modifications in Earth's electrical and magnetic fields observed by the *ISEE 2* and *ISEE 1* spacecraft, positioned much closer to Earth.

An inventive mission analyst at Goddard, Dr. Robert W. Farquhar, realized in 1980 that *ISEE 3* had so large a rocket-fuel reserve that it could become a comet probe as well. With five gravity swing-bys of the Moon, he calculated, *ISEE 3* could be speeded up enough to move fifty million miles (eighty million kilometers) ahead of Earth and also slightly farther out from the Sun. There, on September 11, 1985, an old comet first discovered in France in 1900 by M. Giacobini and then found again in 1931 in Germany by E. Zinner (and thus named Giacobini-Zinner) would come racing by. The largest antennae of JPL's Deep Space Net would just be able to extract *ISEE 3*'s faint radio signals from the background noise at a communications distance fifty times greater than the one originally planned for *ISEE 3*.

So *ISEE 3* was renamed the *International Cometary Explorer (ICE)* in 1982. In August and early September 1985, the worldwide network of the International Halley Watch secured critical last-minute observations of the wavering path of the comet Giacobini-Zinner. On September 8, 1985, one final burst from the *ICE* spacecraft's on-board rocket motor refined its path perfectly in order to pass 5,000 miles (7,800 kilometers) ''behind'' the Giacobini-Zinner nucleus on September 11.

There was no camera, of course; *ISEE 3*'s mission had been to observe the invisible solar wind by means of on-board systems that measured electrical and magnetic fields. But *ICE*'s instruments produced a surprise. The electrical and magnetic effects associated with Giacobini-Zinner's passage were much larger than expected; the comet's interaction with the solar wind took place over a much larger volume of space than had originally been thought. *ICE* detected a thin cloud of molecular debris from the breakup of water molecules evaporating from the comet's icy surface when the comet was still several million miles away. Fragments of water molecules were to be expected; comet nuclei were believed to be mainly homogeneous aggregates of water ice and dust. It was the scale of the interaction that surprised *ICE*'s scientists.

What would the much larger and more volatile comet Halley be like six months later when it crossed paths with the five robotic explorers of the Halley Armada (*VEGA 1, VEGA 2,* Sakigake, Suisei, and Giotto)? JPL's Deep Space Net would also play a crucial part in this great international effort of space exploration. Critical data would be received from Sakigake and Suisei by the DSN stations in Madrid. Even more important, an elegant and closely timed operational interaction (termed Pathfinder) between the DSN and the Soviet VEGA spacecraft was quietly scheduled. This would be needed to capture critical last-minute updates on Halley's dynamic* trajectory so that Giotto could be targeted to pass daringly close to the mysterious nucleus (see figure 19.2).

First, the ground-based International Halley Watch provided the preliminary estimate of Halley's path. Then, on March 6, 1986, as *VEGA 1*

*The jetting of dust and gas from comet nuclei produces significant and unpredictable orbital shifts.

sped through the dusty coma only six thousand miles (10,000 kilometers) from the solid nucleus, the 210-foot dish at the DSN in Australia precisely tracked both Giotto and *VEGA 1*. The latter's cameras simultaneously pinpointed the position of the streaking nucleus relative to the spacecraft. Using high-speed electronic data lines switched through ESA's Space Operations Center, in Darmstadt, West Germany, the three international groups quickly exchanged the tracking and camera data and independent computer orbit runs. VEGA's optical measurements of Halley's position and the DSN's radio location of VEGA and Giotto were mated to give a critical update on Halley's swooping path, completed once every seventy-six years. Giotto's controllers flashed last-minute trajectory corrections up to their bird, which smoothly fired its on-board thrusters as instructed.

Everything worked beautifully. Giotto passed within 370 miles (590

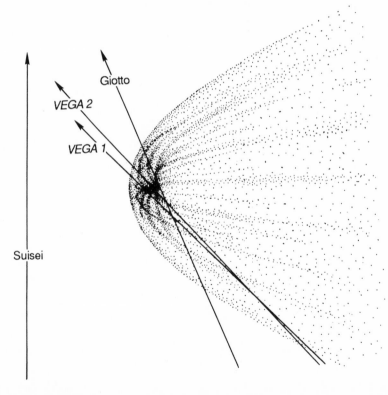

19.2 ENCOUNTER WITH HALLEY • In this diagram the paths of Suisei, Giotto, and the two VEGAs are shown relative to a schematic representation of the coma. Suisei passed within 125,000 miles (200,000 kilometers) sunward of the nucleus on March 8, 1986. On March 6, 1986, *VEGA 1* penetrated to within 6,000 miles (10,000 kilometers) of the nucleus, and on March 9 *VEGA 2* passed within 5,000 miles (8,000 kilometers). Finally, on March 14, Giotto swooped within 370 miles (590 kilometers) of the nucleus.

kilometers) from Halley's nucleus on March 14. To appreciate this feat, consider that Halley covered 370 miles in only eight seconds! The navigation thus had to be good to the second, or Giotto would have come too late. This technical partnership between the United States, Europe, and Russia came about despite the absence of any formal relations between NASA and the Soviet Union. The original U.S.-USSR bilateral space agreements of 1972 (which facilitated, among other endeavors, the Apollo-Soyuz handshake in space in 1975) expired in 1982. Renewal became a casualty of U.S. hostility to the USSR, triggered by the Soviets' suppression of the Solidarity movement in Poland and their invasion of Afghanistan.

Just at its closest approach, Giotto was knocked off its spin—literally—by a gram-sized chunk of something in the coma. About twenty minutes' worth of pictures and other data were lost immediately before communications resumed. The VEGAs (not as well shielded as Giotto) likewise suffered substantial damage to on-board equipment while passing more than ten times farther out. But none of the three spacecraft was permanently incapacitated. Thus, the assessment of risk from collisions within the dusty coma by Soviet and ESA project officials was correct. It is difficult to imagine how either VEGA or Giotto could have achieved significantly more than it did. The same is true for the more modest Japanese effort.

Truly, the world, apart from the United States, gave the Halley encounter its "best shot." And the object found at Halley's nucleus justified this effort. Halley, it turned out, was not the simple "dusty snowball" long imagined by astronomers (and seemingly confirmed by *ICE*'s blind "taste" of Giacobini-Zinner's energetic tail). The cameras of VEGA and Giotto revealed, instead, a far larger, darker, rougher, and more irregularly shaped body. Great jets of gas and dust spewed out from *hot* fissures or caverns within the nucleus (see figure 19.1). VEGA's spectrophotometers sensed heat radiated from portions of Halley's nucleus more intense than that from boiling water. And the jetted material that bombarded Giotto and VEGA proved to be chock-full of carbon and exotic combinations of carbon with hydrogen, nitrogen, and oxygen. These atoms are the fundamental chemical basis of life. Tell-tale signs of carbon monoxide (CO), hydrogen cyanide (HCN), ammonia (NH_3), methane (CH_4), hydrocarbon vapor, and possibly even formaldehyde (H_2CO)—more familiar on Earth as embalming fluid—showed up in data obtained from the encounter.

What is its surface really like? A thick, cohesive dark crust must be present (to explain the hot jets and photographic appearance). That crust appears to be rich in tarry organic compounds. What a bizarre visitor to the inner Solar System!

After thousands of years of human awe and fear, the Halley Armada revealed Halley's nucleus to be a Manhattan-sized, freeze-dried lump of dusty organic broth, probably formed in the rarefied froth around the edges of the Sun's birth. (It may even be contaminated with traces of interstellar debris as well.) Halley's Armada provided a rapid but detailed portraiture of the comet's nucleus, one that transported humankind further back in time—and closer to its chemical origins—than ever before. It may have been the space encounter of the century.

After Secretary General Gorbachev finished yet another speech promoting his far-reaching changes in the Soviet economy, his scientific counselor, the physicist Eugeni Velikhov (also vice-president of the Soviet Academy of Sciences), greeted him as he entered a Kremlin meeting room. Velikhov presented Roald Sagdeev, director of the Institute for Space Research and leader of the world-acclaimed VEGA mission that had encountered Halley's comet a week earlier, and five other men associated with that mission. Gorbachev already knew a lot about VEGA, having the preceding June touted its exploits at Venus as a commendable example of the kind of excellence he wanted to see achieved elsewhere in the Soviet economy.*

Before his guests had even seated themselves, Gorbachev asked, "How could you manage to encounter Halley's comet the last day of the Party Congress?"

Sagdeev replied, "We have the perfect alibi. The orbits of comets are controlled by God."

Gorbachev shot back, "That means God is with us."

When Sagdeev shifted the discussion to the photographs returned by VEGA, Gorbachev was interested and enthusiastic, asking question after question about comets and the VEGA endeavor itself. He reiterated the importance of the international aspect of the mission and said he was so pleased with VEGA's achievements that he promised to mention them to the Politburo (which he subsequently did, according to *Pravda*).

As the three-hour meeting neared its end, he reportedly said, "And what are your future projects? ... If you need me, don't hesitate to let me know."

Ten days later, on March 26, 1986, in an unconscious superpower minuet, six leading American scientists gathered in the Roosevelt Room

*This account is based on informal talks by Sagdeev on July 6, 1986, and December 5, 1987. A fuller version is given in Jacques Blamont, *Vénus dévoilée* (Paris: Editions Odile Jacob, 1987), 350.

LUNCHEON AT THE WHITE HOUSE • Ronald Reagan and the comet expert Fred Whipple exchange information and jokes on this occasion for Reagan to learn about recent discoveries in space science. To Whipple's right is the Caltech professor Ed Stone. Hidden behind him is Tom Krimigis of the Applied Physics Laboratory of Johns Hopkins University. The late Frederick Scarf of TRW is to Whipple's right, followed by Jack Brandt of the NASA Goddard Spaceflight Center. Torrence Johnson of JPL in out of the picture on Brandt's left. Across from Brandt is Jack Svahn, an assistant for policy development; Donald Regan, White House chief of staff, is on Reagan's right; and Al Kingon, secretary of the Cabinet, is on his left. At the far end, next to Kingon, is the acting science adviser John McTague, who organized the meeting held on March 26, 1986. *Official White House photo*

of the White House for lunch with President Ronald Reagan, his chief of staff, Don Regan, and four White House assistants.* The topics were comets and Voyager's encounter with Uranus.

After the usual picture taking and introductions, the eighty-year-old astronomer Fred Whipple, the originator of the "dirty snowball" model of comets, described how interesting and strange comets are. Don Regan promptly asked whether cometary gases had any influence on Earth's environment or atmosphere. Whipple said no but noted that during Halley's visit in 1910 our planet actually passed through the comet's extended tail and a lot of quacks sold "protective" medicinals. The President jokingly commented, "You don't suppose that for those who were not born in 1910, but rather in 1911, the gases might have had some ill effects?"

Whipple then showed the President some Giotto close-ups obtained

---

*This account is based on notes by S. M. Krimigis [16] and on comments by others.

just two weeks earlier. He discussed the fuzzy jets and partly obscured dark nucleus, not easily recognizable even by specialists. Reagan had some difficulty in discerning them. The Caltech professor (and Voyager project scientist) Ed Stone came to the rescue by outlining the inferred nucleus on the photograph, and Whipple produced a sketch diagramming the irregularly shaped object. Then the TRW space scientist Dr. Fred Scarf elicited laughter again by questioning whether the nucleus looked more like a peanut or a potato.

Don Regan wanted to know more about recent close encounters of Earth with comets or other objects. Whipple mentioned the scientifically famous 1908 cometary impact in an uninhabited region of Siberia (which pulsed the Earth's atmosphere like a ten-megaton hydrogen bomb) and several more-recent close asteroid passages. The President observed that if there were an impending collision with an asteroid, all countries would forget their differences and unite to figure out what to do. This, he said, would show that all humans are brothers. Scarf responded that in science we were already working together like brothers.

But this trend of thought stopped short here. No discussions followed regarding the significance of international collaboration in exploring Halley's comet. And mention of the dominant Soviet role in the Halley encounter was avoided by the carefully coached visiting scientists.

Instead, under the portraits of Theodore Roosevelt and Franklin Delano Roosevelt, the discussion turned to the ages of comets and of the universe. Reagan chimed in with another story:

Once there was a professor who was explaining to the students information about the Solar System and about the fact that in a billion years or so the Sun would run out of fuel and everything would grow cold and the Earth would freeze. This was quite impressive to the students, and after some respectful silence, one of the students raised his hand timidly and said, "Professor, tell me, how long did you say that would be," and the Professor said, "Oh, about a billion years," and then the student, "Whew! I thought you said a *million* years!"

After the laughter subsided, Reagan went on in a more serious vein to relate how, as a child growing up in the Midwest, he would go out at night, look at the stars, and wonder (rather like Carl Sagan) whether anyone was living out there. He inquired, "You gentlemen have investigated a lot of things in space. Have you found any evidence that there may be other people out there?" His luncheon guests had found no such evidence. Then Reagan caught them by surprise when he carried their negative results to his own conclusion by remarking, "One can't help but think about creationism.''* His guests politely avoided responding to his conjecture.

The conversation reverted to the here and now of comet science, with new facets provided by the NASA scientist John Brandt and the chief scientist from the Johns Hopkins University Applied Physics Lab, Tom Krimigis. Then Ed Stone eloquently summarized the findings of Voyager at Uranus, emphasizing the novel attribute of Uranus tipped over on its side with its pole of rotation pointing toward the Sun. But the President's interest was visibly waning. So JPL's Dr. Torrence Johnson (another of my former graduate students) identified himself as a longtime ''moon'' person and enthusiastically described the extraordinary moon of Uranus, Miranda (see chapter 11).

Ed Stone proceeded to emphasize the positive cultural benefits of space exploration and its special importance for leading young people to consider scientific careers. The President listened politely. By one o'clock, precisely an hour after Reagan first entered the room, Don Regan was getting restless concerning the next appointment. But his boss insisted on telling one more story, this one ridiculing environmental-impact statements. After a final round of laughs, the President thanked his guests, remarked how wonderful it was to live in an age when all these technical achievements were possible and how wonderful the space program was, and said he hoped his guests would have more and more excitement as time went on.

The group made its way out of the White House in high spirits. John McTague,† the acting science adviser, was exceedingly pleased and complimented the six scientists for being ''a bunch of professional actors,'' adding that ''it couldn't have gone better if it had been scripted.''

It had been showtime at the White House, a fitting conclusion to Washington's superficial and distracted response to Halley's visit. Mark Twain would have enjoyed the jokes but could not have imagined such an ending—much less the plot.

The impartial mirror of space reflected an image of an America profoundly different from Mark Twain's. Indeed, the motives and national self-expectations of even the 1960s and 1970s seemed almost extinct in America's 1986 response to Halley's visit.

What kind of America will Halley's comet encounter the next time, in the year 2061?

---

*That is, the fundamentalist view that intelligent life formed not as a natural consequence of chemical and biological evolution in this stellar system—and presumably on others—but as the result of a singular, miraculous process on Earth alone.

† Jay Keyworth still had not been replaced four months after resigning.

PART SIX

To Mars... Together?

20.1 TOGETHER IN SPACE • The cosmonaut Alexsei A. Leonov shares cramped but weightless conditions with the astronauts Donald K. Slayton and Thomas P. Stafford in mid-July 1975 while the Soviet Soyuz capsule and the U.S. Apollo command module were docked together in low Earth orbit. Leonov and Stafford were both career military officers and later rose to the rank of general.
NASA/Johnson Space Center photo

20.2

Steps toward Mars

U.S.		USSR
	Second NASA/Soviet Agreement	
	Moscow Summit	
	1988	
	First NASA/Soviet Agreement	
	Washington Summit	
	New Space Accord Signed	
	1987	
NASA Tracks Halley Armada		Vega/Giotto Halley Encounter
	1986	
NASA Tracks Vega Balloons		Vega Venus Encounter
	1985	
		USSR Boycotts LA Olympics
	1984	
	1983	
	1972 Space Accords Expire	
	1982	
	1981	USSR Pressures Poland to Supress Solidarity
U.S. Boycotts Moscow Olympics	*1980*	
		USSR Intervenes in Afghanistan
	1979	
	1976	
	Apollo-Soyuz Joint Flight	
	1975	

20. An Emerging Vision

THE MILKY WAY, blazing more luminously than I had ever seen, bisected an inky Caribbean night as the sea gently rocked our cruise ship. On the darkened fantail, Suzanne and I picked out Halley's faint smudge in the sky. It was early April 1986. We were cruising in tropical latitudes between coral-ridden Barbados and volcanic St. Lucia on a cloud-free, moonless night that climaxed the "Halley Comet Cruise," a Planetary Society* educational fund-raiser. At this moment the comet was rushing out past Earth's orbit on its long voyage back to a frigid home between Neptune and Pluto.

Like millions of others, we sensed a bit of our descendants' futures, and our ancestors' pasts, as we watched this famous cosmic metronome. We would not witness its next stroke, in 2061.

I had just seen the Southern Cross for the first time and had discovered what an impoverished sky we northerners live under. Devoid of a local horizon, I felt acrophobic, like a spectator perched high up on narrow bleachers in a cosmic amphitheater. The faint glow from the center of the

*See appendix.

20.3 TO MARS . . . TOGETHER • This black-and-white photograph of an original oil painting by the space artist Michael Carroll depicts the technical configuration for a human flight to Mars that was proposed in 1984 in a study funded by the Planetary Society. A circular aeroshell 167 feet (54 meters) in diameter dominates the center of the scene. The aeroshell carries a large Mars orbiter and landing system and has just been detached from the interplanetary cruise module, which is 280 feet (90 meters) long, carries huge solar panels, and contains cylindrical crew compartments at its end and middle. The artist has displayed prominently a combined logo of "CCCP" and "NASA" to indicate that this is a joint effort to explore Mars. Such a mission might be considered for the second decade of the next century. *Photo courtesy of the Planetary Society*

Galaxy, that mysterious home of black holes and who knows what other monsters of modern physics, seemed to pull me in upon itself.

To make this grand planetarium show complete, the red disk of Mars squeezed in between the Galaxy's brilliant festival of lights and Halley's faint remembrance. Mars was glowing more brightly now than at any other time since 1971. In that year, when Mars and Earth last approached each other so closely, *Mariner 9* had discovered a wondrous new place crowded with giant volcanos and canyons, and crossed by water-scoured dry valleys surviving from a mysterious ancient fluvial period.

In 1971 Mars had been a magic Fabergé egg that scientists working on *Mariner 9* had plucked from a Mariner Christmas stocking. The stocking seemed stuffed with limitless worlds to explore—Venus, Mercury, Jupiter, Saturn, their myriad moons, and even distant Uranus and Neptune.

By 1986 robotic explorers had practically emptied the Christmas stocking, ripping away veils of fantasy and obscurity to expose real worlds, far different from those imagined. Finally, a few weeks before this trip, Giotto and the VEGAs had given us a spectacular first look at the nucleus of a comet as they penetrated Halley's turbulent coma.

I reflected on how Apollo's lunar challenge had triggered an extraordinary outburst of American energy, ingenuity, and high purpose to blaze a trail through the Solar System. Now that trailblazing was nearly completed. Apollo's focusing of America's national purpose had long since dissipated.

"Is it the beginning of the end?" The question gnawed at me, even after I fled JPL in 1982 to distant lands and, for a few years, to distant occupations.* No, I came to realize, it is merely the end of the beginning. In the twenty-first century that glowing red disk will surely be the destination of intrepid representatives of ascendant nations. The first human landings on Mars beckon as the next great milestone in human exploration, now that Earth orbit and the Moon have been visited.

Why Mars rather than nearby Venus? Because Venus's surface not only is hot enough to melt lead but also is blanketed by a poisonous atmosphere pressing down with a force a hundred times greater than does our own. Lonely Mercury, by contrast, is airless, like the Moon. But its surface is seared by solar radiation ten times more intense than that on the Moon. Life for astronauts would be much tougher on Mercury than on the Moon, with little additional reward—scientific, spiritual, or otherwise.

---

*In 1984 and 1985 I went on part-time status at Caltech, opened several small entrepreneurial businesses, and worked with venture capitalists in an attempt to apply information technology to mass education in the United States. In mid-1985 I returned to Caltech to pursue Mars full-time, leaving educational reform to others.

How about Jupiter, the grandiose stage setting for Stanley Kubrick's epochal film *2001?* With good reason Arthur C. Clarke's original script placed the adventurers near Saturn, approaching its peculiarly marked moon Iapetus. Jupiter's radiation belts are so intense that even inanimate electronic components are rapidly fried there. Obviously, humans cannot be "radiation hardened," as were Voyager's and Pioneer's specially prepared electronic brains and nervous systems. So, despite its grand size and diversity of locales, Jupiter will be explored only by robotic descendants of the Galileo mission. The human participation will remain at the other end of the radio link.

Saturn, on the other hand, bared no unusually menacing attributes to the probings of *Pioneer 11* and *Voyagers 1* and *2*. But as a site for human exploration, it suffers from incredible remoteness—a *billion* miles from Earth. It is illuminated by a sun less than one-hundredth as bright as what Earthlings consider normal. Human expeditions to Saturn and Iapetus, or to even more distant Uranus and Neptune, must remain the province of science fiction for the indefinite future.

Mars, by comparison, is hospitable and, in the long run, perhaps even habitable. To be sure, space suits are mandatory dress for traversing its surface. The atmospheric pressure there is at best only 1 percent of that on Earth's surface. And the Martian nights are cold, minus 125 degrees centigrade frequently.

But the daytime sunlight—half as bright as Earth's—warms dark rock and soil (and future astronaut habitats) above the freezing point from time to time. Even compared with the fiercest polar locales on blue, aqueous Earth, it is an arduous environment, but one more benign than that which Apollo astronauts overcame on the Moon. And it is a Club Med site compared with any other locale in the Solar System.

Moreover, Mars's surface, unlike that of its cousins Moon, Venus, and Mercury, is rich in the most crucial resource for human sustenance—water. At the residual frost caps that ring the north and south polar regions of Mars, solid water ice is exposed at the surface. Over much of the temperate region of the planet, permafrost is believed to occur at shallow depths, within easy reach of future drillers, both human and robotic. In addition, the Martian soil probably contains extractable water bound up in hydrated mineral grains. Abundant ice, sunlight, and atmospheric carbon dioxide gas provide the keys to future human subsistence on Mars.

In situ Martian water can replenish crucial vapor lost from life-support systems upon which future explorers will depend. It can also provide oxygen for breathing. By means of sun-powered electrical generators (or miniature nuclear reactors transported from Earth), water vapor can there be disassociated into pure hydrogen and oxygen gases.

Some visionaries have higher aims.[1] They ponder harnessing Mars's ubiquitous carbon dioxide gas to create rocket fuel components like methane (CH_4), oxygen (O_2), and carbon monoxide (CO). Local sunlight or imported nuclear power could thus be transmuted into liquid rocket fuel for powering astronaut-controlled vehicles across Mars's enormous and diversified surface. Martian greenhouses rich in carbon dioxide could provide food.

Bolder visionaries even dream of great space passenger ships, plying the "ocean of space" between Mars and Earth. Locally produced rocket fuel would be loaded aboard at both ends of this kind of fanciful journey. Such a "Mars ferry" would rotate crews and transport critical supplies to a Martian research outpost scientifically analogous to those of Antarctica today, but with hardships and risks far more akin to those experienced by polar explorers on Earth in the days of sailing ships and dogsleds.

Are scientific outposts on Mars a vision? Yes, but one that could be technically feasible for the children of today's children.

An even more fevered dream is colonization. Some visionaries imagine that self-sustaining colonies on Mars can be established. To true believers, all the ills of humanity might be left behind by a brave band of third-millennium pioneers who purchase one-way tickets on the "Mars ferry" in order to raise new generations on Mars. Poignant images of such space-age colonists took life in the 1950s and 1960s in the vivid prose of Robert Heinlein, Ray Bradbury, Arthur Clarke, and dozens of other science fiction writers. In the 1970s and 1980s these appealing themes and characters were presented in living color by such filmmakers as Steven Spielberg, Stanley Kubrick, and George Lucas. For many, such imaginings have now acquired an aura of reality as a plausible, if distant, human prospect.

In fact, there is little in human experience upon which to project extraterrestrial colonization and migration. Sometimes Columbus's voyages to the New World are cited as a historical precedent. But those European endeavors were motivated and financed by parent-country expectations of economic return. The pursuit of individual goals in the New World—scientific curiosity, religious and political freedom, adventure, and personal fulfillment—was possible mainly because of economically motivated kings and parliaments, who remained safe at home awaiting their financial return.

No Martian commodity or process could ever be valuable enough to motivate individual or national investment. Economic gain will not be the engine to finance human travel to Mars (or back to the Moon, for that matter). And there are no heathen souls to be saved there.

The exploration of the Antarctic may offer a more useful precedent.[2] In 1906 Scott and Amundsen raced death and the Antarctic season to be first to reach the South Pole, then the ultimate in terrestrial remoteness. Amundsen (and an emerging Norwegian nationalism) won. Scott (and an overextended Edwardian English pride) lost. For a decade more, Shackleton and others continued to add depth and breadth to that early exploratory process. But World War I soon suffocated such imaginative and daring polar endeavors.

The brief interlude between the two world wars provided an opportunity for experimentation with circumpolar air travel and the long-term survival necessary for sustained human functioning in that loneliest of environments.

Not until after the second global war did the exploration of Antarctica reach its full flowering. World War II reduced the cost and enlarged the capacity of air and sea travel. By 1946 getting to Antarctica—even to the South Pole—was within existing military capabilities. Antarctic exploration became a benign task for an underutilized postwar Navy. Scientists soon began to exploit this capacity on a large scale; their efforts led in 1959 to the International Geophysical Year, which lasted a decade. Energetic and pragmatic scientists brought into being an international program of global geophysical measurement, focused on the Antarctic and its important polar phenomena. In the ensuing decades thousands of scientists from dozens of nations traveled to and around Antarctica on ordinary government scientific programs. The terrain, climate, biology, and processes of the Antarctic were explored in detail, to the benefit of our entire globe.

Along the way a good thing happened. Past claims of sovereignty and future economic and military potentialities yielded to new treaties, providing for an internationalized, peaceful, and environmentally protected Antarctica. Antarctica today is a unique international scientific outpost and laboratory *cooperatively* managed for peaceful purposes by many different nations. The peacefulness of these activities is assured by mutual on-site inspection of facilities.

To be sure, expenditures there are still largely motivated by national self-interest—by prestige, rivalry, and positioning for future scientific and technical development. The issue of resource development is emerging once again. But in a century characterized by unprecedented nationalistic awareness and by dizzying modification of Earth's environment, Antarctica stands out as a glistening counterexample, a refreshing demonstration that international cooperation and scientific management of resources can be a practical part of humankind's future.[3]

The Antarctic experience helped fashion the treaties and popular attitudes governing the disposition of the Moon and planets. When Neil Armstrong placed a plaque on the Moon on July 20, 1969, which read, "We came in peace for mankind," he expressed the spirit of the 1967 treaty that provided for such peaceful, internationally oriented exploration of space. Territorial claims and national sovereignty end at the top of the atmosphere.

The first human flight to Mars is the nearly inevitable extension of current national rivalry, popular yearnings, and evolving space technology. It is, of course, a major step replete with intrinsic risk and unknowns, but it is not as great a technical or political challenge as was Kennedy's commitment to Apollo. In May 1961, imponderables loomed large. How would humans function in space, beyond the safety of quick return from Earth orbit? At that time there was another challenge—the necessity to build rockets far more powerful than ever before. Most of all, time was of the essence then. That is not true now.

For the taking of the great step to Mars, there is ample time. Unlike Kennedy's Apollo counterpunch in response to the Soviet space successes of 1957–61, the next great human leap to a new world need not be a nationalistic space race. It need not stress human and material resources as intensely as did Apollo or compete as furiously with other national and global imperatives.

Next time, we can go to Mars together in a resounding victory of human intelligence and spirit over runaway technological and political change. The U.S.-Soviet rivalry for world leadership can evolve from unrelieved military confrontation to sophisticated competition to lead and facilitate international cooperation. If the United States and the USSR decide to cooperate in the long-term exploration of Mars—rather than race one another—their leaders can also choose to do so in measured, affordable steps.

That glowing red disk of Mars thus warranted prominence in the magic sky surrounding our Caribbean cruise ship. It is the next destination.

Long before human travel to Mars was technologically possible, visionaries dreamed of it. The first political consideration of it, though, did not occur until 1969. Apollo's successful lunar achievement then demanded discussion of a new national purpose and direction for Americans in space. Vice-President Spiro Agnew played Nixon's stalking-horse by advocating a superficially conceived human flight to Mars as the next round of U.S.-USSR space competition. A race to Mars would follow the race to the Moon. Agnew proposed lift-off in August 1981. That trial balloon popped, leaving only "Shuttle for Shuttle's sake" as NASA's future program.

By the early 1980s, however, Mars had again attracted attention, this time from serious professionals who wanted to pick up where NASA left off after Agnew's premature and ill-informed advocacy. But for over a decade NASA's Shuttle difficulties had prevented any official thinking about Americans' once again going out beyond Earth orbit. So the newly formed Planetary Society stepped in to play a key nongovernmental role in defining a Mars-oriented future. Carl Sagan and I founded the Planetary Society in late 1979 to provide a vehicle for the broad public interest in planetary exploration and the search for extraterrestrial life (described fully in the appendix). In 1983 the Mars Institute, a loose network of enthusiasts, working mostly in universities or on their own, was partly funded by the Planetary Society in order to mobilize renewed interest and study. In June 1984 the society brought together American, Soviet, and European scientists in a nongovernmental setting in Graz, Austria, to formulate possibilities for cooperative planetary exploration. In 1984 it commissioned the first professional analysis of a manned Mars mission since Agnew's days.

Even as public interest in Mars's future began to coalesce under the Planetary Society's leadership, important new scientific insights about Mars' aqueous past sprang up. Since the end of the Viking mission in 1979, specialists at the U.S. Geological Survey in Flagstaff, Arizona, had been patiently improving their computer tricks to enhance the highest-resolution Viking images.

As with each previous increase in image quality, startling new facts about Mars's surface emerged, especially for areas within the giant valley system of Valles Marineris. The new pictures showed vast thicknesses of layered sediments blanketing portions of the valley floor, looking every bit like *lake*-bottom deposits.[4] Lakes on Mars? Much older surface features are quite uneroded. Their pristine character rules out completely the possibility of an ancient aqueous atmosphere on Mars. How could there have been giant standing bodies of water hundreds, even thousands, of miles long and permanent enough for thick sedimentary sequences to accumulate on their bottoms without vast amounts of water evaporating into the atmosphere and eventually causing Earth-like erosions over the whole planet?

Antarctica provided clues to that paradox. Prescient scientists at NASA's Ames Research Center studied ice-covered lakes there and learned that sedimentation, even algae growth, occurs in the permanently ice-covered water bodies.[5] Thus, an exciting new geological episode in Mars's history has been recognized, one that again opens up speculation about ancient life there.

Surprises about Mars continue. What new twists and turns await camera-equipped mobile surface robots in the future, as well as higher-resolution imaging from orbit?

The leaders of the Planetary Society—President Carl Sagan, I as vice-president, and Executive Director Louis Friedman—organized a lively debate for January 12, 1985, in Washington. During the morning session our cosponsor, the venerable American Academy of Arts and Sciences, analyzed the pros and cons of building Reagan's Strategic Defense Initiative, or "Star Wars." That afternoon I chaired the Planetary Society's program—a debate on the theme "Potential Effect of Space Weapons on Civilian Uses of Space." Carl and the MIT physics professor Philip Morrison were pitted against a very able Defense Department pair—Robert Cooper, fresh from heading the Pentagon's Advanced Research (and former director of NASA's Goddard Space Flight Center), and the Strategic Defense Initiative's chief, Lieutenant General James Abrahamson, a former manager of NASA's Shuttle program. Later we added Roald Sagdeev, taking advantage of his presence in the United States. This caused Abrahamson to send in his speech with Dr. Greg Canavan, of Los Alamos, who served as a last-minute substitute, presumably to avoid the appearance of an official dialogue between Abrahamson and Sagdeev.

Morrison led off as follows:[6]

There are two major dynamic and growing forces that run as strong threads throughout all our centuries. They are the rise of science and technology, and the rise of the nation-state. The greatest conflict in the world, probably more enduring than that between war and peace, or between the several states, or even between the rich and the poor, is the conflict between the steady and cumulative growth of these two institutions. I don't think that we will long endure as a complicated, world-inhabiting species if we don't, within some proximate future, find some way to moderate the impact of the simultaneous rise of these two forces.

... I think the clearest problem is that we need to moderate, to temper, to avoid establishing another permanent nation-versus-nation battlefield in space. If we don't do that, we will not be able to exploit that domain in the way promised by enthusiasts for civil uses of space.

Cooper countered with the military view that space is a place, nothing more, nothing less:

... Those of us who have grown up in the space era look at space with an emotional eye, and it has an extraordinary allure for us. In the Pentagon, you can hear extremists caught up in this allure say that we should seize the high ground of space to keep the nation secure. On the other hand, in the civil community, you hear people calling for making space a sanctuary from military operations— meaning not only a refuge and an asylum, but also a sacred or holy place....

But when you get down to business, these dramatic appeals are quite unrealistic and not helpful as guiding principles for any of our space endeavors, whether they be scientific activities, space exploration, commerce or military operations.

Let me tell you that, from a military point of view, there is absolutely nothing special or dramatic about space. It is just another environment in which human activities will take place, and one of those activities is bound to be military.

Abrahamson, through Canavan, sounded the familiar arguments about technological "spin-offs": SDI's vast array of new technologies will permit new space exploits to be launched.

At this point Carl Sagan, responding to Sagdeev's presence and to political expectations heightened by the first meetings between Shultz and Gromyko then under way in Geneva, countered with a powerful appeal to hope and optimism:[7]

If you are interested in pumping the economy, there are far better ways to do it. Think of the major cooperative programs which are fully within our technological capability and which could be done at a tiny fraction of the cost of "Star Wars" technology. Think about a joint US / Soviet manned (and womanned) mission to Mars.

As I moderated the questions afterward, I could *feel* the presence of a big new idea born of fear and hope. Using the moderator's privilege of making the final statement, I summarized,

... The discussion has focused on the central issue: Is the future a linear extrapolation of the past, which according to Philip Morrison's view is a very uncertain future indeed? There is a sense that we can, therefore we must, try to change that future. Space is viewed as the next frontier for that change... We should be looking toward a future in which humans will go to Mars and do many other exciting things. That will be the hallmark of our actions in space.

Something new had been added to John Kennedy's dreams. The national imperative in space had been redefined to fit the nuclear world of the 1980s.

Five months later Carl, Lou, and I were back in Washington in the same auditorium of the National Academy of Sciences. This time we had teamed the Planetary Society with the American Institute of Aeronautics and Astronautics (AIAA), the main professional organization of aerospace engineers (and their companies). It provided a conservative balance to our January partnership with the mildly liberal American Academy of Arts and Sciences. But now we directly targeted the subject: "Steps to Mars." Detailed engineering and scientific discussions analyzed the voyage to Mars and the measurements to be made there. Robotic and human-piloted techniques received equal billing.

Noontime brought a surprise. The retiring president of the AIAA, Dr. John McLucas (a former secretary of the Air Force, a former administrator of the Federal Aviation Administration, and a deeply committed internationalist), chaired a moving reunion of the two Soviet cosmonauts and the three U.S. astronauts who had joined hands in space in 1975, during the Apollo-Soyuz mission. McLucas was eager to exploit the fact that our meeting date commemorated the tenth anniversary of that rendezvous, even as it commemorated the sixteenth anniversary of the Apollo landings on the Moon and the ninth anniversary of *Viking 1*'s touchdown on Mars.

I was skeptical and a little concerned about this mixing of an unproven political initiative—Russian and American—with the technically bold idea of humans' going to Mars, especially in Washington, D.C., which had been hostile for so long to both novelties. But it turned out to be a very successful occasion. Sagan later wrote in *Parade,*[8]

In the darkened auditorium of the National Academy of Sciences in Washington, D.C., the five of them—veterans of many space missions—reminisced about the silent movie being projected on the big front screen. With an easy, self-mocking humor, they described the design of the compatible docking module, the separate launches from Cape Canaveral, Fla., and Tyuratam in the USSR, the dangerous rendezvous, the triumphant crawling through airlocks to visit one another, the exchange of gifts, the camaraderie and their separate returns to Earth. Occasionally, a little shyly, they would put an arm around each other. Many in the audience were struck by the mutual affection and respect of Lt. Gen. Thomas Stafford, U.S. Air Force, former commander at Edwards Air Force Base, where high-performance aircraft are tested, and Maj. Gen. Alexie Leonov, Soviet Air Force, the first human to walk in space. As the film ended and the lights came on, there arose from the sedate audience of engineers and scientists a sound I have rarely heard—an ovation of such a timbre and intensity that you knew something deeply felt had been touched in the hard-bitten and tough-minded audience....

Now, here they were, the veterans of that contact between alien civilizations, describing a mission whose accomplishments, apart from some worthwhile science, were chiefly in the cause of human understanding. There was a hunger in that audience—as there is throughout the world—a longing for the two nations to do something together for a change, something on behalf of the human species. Our powers are so great and our accomplishments so feeble. Think of what we could do together. As the five astronauts and cosmonauts were given relief maps of the Kasei Valles region of Mars, you heard another stirring ovation, and again the thought arose unbidden: Maybe it was possible after all.

That afternoon I moderated a panel that faced head-on the key question "Why go to Mars?" Once again Carl challenged East and West to rise to the emerging opportunity. But now Harrison ("Jack") Schmitt, the first scientist and one of the last two humans to reach the Moon, added support from a more nationalistic point of view. He challenged Americans to commit themselves to an eventual trip to Mars.

> ... In addition to the questions of why, how and when we should go to Mars, there will be the question of whether the nations of the world can find it in their interests to work together on this truly magnificent endeavor. The answer to this last question should be "Yes" for both the United States and the Soviet Union, I hope in cooperation with as many other nations as wish to and can participate. However, that answer *will* be "yes" only if the United States makes an unequivocal commitment to go to Mars under any circumstances.

Next came support from Dr. Sally Ride, a two-time veteran of shuttle flights and America's first woman into space. Ride is a broadly respected and thoughtful person. (Six months later, she was a key member of the Rogers Commission investigating the *Challenger* disaster, and following that she created NASA's Office of Exploration.) With characteristic incisiveness and humor she raised the following question:

> Why should we send people to Mars? We can't justify it with planetary science alone; we'd get a better return on our scientific dollar by sending unwomanned probes. Unfortunately, that philosophy puts me out of a job. We also can't justify it with assured economic benefits. . . .
>
> We will be going to Mars; the questions are, "When and with whom?" There are obviously pros and cons to joint US-Soviet expeditions, but astronauts and cosmonauts do share a common ground. Anything that cultivates that common ground can only improve the world situation.
>
> I'd like to leave you with a little bit of *Apollo 17* history. When Jack Schmitt stepped out on the lunar surface, he offered a challenge to the next generation—to leave their footprints on other, more distant worlds. I think we should take him up on that challenge.

International support came from Dr. Robert Bonnet, head scientist for the European Space Agency, who said,

> I strongly believe that manned exploration of Mars should be conducted on a broad international basis, for two reasons. First, the project could cost $40 billion, a prohibitive cost for a single country or organization. The second reason is, of course, political. The two superpowers, the US and USSR, should collaborate. As long as they have to peacefully cohabit on this planet, why not do it constructively? A team of astro-cosmonauts in space for several years would offer a means of political stability.
>
> Europe wants to participate in this venture.

Finally, NASA Administrator James Beggs brought official recognition and personal enthusiasm. He remarked,

The time will come within the lifetimes of most of the people in this audience when human beings will go to Mars. They will be the precursors of a permanent establishment on Mars and, perhaps, will allow us to look forward to even more interesting and ambitious ventures in space. . . .

Consider George Bernard Shaw's line from *Back to Methuselah:* "Some men see things as they are and say, why? I dream things that never were and say, why not?" Why not go to Mars to advance the human presence in space? Why not go to Mars and build a gateway to the asteroid belt and the outer planets? Why not go to Mars to use human judgment, abilities and intelligence to explore an exciting new world? Why not go to Mars to recognize, to describe, to organize, to correlate, and to solve problems as only human beings can solve them? Why not go to Mars to better understand ourselves, Earth, its ice ages, atmospheric changes and climate? In T. S. Eliot's words, "The end of all our exploring will be to arrive at where we started, and to know the place for the first time."

. . . The immensely challenging program could be a strong force for peace. It could redirect creative human efforts from dealing with armed conflict to planning and carrying out a peaceful, stimulating and ultimately more valuable program of unprecedented scope and imagination. Ironically, perhaps Mars, the "war god," could become a powerful instrument for peace.

Humans to Mars? Indeed, why not? "To Mars ... Together" was spreading like wildfire. But the total cost, over decades, would be huge—perhaps as much as that of Apollo. Is it worth it? Fortunately, there is plenty of time, and, if done cooperatively, a journey to Mars will provide a means of avoiding a disruptive and expensive space race.

A deeper, more difficult challenge to the idea of going to Mars—together—arises immediately. Is long-term space collaboration between the Soviets and the United States practical?[9] How could it survive the inevitable conflicts on Earth and periods of cool relations such as those of the 1970s and 1980s? Sally Ride jumped in with the answer to that question:

You have to assume, going into a joint venture, there's always the possibility that your partners may pull out. But there are many ways you can ameliorate the situation by initial negotiations. For example, instead of a group splitting the launcher equally, one country provides the launcher, another the habitation module, another something else, so that you're not left with a system that is defunct if one group pulls out. You will have the option of building your own orbital transfer vehicle or aerobaking technique or whatever without the mission partner.

On January 7, 1986, the Sunday *New York Times* endorsed "To Mars ... Together"—a significant step, given the *Times*'s previous refusal to endorse large space endeavors. On January 27, 1986, immediately following Voyager's exploits at Uranus, the paper again emphasized NASA's need for a clear goal and raised the possibility that "To Mars ... Together" could represent it. (The *Times* made this point seven more times subsequently.)[10]

Less than twenty-four hours later, *Challenger*'s televised explosion over Florida changed all hopes and presumptions about America's future in space.

Two months had passed, and divers were still plucking *Challenger*'s wreckage from the floor of the Atlantic, when our cruise ship pulled into Ft. Lauderdale, two hundred miles south of Cape Canaveral. It was a warm and clear morning as we finished our brief encounter with luxury and the Caribbean. Now reality had to be faced once more.

Was "To Mars ... Together" also a casualty of that disaster, or could *Challenger*'s grievous loss generate a new sense of American purpose in space?

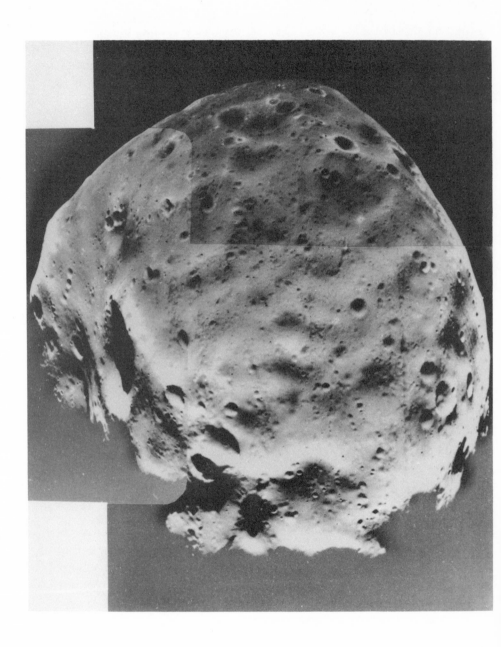

21.1 PHOBOS • *Viking Orbiter 1* took the high-resolution pictures that make up this mosaic of Mars's larger moonlet. Phobos is one of the darkest objects in the Solar System, reflecting only a tiny percentage of the sunlight falling on it. Phobos is less than twelve miles (twenty kilometers) in maximum length.

21. Return to Mars, via Moscow

"WE HAVE CROSSED THE BORDER of the Soviet Union. No pictures are permitted over Soviet territory."

The limpid tones of an Air France steward's voice belied the chilling reality behind his words. It had been eighteen years since my only previous trip to Russia. Yet now, in November 1986, I had the same claustrophobic feeling that had then made me vow never to return.

The purpose of my trip in October 1968 was to report at a scientific conference on the then novel idea of carbon dioxide frost on Mars. Russian tanks had just crushed the Czechoslovakian move toward greater independence, and Nikita Khrushchev's belligerent rattling of Russia's new space rockets a decade earlier still lingered in my memory. Nevertheless, I looked forward to a firsthand look at our World War II allies and postwar adversaries.

Before leaving for Russia in 1968, I had been exposed to the usual stories of KGB attempts to entrap Americans. Since I was a member of government advisory committees with access to American government secrets, such tales were less easy to dismiss for me than they might have been for traveling American college students who sometimes illegally sold their clothes and records at black-market prices to pay expenses. In fact, specific accounts of harassment of American aerospace specialists were popular topics at cocktail parties in southern California. As a precaution, another American scientist and I had arranged to share a double room at Moscow's giant Hotel Rossia, right next to Red Square and Lenin's Tomb.

So it was a bit disconcerting to find ourselves split up into single rooms at the Rossia. Was this intentional? The answer seemed evident in a midnight phone call.

"A rendezvous, monsieur? A little fun? Are you interested?"

This would-be temptress could never have gotten the phone extension of an English-speaking guest without active help from the hotel management. Obviously, she worked with the government's blessing, probably probing for personal (and blackmailable) weaknesses in guests. I mumbled a rejection, slammed down the telephone, and tossed through a sleepless night.

It was a relief to leave dour and threatening Moscow the following morning for the more amiable environment of Kiev and our scientific conference. There, in a tired-looking college lecture hall, I presented the new idea of carbon dioxide frost caps to an unresponsive audience, many of whom could not understand me, because the promised translation service never materialized. The meetings were dull and stiff, bereft of either new ideas or new data.

By midafternoon I was once again depressed, and somehow angry. Russia was not what I had expected. Half a century after the Russian Revolution, some political moderation and diversity should have developed. Yet the same old autocracy remained, just managed more efficiently by the new, nonhereditary power managers. The all-pervasive KGB had replaced the czar's Okhrana and Lenin's Cheka: 1917 had really been an economic revolution, not a political one.

I couldn't wait to get out of that place.

By November 1986, when the Air France flight returned me to Russia, U.S.-Soviet rivalry no longer powered America toward historic space achievement. Apollo, the Mariners, and Viking were fading memories, part of an earlier heroic—and tempestuous—period of American history. And I, like America, was older, more aware of the world. Racing Russia to the Moon, or anywhere else, no longer offered us a shiny pathway to the future. Instead, somehow, U.S. and Soviet energies must be fused into peaceful *cooperative* accomplishments. The second golden age of planetary exploration would not repeat the American domination of the Apollo era. But there could be a magnificent new era of international discovery and adventure, fueled by enlightened U.S.-Soviet collaboration.

My optimism for "To Mars ... Together" was linked to my acquaintance with Roald Sagdeev. Born in 1932, only one year after me, Sagdeev, like all Russians his age, grew up amid the horror of World War II. A brilliant scientist and successful bureaucrat within the Soviet system, Sagdeev had been elected to the Soviet Academy of Sciences at thirty-six and was selected to run the Institute for Space Research (or IKI) at forty. He had become, for Americans, the best-known Soviet exponent of strategic-arms limits and space cooperation. Sagdeev had another distinction. He somehow managed to put a limited form of *glasnost* into practice years before Gorbachev came along. Sagdeev's 1986 VEGA Halley encounter had brought together scientists from eight nations into a genuine partnership (see chapter 19).

I came to know Sagdeev through our Planetary Society activities in 1985 and early 1986 in Washington. Indeed, he became a member of the society's board of advisers, an unusually bold affiliation for a Soviet official. Sagdeev moved easily in the West, overflowing with colloquial humor and slang. He displayed both his own power and a remarkable knowledge about America and Americans.

After my return to full-time teaching and research at Caltech, in 1985, I had resolved to concentrate totally on Mars. But how to rebuild an active research program that would create opportunities for the top-flight graduate students Caltech attracts? NASA had little new activity. When Sagdeev offered me a chance to participate in the planning and execution of a new Soviet mission to be launched toward Mars in July 1988, I responded enthusiastically.

The "Phobos '88" mission would send two identical large spacecraft to orbit Mars and thence to maneuver close to Mars's diminutive and mysteriously dark satellite Phobos. Phobos is the darkest object known in the Solar System and may contain "fossil" chemicals from the origin of the Solar System. It has also been suggested by some scientists in the United States and the Soviet Union as the site for an eventual human outpost at Mars. The mission was broad, calling for a close-up observation of Phobos and an instrument package dropped directly on the surface. New observations of Mars would be garnered along the way, as would observations of the Sun and of interplanetary electrical and magnetic properties.

Here was an opportunity to get back into "hands-on" space exploration and to further the vision of "To Mars ... together" in a practical way.

I was to rendezvous in Moscow with three other American scientists who were also seizing Sagdeev's proffered Phobos '88 opportunity. This time we entered Russia as private citizens, whereas in 1968 I had reported our new Mariner results with NASA's blessing. Now there was no NASA approval, not even travel support.

The era of U.S.-Soviet space collaboration inaugurated by Nixon and Brezhnev in 1972, which had resulted in the Apollo-Soyuz joint flight and other cooperative space endeavors, ended in hostility (see figures 2 and 4 in the Prologue). Cold war tensions exacerbated by the Soviet suppression of the Solidarity movement in Poland squelched official U.S. government enthusiasm for any public space cooperation with the Soviets. If President Carter could direct U.S. athletes to boycott the summer Olympics in Moscow in 1980, he could hardly condone new space partnerships with these same Russians.

By 1982 the political environment had deteriorated further. A very hostile Reagan administration rebuffed Soviet proposals for renewal of the 1972 bilateral accords, and with their expiration the unvarnished space rivalry of the 1960s seemed destined to begin again.

Well, not quite. In 1972 America had an active space program, and its astronauts still walked on the Moon. In 1982 the Soviets, in contrast, had for a decade steadily built up a reliable and effective planetary exploration capability focused mainly on Venus and had methodically established a continuing manned presence in low Earth orbit. This mattered little to the White House. There would be minimal U.S. cooperation in space with "the evil empire" until Reagan and Gorbachev rediscovered détente.

"We are preparing to land at Moscow's Sheremetyevo Airport," echoed the mellow voice of the steward. Layers of dark, cold overcast obscured the runways until the last minute, adding to my unease at returning to Russia in a role neither sanctioned nor encouraged by my own government.

As we left the comforts of Air France, I realized that the terminal was new. It was in fact built for the 1980 Olympics, just as my city, Los Angeles, expanded its international air facilities for the 1984 Olympics.

Then I encountered a scowling young border guard—just as I had done eighteen years earlier. From within a sentry box that framed his military hat and uniform, he scrutinized each passenger's passport and visa. Repeatedly, he checked the photograph against the features before him, including an overhead view acquired through a tilted mirror system.

"Are you Bruce Murray?" he asked me coldly.

"Yes," I said, wondering who else he imagined I might be. In order for me to get the business visa he was now scrutinizing, Sagdeev's institute had had to issue the invitation and send another official transmission through Intercosmos, the Soviet government agency regulating all space relations with foreigners, to the Soviet embassy in the United States.

A few more moments of staring and checking, and I was waved on.

"Things don't seem to have changed much here," I thought angrily.

Suddenly, there appeared a smiling, bearded face. My friend and scientific colleague Slava Linkin greeted me warmly. Slava (short for Viatcheslav) produced a piece of paper that spared my luggage from a meticulous and time-consuming customs inspection.

That life in Moscow had improved somewhat between 1968 and 1986 was evident even on the ride to the hotel. There were more shops and more goods, although the selection was still quite limited, by U.S. or European standards. On a subsequent trip, Linkin invited me to a family dinner at his home, a gesture as much in contrast to the frozen, police state environment of 1968 as was his ownership in 1986 of the car in which we went sightseeing. Moscow was changing slowly on its own, even as Gorbachev proposed revolutionary changes in many aspects of Soviet life.

At an informal family dinner, Sagdeev announced that he would miss the Phobos '88 meetings because of a higher calling. He was off to India the next day as part of Gorbachev's official party. The dinner in his smallish, comfortably furnished apartment showed no trace of the pressure he must have been experiencing. His wife, working daughter, and married son conversed easily with me in excellent English, full of natural courtesy and poise as well as obvious affection for one another. A nice family.

His impending travel and my eye-stinging jet lag notwithstanding, we planned a further expansion of cooperation. First, we worked out an agreement between Caltech and IKI to facilitate frequent staff visits and collaborative research.* Professor James Head, the other dinner guest, had pioneered a similar arrangement between Brown University and another Moscow research organization, the Vernadskii Institute. Now there would be a new nongovernmental link to Moscow, permitting planetary scientists and astronomers at Caltech to work on joint space research with their counterparts at IKI.

The second topic Sagdeev and I discussed intrigued me even more,— ballooning on Mars. Just the preceding summer I had led a small group of Caltech students in a free-ranging study concerning future mobility on Mars. Mars, after all, is equal in surface area to the entire land area of Earth. Exploring it is akin to exploring all of Earth's continents combined. How would we learn what Mars is really like—at the scale of a human explorer or geologist—unless our robotic surrogates could somehow move swiftly across that rocky desert surface? Where would the motive force come from? Our answer was, Use Mars's own natural power system, its wind.

Jacques Blamont had followed this same line of reasoning five years earlier. His solution then involved a giant, lightweight rolling ball that could be blown across the Martian surface carrying a small scientific payload. But a fundamental flaw seemed inherent in the concept. Just like a tumbleweed on Earth, the ball on Mars would sooner or later be trapped by natural obstacles in the terrain—a steep-sided Martian crater, for example. Its journey might end prematurely. Nevertheless, Blamont's "Mars ball" so stimulated students at the University of Arizona that they tried wheeled and powered versions in the nearby desert.

* Sagdeev and Caltech's president, Marvin Goldberger, signed the agreement in February 1987, and substantial visiting began that year. Each side pays the other's travel and living costs in the host country. In 1987 Caltech received a grant from the Greve Foundation of New York City to help defray its share of these travel costs.

JACQUES BLAMONT • Blamont is a pioneer in both space exploration and international space endeavors. He played a major role in the birth and development of the French space program (now the most significant in the world after those of the United States and the Soviet Union). He began collaboration with NASA in the 1960s and with the Soviet space science program in the 1970s. These have led to important international collaborations in the exploration of Halley's comet and in that of Venus and Mars by balloons.

Somehow, the buoyant device must be lifted off Mars's surface from time to time, to use the wind and overcome obstacles. But how? Informal discussions between French and Soviet scientists pointed to a novel answer—a *dual* balloon (see figure 21.2) composed of one sealed helium balloon, like those the Soviets and French used on Venus in 1985, and one solar "hot air" balloon open at the bottom. The envelope of the solar balloon would be dark and would thus be heated by sunlight during the daytime. The additional buoyancy gained from the solar "hot air" balloon during the Martian day would enable both balloons, tied together, to lift a small payload off the surface and then settle back down again each evening, having been blown perhaps hundreds of miles by the winds on Mars. The next day would bring another voyage to another site, and so on, for perhaps ten days, until the batteries gave out. In fact, Blamont and Slava Linkin were already proposing this concept for the next Soviet Mars mission, then aimed at 1992 (later deferred to 1994).

What an elegant idea! My students and I confirmed Jacques's calculations and went beyond them. By the end of the summer of 1986, we had attracted the participation of some imaginative JPL engineers who were likewise fascinated by the idea of moving about on Mars. An informal collaboration spontaneously sprang up among the Caltech campus, JPL, The Planetary Society, and, through Blamont, the French National Space Agency (CNES).[11] Representatives from Utah State University, the Uni-

THE MARS BALLOON

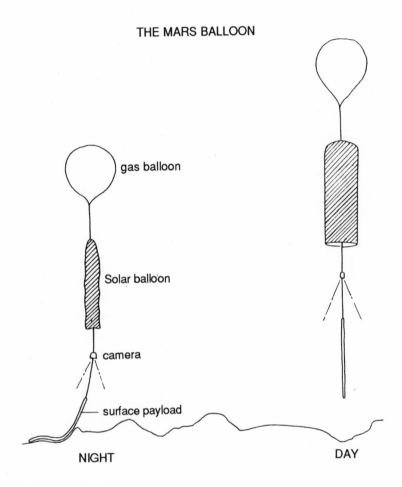

gas balloon

Solar balloon

camera

surface payload

NIGHT

DAY

21.2 BALLOONING ON MARS • The principle of the Mars dual balloon (under consideration by the French and the Soviets for inclusion in a Mars mission planned for 1994) is illustrated here. The gas balloon, a sealed envelope containing helium, is perhaps sixty feet (twenty meters) in diameter; it has enough net buoyancy in the rarefied Martian atmosphere to support the weight of a second, solar balloon and a miniature scientific payload. At night the dual balloon drifts along the surface under the influence of surface winds, ballasted by an easily sliding surface payload. In the morning, sunlight striking the dark covering of the solar balloon (open at the bottom) heats the atmospheric carbon dioxide within it. The warm, less dense carbon dioxide fills the balloon to capacity. The additional buoyancy produced suffices to lift the surface payload and propel the balloon up to an altitude of perhaps four miles (six kilometers). The brisk Martian winds normally prevailing at such heights would be expected to transport the balloon up to hundreds of miles by late afternoon, when decreasing solar heating would cause the solar balloon to descend. The dual balloon would then return to the surface for the night. This process might continue for ten to fourteen days, governed by the lifetimes of the battery and of the helium balloon. U.S. scientists and engineers have played an important, if informal, role in the development of this innovative concept. *Drawing by Eric J. Gaidos*

ROALD SAGDEEV • Sagdeev was a child chess prodigy in Moscow, where he was born in 1932, and went on to become one of the outstanding physicists of the USSR. He was elected a member of the powerful Soviet Academy of Science in 1968, and became head of the Institute for Space Research in Moscow in 1973. Under his leadership, the VEGA mission to Venus and to Halley's comet and the Phobos '88 mission to Mars and its moonlet Phobos have made use of an unprecedented degree of East-West cooperation. He has also been a conspicuous figure in USSR / U.S. arms control discussions. In 1986 he was elected to the Supreme Soviet, where he has been a vigorous proponent of *glasnost* and *perestroika,* which he also supported at the June 1988 Party Congress. *Photo by the author*

versity of British Columbia, the Ball Aerospace Corporation, and the Titan Systems Corporation soon joined the enthusiastic band.

So when Sagdeev learned of our efforts after dinner at his apartment, he was keenly interested. The prospect of a *trilateral* Mars balloon project as part of the next Soviet Mars mission was just the kind of expansion of international cooperation in exploring Mars that he wanted. "Please continue," he urged. "Let's make this thing happen."

The next morning the official Phobos '88 meetings began at IKI, which is housed mainly in a massive, thirteen-story building in the southwest section of the city. Earlier in 1986 IKI had become *the* center for international participation for eight countries in the VEGA comet mission. Now, to ready the 1988 flight to Mars and its moonlet Phobos, Sagdeev and his colleagues took an even bigger step. Some fifty scientists from Austria, Bulgaria, Hungary, East Germany, Ireland, Poland, Finland, West Germany, Switzerland, France, Czechoslovakia, and Sweden were banded together in small groups with Soviet colleagues to develop and supply over thirty scientific instruments for each of two large new Phobos '88 spacecraft. Great Britain and the European Space Agency were represented by high-level observers. Because of restrictions on the export of U.S. hardware to Russia, the American participants could not supply instruments or components. Instead, we would use our experience with Mars and Phobos to help coordinate scientific plans from various instruments to yield maximum return. Sagdeev used the American term *interdisciplinary scientist* to designate our role.

The act of watching European scientists working with their Soviet counterparts drove home the political wisdom of Sagdeev's *glasnost* in space. While I had no doubt of his deep commitment to a fair and balanced Soviet-U.S. cooperation in exploring the planets, he was nevertheless in a

"win-win" position. If the United States continued to reject Soviet overtures for cooperative Mars exploration, the USSR would receive Western
European participation by default. The American position was weak. In
fact, even Britain and ESA, America's strongest allies in Europe, were
already supplying exotic new X- and γ-ray instrumentation for planned
Soviet astrophysics missions. Only Americans were excluded—by their
government. I wished that people from the White House had been sitting
here; then they might have realized that as long as we continued to view
space as a domestic football, even as a giant WPA in the sky, rather than
as an instrument of American foreign policy, the Soviet Union would
steadily increase its influence over Western Europe. And surely, I thought
moodily, there were others here in the USSR who preferred Soviet dominance in space over Sagdeev's vision of cooperation with the United States.

The meetings at IKI were businesslike and even-handed. No hint of
Big Brother showed through. An excellent simultaneous translation (Russian, English, French) facilitated the free exchange of ideas. I had never
seen anything like this in the United States except at the UN meetings in
New York. European and Japanese scientists who participate in U.S. space
missions do so in English. No accommodation is made for their language
needs. And NASA is no different from the rest of the United States.
Egalitarian relations are hard for Americans to develop; paternalism is
easier. It is hard to give up the client-state attitudes we inherited from the
postwar period. It is easy just to assert that English is the international
language and ignore others' needs.

Language differences aside, the key players at the Phobos '88 project
meeting could have come straight out of a large project meeting at JPL.
IKI was represented by its deputy director for engineering, V. M. Balebanov.
He cochaired the meeting with Dr. Roald Kremnev, the newly appointed
director of the Babakin Center, a closed facility that develops and supplies
planetary spacecraft, for which IKI provides instruments. It quickly became clear that the new Phobos spacecraft was overweight—an almost
universal difficulty of new space systems, regardless of the nationality of
their designers. Kremnev, representing Babakin and holding responsibility
for the spacecraft component of the mission, drew attention to the growth
in weight estimates of many of the thirty scientific instruments being overseen by IKI. Balebanov quickly countered by noting that there had been a
parallel increase in weight estimates in certain subsystems of the spacecraft
itself. He suggested that Kremnev might wish to use some of "his" weight
reserves to effect a balance.

Because of the simultaneous translation flowing into my ears from the
remote headset, I could follow this verbal Ping-Pong match as it was
played. Such conflict between the spacecraft (the truck) and the science

instruments (the cargo) was the hallmark of every scientific mission I had known. The Russians were just like us. John Casani and his JPL team would size up the situation in an instant, I thought, smiling to myself.

One big difference did surprise me. Who was in charge? At an American meeting there would have been a single chairman. In an American project the lines of authority are normally well defined. In fact, it is a truism in American aerospace endeavors that divided lines of responsibility always lead to problems. Where does the buck stop in Russia?

I gradually realized that neither IKI nor Babakin (nor the launch vehicle organization nor the tracking organization) reports to the same organizational boss, as they do in NASA. With a JPL planetary mission the spacecraft, the scientific instruments, and the deep-space tracking facilities (the DSN) all fall under the purview of the JPL director. In the Soviet Union those functions are coordinated behind closed doors at a higher level by committee. Thus, Babakin could plan successive generations of planetary spacecraft, while IKI (and other scientific institutes like Vernadskii) could develop scientific instruments for the spacecraft. Similarly, the tracking stations served many customers in the Soviet Union. Planetary missions were only a minor consideration in their overall work load. The Proton launch vehicles were allocated for planetary as well as many other purposes by some high-level committee. All four elements probably received fairly stable budgets, which were not tied to specific missions.*

There is great strength in this system: a steady flow of launch vehicles, spacecraft, and scientific experiments—with frequent planetary missions as the result. But there is a flip side as well. With no single organization in charge of maximizing overall mission achievement, conspicuous imbalances cropped up in the Phobos '88 mission. For example, the spacecraft had no scan platform, so the entire spacecraft had to be turned on gyros to point properly for *each* photographic or other remote-sensing observation. Nor did the spacecraft have good radio communications. The data rate to Earth is at best only 16,000 bits per second, barely equal to the one we used at Mars in 1969. Yet the Soviet VEGA mission in 1986 had both a scan platform and a fourfold higher data rate than Phobos. The proper spacecraft technology had once been available in the USSR. Why not again? The answer seemed to be that, without a single boss, there was no way to optimize the technical capacities of each different element.†

I began to realize that I had been making a typical mistake: assuming

---

*Under *perestroika,* the USSR is moving away from stable funding to project-by-project funding. Major changes in the character of the Soviet program seem likely.
†This divided responsibility may have contributed indirectly to the failure in flight of both *Phobos* spacecraft.

that because the Russians had made achievements in space comparable to ours—such as exploring the planets—they must be organized as we are. But there is no NASA in Russia. The manned space activities there are promoted and carried out by an organization totally different from those that carry out space science. Their manned-flight organization has close ties to the military (just as Hollywood ceaselessly suggests that our manned space program is linked to the Air Force). In contrast, the Soviet Academy of Sciences—a major branch of the government, with more than 100,000 employees—leads most of the scientific space activities. It is as if in the United States the National Science Foundation and other governmental scientific organizations ran a separate, automated scientific space program. Manned and unmanned groups in Russia both use the big Proton launch vehicle, but they otherwise have little contact with one another.

Had the United States been organized this way in the post-Apollo period, there would have been no "Shuttle only" policy to put all robotic missions onto a man-piloted Shuttle. In fact, the Russians' current strength in civilian space is due largely to their separation of the manned space program from their scientific unmanned satellites and to their pragmatic use of expendable rockets for both. Thus, the Russians now compete effectively with us despite their lagging overall technological capability.[12]

I was still pondering these differences a week later as I entered the crowded cabin of a Lufthansa airliner to leave Moscow. Suddenly, I caught sight of an *International Herald Tribune* in the arms of a waiting flight attendant. Its pages seemed to me to be printed in pure gold—freely available information about the world! Moscow was better, much better, than it had been in 1968, but it was still a closed society, deeply troubled by Gorbachev's attempt to impose *glasnost* and *perestroika* from the top.

It felt good to be going home. But this time I knew I would be back. I had friends and scientific colleagues in Moscow, and we had important work to do—together.

22.1 VOYAGER AT NEPTUNE • In this artist's rendition of the encounter of *Voyager 2* with Neptune on August 25, 1989, the spacecraft is passing above the planet's equatorial plane, in which peculiar partial rings (called arcs) occur. Soon it will almost graze the northern polar regions of Neptune's thick methane and hydrogen atmosphere and then quickly swoop close to Triton, its large moon. Triton, like Earth and Saturn's Titan, appears to have a nitrogen-rich atmosphere. Following the Neptune encounter, *Voyager 2,* like its twin, *Voyager 1,* will continue to probe ever farther from the Sun. The Voyagers may continue to radio their findings to Earth for another decade or more. *NASA/JPL illustration*

22. A Last, Best Chance for America

DALE MYERS WAS really steamed up.

A longtime acquaintance who had once lived in Pasadena, Dale was now deputy administrator of NASA. It was May 14, 1987. We sat in the same corner office on the top floor of the NASA headquarters building in Washington where, in earlier years, I had faced Hans Mark, and before him, Alan Lovelace. The person was new, but the issue was the same—planetary exploration.

This deputy administrator's tone, however, was angrier than that of his predecessors. I did not take offense, because I knew what was bothering my frustrated and dispirited friend. I had sought this meeting to try to recover a constructive relationship.

The problem erupted without warning in August of 1986, when NASA moved to delay the small but highly regarded Mars Observer mission, America's only funded Mars program. Originally scheduled for launch in 1990, Mars Observer now would be rescheduled for 1992. The stated reason was that the Space Shuttle was slipping still further behind and could not be counted on for a launch in 1990. Once again America's journey into space was sputtering to a stop because NASA could not see beyond the Shuttle. Once again the obvious solution—an expendable launch vehicle—was being bypassed in favor of delay.

I was stunned and dismayed at this news. Here was the one, small American entry in the Mars derby in the face of accelerating Soviet efforts, the one NASA program that could be looked upon as a stepping-stone to human exploration of Mars. Congress had easily approved Mars Observer in 1985 and had already appropriated funds for the coming year, 1987. Nevertheless, NASA chose to delay rather than switch Mars Observer to an available expendable launcher.

When this decision percolated up NASA's chain of command in August 1986, James Fletcher had been NASA administrator for only three months. The angry man across from me now, Dale Myers, had been Fletcher's close associate at NASA in the early seventies and was a highly regarded technical manager in government and industry circles. But he had not yet been able to rejoin Fletcher at NASA. Reagan still had to find a home for William Graham, the deputy administrator whom Fletcher inherited. After some hesitation (and to the disappointment of the American scientific and technological community), the White House solution for the "Bill Graham problem" was to kick him upstairs and appoint him the new science adviser to the President.

Fletcher's highest priority for his second tour as NASA administrator was to "get the Shuttle flying again." He also moved swiftly to buttress the flagging Space Station program. But despite some conciliatory words about the need for a "mixed fleet," Fletcher, in fact, put new expendable launchers for NASA well behind plans for the Shuttle and the Space Station. This low priority ignored the urgings of the National Academy of Sciences, key congressional committees, most professional societies, and NASA's own advisory committee. Fletcher's priority should have been to get *America* flying again, not just the Shuttle.

Fletcher evidently would not even consider using an expendable launcher for Mars Observer when he approved delaying that mission for two years. Had he done so, he would have discovered that Mars Observer could be launched with an existing Titan rocket. The Titan 34D (or Titan 3) had long been used by the military and—after *Challenger*—was offered by the Martin Marietta Corporation as a commercial product for NASA and the communications satellite industry.

Instead, Fletcher's staff took the $50 million earmarked for the development of Mars Observer spacecraft in 1987 and retargeted the money to augment other space-science activities, especially those supporting the Space Station. Consequently, even though NASA's proposed new budget jumped up 12 percent overall, the funds for planetary exploration dropped 15 percent.

To make the whole, sorry mess even worse, Fletcher's aides grossly underestimated the future costs of delaying Mars Observer (partly because they acted swiftly and in secret, without consultation of key project personnel at JPL).

What could I do? What could the Planetary Society do? Friedman and Sagan were as dismayed as I was. In addition, we raised a cry in the news media and elsewhere. Our actions had an effect. Fletcher soon stated, "We have decided ... to continue working on the Mars Observer on schedule to provide launch readiness for the 1990 opportunity.'' The head of NASA's Office of Space Science and Applications backed him up, saying, ''We have reconsidered the proposed delay and now still are working on the planned 1990 launch for Mars Observer.''

Our action had created some time to help NASA find and finance a Titan 3 expendable launcher. And indeed there was one to be had, for $125 million to $150 million. This amount was no more than the increased cost of the spacecraft if the launch were actually delayed from 1990 to 1992.* Working with the Martin Marietta Company, RCA (the spacecraft's prime contractor), and key Air Force officials, we helped find a way for NASA to reserve a Titan 3 for the 1990 launch and defer any payment until 1988.

But we had not appreciated how little Fletcher and his aides cared about going to Mars, about using expendable launch vehicles, or about competing or cooperating with the Russians. They were fixated on the Shuttle and the Space Station. The basic NASA plans of August 1986 were never altered, pious statements in September notwithstanding. The budget plan submitted by NASA to the OMB in September wiped out most Mars Observer funding for the following two years. This unhappy picture became public on January 2, 1987, in a letter from NASA to the relevant congressional committees notifying them of NASA's intention to change the current operating plan.

''Why not use the Titan expendable launch vehicle?'' I asked Dale Myers, now Fletcher's deputy, when I phoned him in early January.

''Where am I going to get the $150 million?'' he shouted back angrily. (In NASA's *Alice in Wonderland* accounting, a Shuttle ride is ''free,'' but an expendable launch must be explicitly budgeted.) The obvious answer to me and, I presumed, to him was to take it from the same place where NASA would have to find the additional money to pay for the delayed development of the Mars Observer spacecraft.† However, not wishing to challenge him further, I pointed out that there were 100,000 dues-paying members of the Planetary Society,‡ the world's largest space-interest group by far. Perhaps their intense interest, reflective of the broader public interest in Mars exploration, could prove effective.

---

*All the scientists had been selected and were at work, along with fully staffed project groups at JPL and RCA, the prime contractor for the spacecraft. It is not feasible to disband such groups and then expect to pick up two years later where one left off. To further complicate matters, the contract with RCA was for a fixed price and would have to be completely renegotiated if NASA chose to delay the mission.

†In fact, Mars Observer is now scheduled to go on a Titan expendable launcher in 1992. However, the financial catastrophe caused by NASA's slipping the mission from 1990 led to the dropping of a key scientific instrument from the payload.

‡Now 127,000.

"What should we do?" I asked.

"Take a deep breath and wait a few months," he answered.

A few months is too long to hold your breath. Sagan, Friedman, and I launched the Planetary Society on a major advocacy campaign. Our members were informed of the situation and its implications. They were urged to write their congressman and NASA Administrator Fletcher expressing their personal concerns about the delaying of Mars Observer. The results were astounding. More than 10,000 of our members sent off some 25,000 letters. Fletcher received more than 5,000, as we learned from supporters in NASA headquarters who stacked up the bales of unanswered mail.

The response in Congress was almost unprecedented for a minor budget issue. Nearly every congressman—conservative and liberal, Republican and Democrat—received Mars Observer appeals from constituents, young and old. The staffs of the House and Senate Space Committees enjoyed sudden popularity; requests for information flooded in from congressmen and senators who had never before demonstrated any interest in space.

Fletcher was also besieged by personal phone calls from members of Congress, some of whom were key to his "important" budget items like the Shuttle and the Space Station. But it was to no avail. NASA's "ship of diplomacy" was too mired in the ice floes of the budget process and too hemmed in by its own conflicting statements regarding the reasons for Mars Observer's delay. It could not extricate itself from what even NASA insiders recognized as a concatenation of mistakes. Fletcher held firm. On April 15, 1987, NASA ordered JPL to stop work on readying the Mars Observer spacecraft for a 1990 launch. (On the same day Secretary of State George Shultz and Foreign Minister Eduard Shevardnadze signed the new bilateral accord to cooperate in space, which specifically cited Mars missions.)

Although we had lost the battle to preserve a Mars Observer 1990 launch, "To Mars ... Together" gathered momentum from this episode. Congress took notice of popular support for a small planetary mission, one without support from major aerospace firms, big unions, or powerful geographical lobbies. If it was obvious that NASA had blundered in delaying Mars Observer, it became obvious as well that "To Mars ... Together" was a potent new political concept.

The Mars Observer battle had one major negative consequence. Fletcher and his loyal deputy Myers were furious that we had created such a big problem for them. All this led to my effort to calm the waters in May 1987 and to search with Myers for some focus for NASA that would appeal to the Planetary Society.

"How can we help, Dale?" I asked, hoping he would use that opening to outline an area of future cooperation. "We're 100,000 dues-paying members, a real Mars constituency. And you've got Sally Ride studying four long-range options for America in space—two of them involving Mars.* Can't we work together?"

"We may not choose the Mars options, Bruce," he replied gruffly. "If we do, we'll contact you."

I could hardly believe my ears. This likable and talented technocrat, second in command of an agency spending $10 billion of the "discretionary" part of the civilian federal budget, was dismissing an opportunity to rebuild its tottering popular support.

My time was up and I had failed to find some common ground upon which we could build anew. How sad.

As I walked out of Myers's office, past the portraits of the previous administrators of NASA, I suddenly realized that he considered us a pressure group, not a constituency. But if we were not a NASA constituency, who *was?* Certainly not the openly critical National Academy of Sciences, or the enemy-filled White House, or Congress with all its special interests.

As I got off the elevator, it hit me: "NASA's constituency is the aerospace industry! Just what my cynical university friends have been saying for years."

Suddenly I felt naive. I had dealt with top NASA managers for all these years on the assumption that they worried as much as I did about leading the country grandly into space. If we shared that goal, we must always be constituents of one another.

But the facts suggested otherwise. On one point NASA had been consistent. Top brass at NASA always aimed to maximize NASA's annual budget in the short term. NASA seemed to believe that the support of the aerospace industry was the surest means to push NASA's budget requests through the OMB and Congress. The public was at best an indirect, distant ally. The drive toward historic achievement often gave way at budget times to expediency and to short-term institutional protectionism.

That was the real key for these men—next year's NASA budget, not

---

*The other two options were a lunar base and a focus on Planet Earth. The latter involves deployment of many unmanned Earth satellites to monitor the global environment. In a major policy paper written in late 1985 (mentioned in the prologue), I argued that an *international* mission to Planet Earth should be a major U.S. policy objective. The United States, in conjunction with most of the nations of the world, would help lead a global program of space and surface measurements, in order to chronicle natural and man-made environmental change. "To Mars ... Together" and an international mission to Planet Earth were the twin themes I proposed.[13]

the achievements in space in later years that the money was supposed to provide. Private dreams notwithstanding, each year the mediocre process was repeated, because there were no exploratory achievements to which NASA was committed.

Just five months earlier, hopes had risen when the astronaut Sally Ride came to NASA headquarters to draft an agency position on "leadership and America's future in space."[14] Fletcher finally seemed to grasp the need to justify the Space Station as part of NASA's longer-term endeavor. Sagan, Friedman, and I had for two years spoken of the "lack of presidential leadership" and argued that "NASA needs a goal." These views, now heard everywhere, seemed to prompt a response from NASA. Quickly, Ride blended NASA's own thoughts with the dreams of the National Commission on Space in early 1986[15] and with the sobering realism of the Rogers commission, which investigated the *Challenger* disaster and on which she served. She singled out four possible long-range initiatives, each a "bold, aggressive proposal which, if adopted, would restore the United States to a position of leadership in a particular sphere of space activity": "(1) Mission to Planet Earth, (2) Exploration of the Solar System, (3) Outpost on the Moon, (4) Humans to Mars."

By April 1987 rumors ran through NASA headquarters about just when Fletcher was scheduled to see the President concerning a future initiative— and whether he would push for Mars.

But once again NASA's past reached out to limit its future. Reagan's White House got hopping mad at NASA and cut off Fletcher's access to the President, because of a huge increase in the estimated cost of the Space Station. In 1984 Reagan had approved a Space Station that then NASA leaders Jim Beggs and Hans Mark priced at $8 billion. However, Jim Fletcher, when called back to serve as NASA's administrator a second time, in mid-1986, wisely had his new team "reprice" the Space Station. This doubled the estimate, to $16 billion (in 1984 dollars), which was bad news to a White House trying to pin the federal deficit on "congressional high spenders."

Worse yet, there was still plenty of suspicion that the $16 billion price was low. To underline its lack of confidence in NASA, the White House requested an independent cost estimate for the Space Station from a panel of the National Research Council, an arm of the national academies of sciences and of engineering. That NRC panel, chaired by Robert Seamans, NASA's deputy administrator in the Apollo era, practically re-doubled the estimate, projecting a cost of as much as $30 billion, by including more necessary items.[16] Furthermore, the panel called the Shuttle only "marginally adequate'' for ferrying the Station's components into orbit. The Shuttle would have to be upgraded in order to do the Space Station job Fletcher wanted.*

Fletcher's reluctance to face the Shuttle's and the Space Station's problems realistically cost NASA access to the Oval Office. Now Fletcher could not even get to see his boss regarding NASA's future. The agency's standard post-*Challenger* solution to every problem (''Send more money'') was one the White House did not want to hear anymore.

Meanwhile, the State Department seemed to think ''To Mars ... Together'' was a good option for the President. In October 1986, four years after the preceding agreement had expired, it initialed a bilateral accord with the USSR concerning space cooperation. The State Department wanted a new space agreement ready for a ceremonial signing at the still pending second Gorbachev-Reagan summit. Because of protracted delays in the summit, Secretary of State Shultz and Foreign Minister Shevardnadze signed the agreement at a separate meeting on April 15, 1987.

But optimism about progress in U.S.-USSR space relations did not extend to NASA. Its top leaders soon made it clear that they did not expect substantive outcomes during Reagan's tenure. Instead, their objective was simply to tidy up the bureaucratic process for the next President. The first meeting of the newly formed U.S.-USSR Joint Working Group was not even scheduled to take place until late 1987. And in keeping with the emphasis on form over substance, Sam Keller, an old-line NASA bureaucrat, was designated to lead the NASA negotiating team to Moscow. Keller was an administrative operative, not a technical leader or a policymaker. He was the bureaucratic hatchet man who had delayed Mars Observer in late 1986 and early 1987. I had expected Fletcher to designate a higher-ranking NASA official. However, Fletcher not only put Keller in charge but even distanced himself from the whole business.

It was still ''back to the future'' in Washington in May 1987.

In contrast, there was good news in California. Our unofficial ''trilateral'' project for a Mars balloon was progressing. In early September 1987 we finished key field tests on fresh, rough lava flows in the Mojave Desert.

---

*The NRC panel was not the first to question the Shuttle's capacity to carry pieces of the Space Station into orbit. In late 1986 some of Fletcher's own staff had recommended use of a new, ''heavy lift'' unmanned launcher—like the one the Air Force is now developing—to ferry big chunks of the Space Station into orbit. The bigger the ''chunks'' ferried, the less dangerous and costly the assembly of pieces in orbit by astronauts. But Fletcher rejected such effective and practical suggestions, just as he rejected an expendable launch vehicle to hold Mars Observer on schedule for a 1990 launch. No, the Space Station would use the Shuttle only.

The videotapes of these tests showed dramatically why our new payload design worked in "Mars-like" territory, and why the initial "gondola" concept would not work on Mars.

A great opportunity arose for us to present these new Mars balloon results to the Russians and the French at a meeting in Moscow. Roald Sagdeev wanted Carl Sagan, Lou Friedman, and me (and many other American space figures) to come to Moscow for a thirtieth-anniversary commemoration of Sputnik's opening of the space age on October 4, 1958. He would even provide free travel.

In Moscow we found ourselves part of a group of about four hundred foreigners treated to VIP visits to the ballet and to the cosmonaut base "Star City," in addition to three days in large business sessions, where we mingled with hundreds of Soviet space officials and scientists. Flawless simultaneous translations aided communication.

The primary objective of this Soviet public relations effort was apparently to demonstrate to the Kremlin that visible Soviet space achievements, involving open international participation, could earn the USSR respect in all quarters. The heads of the Japanese space program and all the European ones showed up. But from NASA there was just one, lone fourth-level official, Sam Keller.

I almost fell off my seat during Keller's speech. Sam made a strident appeal to move ahead in exploring Mars, and he effusively complimented the Planetary Society for its promotional efforts! What was going on? Was something happening behind the doors in official Washington?

"If Sam Keller ends up wearing a white hat in this game, I'll really have to rearrange my prejudices," I thought. "But not to worry—old bureaucratic leopards don't change their spots."

I was wrong. The prospect of a second Reagan-Gorbachev summit three months hence had started juices flowing at the State Department and at the National Security Council, if not at the top of NASA. So in early December 1987, as Gorbachev openly endorsed "To Mars ... Together" in Washington, Sam Keller faced Soviet space negotiators in Moscow. They hammered out an agreement legitimizing and endorsing a wide range of American participation, including my own, in the Phobos '88 mission. In return, ten Soviet scientists would participate in the now delayed American Mars Observer mission. The results were revealed to the public. It was a small, but real, start toward "To Mars ... Together."

NASA seemed to be stirring at last, awakening from its Shuttle sleepwalk.

Meanwhile, popular interest in "To Mars ... Together" (by now the title of a *New York Times* editorial and a PBS documentary) continued to grow, fueled by the visceral sense of many Americans that we should strive to lead in space again and that cooperating with the Russians and other countries was a fine way to go. Even in Congress, enthusiasm for "To Mars ... Together" began to spread from a few visionaries like Senator Spark Matsunaga of Hawaii* to the middle-of-the-road chairman of NASA's authorizing committee in the House, Robert Roe of New Jersey.

The end of May 1988 brought President Reagan to Moscow and to his third summit with General Secretary Gorbachev. Our expectations had been heightened by Gorbachev's recent revelation that he intended to raise the subject of joint U.S.-USSR manned Mars exploration with Reagan.[17]

However, neither the primary objective, a new agreement to reduce inventories of long-range missiles, nor a major agreement to jointly explore Mars emerged. It would be hard to imagine how the latter could happen without the former. But three smaller steps toward space cooperation were taken. Each side agreed to fly space instruments on the other's spacecraft; this is an important escalation in the degree of cooperation. In principle, some American instruments could now even be carried to Mars's surface as part of the next large Soviet endeavor, scheduled for launch in 1994.

Second, the United States agreed to carry French-Soviet radio equipment aboard the U.S. Mars Observer spacecraft and use it to relay to Earth close-up pictures of Mars's surface radioed from the French-Soviet balloons that are to be delivered as part of the Soviet Mars mission in 1994. (This was another of Jacques Blamont's good ideas.)

There was also a practical step toward joint robotic exploration of Mars. Soviet and American scientists and planners are all in accord that the automated return to Earth of rock samples of Mars's surface is the next major milestone. Both groups are independently studying this complex and expensive endeavor. In Moscow in May 1988 the United States and the USSR agreed to exchange their national studies. This is a necessary precursor to any consideration of a possible cooperative mission to return samples from Mars.

Not bad for an otherwise "ho-hum" summit. No initiatives, but meaningful diplomatic progress on both "Mars" and "Together." The modest momentum developed in 1988 within the federal bureaucracy will now carry over into the next presidency. We can only hope that the enthusiasm of the Soviets will continue, despite the sluggish U.S. response to their bold overtures.

---

*Author of *The Mars Project: Journeys beyond the Cold War* (New York: Hill and Wang, 1986).

My former nemesis Sam Keller may prove to have been the most successful official in post-*Challenger* NASA. In comparison, Fletcher's "robust" Shuttle goal remains as distant as ever, and his Space Station initiative is bogged down, awaiting a renewed presidential commitment. Originally meant to commemorate the five-hundredth anniversary of Columbus's first voyage, the deployment of the Space Station has now been put off to 1996. it has slipped one year per year for four years. And the technical problems have not even started to crop up yet. Shades of the Shuttle!

In parallel with these governmental stirrings, the Planetary Society continued to stitch together an ever broader endorsement through its "Mars Declaration." This approach was originally suggested by the Planetary Society adviser John Gardner, former secretary of HEW and the founder of Common Cause.* Sagan, Friedman, Thomas Paine (now a society director), and I have therefore hammered out a manifesto. "The Mars Declaration" reads as follows:

Mars is the world next door, the nearest planet on which human explorers could safely land. Although it is sometimes as warm as a New England October, Mars is a chilly place, so cold that some of its thin carbon dioxide atmosphere freezes out at the winter pole. There are pink skies, fields of boulders, sand dunes, vast extinct volcanos that dwarf anything on Earth, a great canyon that would cross most of the United States, sandstorms that sometimes reach half the speed of sound, strange bright and dark markings on the surface, hundreds of ancient river valleys, mountains shaped like pyramids and many other mysteries.

Mars is a storehouse of scientific information—important in its own right but also for the light it may cast on the origins of life and on safeguarding the environment of the Earth. If Mars once had abundant liquid water, what happened to it? How did a once Earthlike world become so parched, frigid and comparatively airless? Is there something important on Mars that we need to know about our own fragile world?

The prospect of human exploration of Mars is ecumenical—remarkable for the diversity of supporting opinion it embraces. It is being advocated on many grounds:

• As a potential scientific bonanza—for example, on climatic change, on the search for present or past life, on the understanding of enigmatic Martian landforms, and on the application of new knowledge to understanding our own planet

• As a means, through robotic precursor and support missions to Mars, of reviving a stagnant U.S. planetary program

• As providing a coherent focus and sense of purpose to a dispirited NASA

*After reading Gardner's *Self-Renewal* in 1975, I recruited him to serve on the newly created JPL advisory committee. We became close friends, pursuing educational innovations together for several years after I left JPL.

for many future research and development activities on an appropriate timescale and with affordable costs

- As giving a crisp and unambiguous purpose to the U.S. space station—needed for in-orbit assembly of the interplanetary transfer vehicle or vehicles, and for study of long-duration life support for space travelers
- As the next great human adventure, able to excite and inspire people of all ages the world over
- As an aperture to enhanced national prestige and technological development
- As a realistic and possibly unique opportunity for the United States and the Soviet Union to work together in the spotlight of world public opinion, and with other nations, on behalf of the human species
- As a model and stimulant for mutually advantageous U.S.-Soviet cooperation here on Earth
- As a means for economic reconversion of the aerospace industry if and when massive reductions in strategic weapons—long promised by the United States and the Soviet Union—are implemented
- As a worthy application of the traditional military virtues of organization and valor to great expeditions of discovery
- As a step towards the long-term objective of establishing humanity as a multi-planet species
- Or simply as the obvious response to a deeply felt perception of the future calling.

Advances in technology now make feasible a systematic process of exploration and discovery on the planet Mars—beginning with robot roving vehicles and sample return missions and culminating in the first footfall of human beings on another planet. The cost would be no greater than that of a single major strategic weapons system, and if shared among two or more nations, the cost to each nation would be still less. No major additional technological advances seem to be required, and the step from today to the first landing of humans on Mars appears to be technologically easier than the step from President John F. Kennedy's announcement of the *Apollo* program on May 25, 1961, to the first landing of humans on the Moon on July 20, 1969.

We represent a wide diversity of backgrounds in the fields of science, technology, religion, the arts, politics and government. Few of us adhere to every one of the arguments listed above, but we share a common vision of Mars as a historic, constructive objective for the technological ambitions of the human species over the next few decades. We endorse the goal of human exploration of Mars and urge that initial steps toward its implementation be taken throughout the world.

Signers of this document include former President Carter and six former Cabinet officers of both parties; every NASA administrator except James Fletcher; a former commander of NATO and half a dozen other retired generals and admirals; a dozen former diplomats and high-ranking specialists in international relations; fifty astronauts and former astronauts, including the first man on the Moon (Neil Armstrong) and one of the last (Jack Schmitt); twenty-two Nobel laureates; twenty-four college presidents; and a full assemblage of business, religious, and entertainment figures.

As I write, it is two and a half years since *Challenger*'s fiery blast shattered America's space fantasies. Americans are beginning to confront their disquieting image in the dispassionate mirror of space. A budding political consensus backing a new national policy for space stirs in Congress. News-media interest in and support for "To Mars ... Together" is everywhere. Even NASA fitfully fashions plans to lift us out of our present malaise.

While Americans ponder their future under a new President, our disciplined robotic explorer *Voyager 2* closes in on Neptune, to complete its voyage of discovery. *Voyager 2* is a faint echo of John Kennedy's May 1961 Apollo speech—a last hurrah for that fantastic moment when Kennedy unleashed American technology in the service of the world's dreams and aspirations.

We Americans are lucky. We still have a last, best chance to journey into space, this time outward toward Mars, traveling in good company.

"To Mars ... Together" is a feasible political option. The next President could seize this means of signaling to a skeptical world that the United States is ready to play a new and more sophisticated role in world affairs, while retaining its frontier enthusiasm. He could respond substantively to Gorbachev's still unanswered challenge to jointly explore Mars. That next President could propose that the two countries cooperatively undertake the three critical tasks that must be accomplished before any human can safely travel to Mars's surface:

1) Overcome Weightlessness • "Astro/cosmonauts" must arrive at Mars, after an interplanetary flight of almost a year, physically and mentally capable of handling the rigors of landing on and exploring its surface. Neither the United States nor the Soviet Union now knows how to do that. The means must be found through long-term human experimentation in Earth orbit. Mixed U.S. and Soviet crews could together develop practical means to deal with protracted weightlessness. It may even be necessary to experiment with large spinning spacecraft (like that in *2001*), which provide modest amounts of artificial gravity to keep the body's functions balanced. Highly instrumented Shuttle flights, long stays on the Mir station, and use of the redesigned U.S. Space Station together could provide the necessary research facilities. The U.S. Space Station program would be saved from mediocrity—or worse—by being transformed into the critical step for Americans to go to Mars. The current concept would have to be redirected to emphasize biomedical research and physiological experimentation, instead of on microgravity research and applications, the current focus. In any case, microgravity research is better carried out on a human *tended* space platform than on one continuously occupied by humans (whose presence is intrinsically incompatible with the maintenance of a true microgravity environment). The purpose of American space flight would now justify the intrinsic risk involved.

*2) **Build a Space Habitat*** • Reliable and self-contained ecological habitats will be necessary to house and sustain flight crews for up to a year, on the way to and from Mars. The Apollo missions were away from Earth orbit for only two weeks' duration at most. Much of the preliminary testing and development could be done jointly on Earth, using human subjects in closed systems. The final demonstration of the proposed Mars voyage habitat, however, will require flight in Earth orbit for a year or longer. This could also be carried out in common, providing a unique experience with joint mission operations and, especially, with integrated mission authority and responsibility. If there is a problem, who decides what to do? Who is in charge of this joint mission?

*3) **Learn More about Mars*** • Robotic acquisition of more close-up and in situ information about Mars's surface is necessary to assure the safe and effective landing and departure of humans, who will also need to ride novel ''cars'' over that rocky, unknown surface. If they are to explore effectively, their space suits and habitats must withstand the ill effects of Martian dust for many months. Landing and lift-off must not be vulnerable to Martian windstorms. The engineering studies of Mars sample-return missions already under way separately in the United States and the Soviet Union can provide the initial technical framework for a joint planning team. Step-by-step cooperative robotic missions could then be designed to gather the needed environmental information in the late 1990s.

These are challenging tasks, at least a decade's worth, but not so demanding or costly as to distort seriously either country's current space activities or budgets. These three primary tasks would have to be accomplished before either nation could commit itself to a serious manned Mars schedule. There would be cost savings overall if the tasks were done jointly.

Most important, the two wary superpowers would gain detailed experience in trying to work together in space. On the basis of that experience, another President, and probably another General Secretary, could decide at about the turn of the century whether or not to commit his country to a joint manned mission to Mars. That great next step could perhaps be targeted for the second decade of the next millennium and structured to keep annual costs for each side manageable. Europe and perhaps Japan would be invited to play an important supporting role, further committing the United States and the Soviet Union to stay the course.

Finally, the American president could do his Soviet counterpart one better. He could unilaterally direct a portion of America's unequaled technological and scientific capacities to help the world establish an *international* mission to Planet Earth. Through multilateral agreements, and with technical support from the industrialized nations, many nations would band together to monitor Earth's global environment, from space and from stations on its surface. U.S. leadership—not dominance—is a prerequisite to preparing to meet the looming twenty-first-century challenge of cooperative, international management of Earth's environment. By managing Earth, humankind can avoid global ecological disaster while still increasing global productivity to meet the world's rising expectations.[18]

The next president could refocus U.S. civilian space on the twin themes of "To Mars ... Together" and an international mission to Planet Earth. That has been my personal dream since 1985, when I wrote,

The civilian space program can once again be used to achieve national goals. To develop new means of global management through space observations, and to join with the Soviets in going to Mars, would be bold new ways to use space for international leadership. They would also constitute a particularly good use of NASA. These new programs would be scientifically and technologically challenging, they would be intrinsically open and international, and they would involve the highest level of adventure. Most of all, they are all about peace and hope and imagination—the stuff upon which NASA has flourished. A second golden age of civilian space can be achieved. Space in the service of priority human needs could become the symbol of an enlightened and vigorous twenty-first century for the United States.[19]

Presidential leadership will be needed to build an enduring partnership with Congress, which must consistently support and fund a step-by-step approach to Mars, cooperation with the Soviets, and an international mission to Planet Earth.

Not least, leadership will be needed to resuscitate a NASA grievously injured by presidential inattention and by the failed tactics and policies of top NASA officials. Renewed space exploration and an international mission to Planet Earth will require NASA to be combined eventually with other technical elements of the federal government in a Cabinet-rank organization.[20]

Challenging? Yes. Feasible? Yes, if the President perceives that the U.S. national interest would be well served by "To Mars ... Together" and by U.S. leadership of global monitoring. And if he possesses the insight, staff, and political skills necessary to lead.

Will America grab that second brass ring on the carousel of destiny? I don't know. We have the capacity for renewed greatness in space achievements, as well as the capacity to fall back to sleep among our dreams and fantasies, hoping the alarm clock never rings.

Either way, space exploration has brought out the finest in us. Our achievements in the last three decades, at sites ranging from Mercury to Neptune, are a luminous legacy for the future, with virtually no bad side effects—a lasting symbol of American optimism and belief in the goodness of knowledge. Our artifacts are safe on the surfaces of the Moon, Mars, and Venus and in silent planet-crossing orbits. They are among the first acquisitions of Earth's cosmic museum. Future archaeologists will admire this initial phase of humanity's reach outward and its cultural influence.

Humankind *will* go on to leave its traces on Mars, on Halley's comet, perhaps even on tiny Miranda. Let us hope that those traces will record how cultures once at odds later shepherded Planet Earth from adolescence to maturity so that humankind's enormous technological creativity could flower in a golden age, bountiful beyond anything we can now imagine.

APPENDIX About the Planetary Society

THE PLANETARY SOCIETY is a public-interest organization dedicated to planetary exploration and the search for extraterrestrial life. It has played a leading role in the support of promising research about other worlds, in public education, and in fostering the kind of political climate that makes such exploration possible. Organized as a nonprofit corporation, with a federal tax-exempt designation of 501(c)3, it solicits dues and tax-deductible contributions from its 125,000 members, who support most of its expenditures of four million dollars per year. Modest foundation support and minor income from society-related sales provide the remainder. The society does not accept contributions from the aerospace industry or from NASA.

Its headquarters are in Pasadena, California,* close to the center of the American planetary program, the Jet Propulsion Laboratory. Its offices fill most of the rooms in a turn-of-the-century California bungalow, where equipment, supplies, and people are stacked wall to wall. On any given day, the casual visitor might find television producers, computer-graphics artists, telescope manufacturers, science teachers, architects, students, or news reporters stepping around busy volunteers to discuss business with the staff. Society members from around the world drop by the headquarters, a crossroads for those fascinated by other worlds. A steady stream of scientists and engineers who have led the exploration of the planets, and who invent ways to make new exploration possible, passes through to discuss new projects, exciting international endeavors, magazine articles in preparation, or public-education events.

The Planetary Society was founded in late 1979 by Carl Sagan and myself. We sought to prove that there was great support among the general public for continuing the exploration of the planets that had begun so auspiciously during the

*65 N. Catalina Street, Pasadena, CA 91106

Apollo era of the late 1960s and early 1970s. To help us create and run the proposed organization, we turned to an accomplished aerospace engineer, Dr. Louis Friedman, who had just returned from a year as a staff member of a congressional science committee. None of us was experienced in the highly specialized work of public-interest groups, but, to get the organization off the ground, we quickly learned the sometimes arcane ways of direct mail. With the help of Harry Ashmore and Peter Tagger, two experienced hands in membership development, we launched our first campaign. In its first year the Planetary Society grew faster than any other membership organization in American history, except for Common Cause after the Watergate scandal.

Sagan, Friedman, and I continue to run the Planetary Society, talking almost daily. We are assisted by a distinguished board of directors, made up of Joseph Ryan, of the O'Melveny and Myers law firm; Henry Tanner, corporate secretary of Caltech; and Thomas Paine, former administrator of NASA. Friedman, as executive director, is a full-time, paid employee. However, neither Sagan nor I (nor any other members of the board) have ever received salary or other financial compensation from the society. On the contrary, Sagan assigned his royalties from the book *Murmurs of Earth* to the society, as I did with *The Planets,* a *Scientific American* reader that I edited. From time to time we also donate to the society honoraria we receive for speaking and writing about planetary exploration.

When it is needed, we can turn for advice to a distinguished and diverse board of advisers. The writers Isaac Asimov, Ray Bradbury, Arthur C. Clarke, and James Michener have contributed their talents. Two former astronauts, Sally Ride and Harrison Schmitt, provide firsthand knowledge of space travel. The Planetary Society's international scope is evidenced by the presence of the advisers Jacques Blamont of France, Cornelis de Jager of the Netherlands, Garry Hunt of Great Britain, and Roald Sagdeev of the Soviet Union. The distinguished scientists Frank Drake, Lee DuBridge, Philip Morrison, and James Van Allen supply indepth expertise when necessary. Many other dedicated people have also helped us as advisers.

The Planetary Society has stayed close to its roots in the planetary-science and aerospace-engineering communities. It offers its members a chance to communicate and mingle with the people who send probes to the planets, study Earth's neighborhood in space, and search for extraterrestrial intelligence.

The people who have accepted the society's invitation to join are bound together by their fascination with other worlds and with the possibility that evidence of the existence of other civilizations will someday be discovered. The membership comprises 125,000 individuals, including 8,000 from Canada and 4,000 from other nations. According to an informal poll, the members are predominantly male and from middle- to upper-income households; some 88 percent are college educated. Politically, they consider themselves neither conservative nor liberal, and only 36 percent belong to other scientific-interest groups. They all share a belief that humanity should explore the Solar System and possibly regions beyond it, and they want the space-faring nations to get on with it.

The most visible manifestation of the Planetary Society is its bimonthly magazine, the *Planetary Report*. It serves as the chief means of communication between the scientific community and the society's members. The articles are written, for the most part, by leading scientists and engineers who actually do the work of exploration. The *Report* started as a small, 16-page, brochure-type publication, with an occasional 24-page special issue. Now it is a respectable 32 pages, containing as much copy and artwork as much larger publications do. The society's board of directors has kept the *Report* free of advertising. Its circulation is slightly larger than the society's membership, since the *Report* is distributed to schools and libraries around the world. It has become an important educational medium in its own right.

A typical issue of the magazine might include a report on the latest results from a planetary encounter, newly reprocessed images of other worlds from old missions, reports on society-supported research, scientific speculation on the evolution of solar systems, plans for future missions, and an essay on what it all means. Special issues have covered in detail fascinating topics in the planetary sciences, such as life in the universe, the return of Halley's comet, the Voyager encounters with Uranus, humans on other worlds, the tenth anniversary of the Viking mission, and planetary catastrophes. The scientists and engineers contribute the articles gratis to share their work with as wide an audience as possible and to help the society ensure that their work can go on. The *Report* also informs members about society activities, political machinations that affect its goals, recently published books and magazines, and the ideas and opinions of other members.

The *Report* regularly updates its readers on the research projects they help make possible. One program with concrete results has been the Planet-Crossing Asteroid Search (PCAS) of the JPL scientist Eleanor Helin. Over half of the recent discoveries of these strange little objects have been made under the auspices of this program. The two most accessible near-Earth asteroids, 1982 DB and 1982 XB, were among the first objects found with Planetary Society support. These are the best asteroid candidates for future spacecraft missions because their orbits make them the easiest to reach for a future rendezvous. Another recent result was the discovery of the comet P/Helin 1987w, a short-period comet orbiting the Sun every 14.5 years. With the Planetary Society's continuing support, and that of NASA and the World Space Foundation, this important work will continue.

The largest society-aided research program is Project META (for Megachannel Extraterrestrial Assay), the most powerful and comprehensive radio search for signals of extraterrestrial origin now under way anywhere in the world. META scans the skies from Harvard University's Oak Ridge Observatory, in Massachusetts, looking for possible radio beacons from other technological civilizations. META grew out of the Harvard Professor Paul Horowitz's ingenious idea for a portable multichannel spectrum analyzer, called Suitcase SETI (for Search for Extraterrestrial Intelligence), that could be carried from radio telescope to radio telescope to sift through the rain of radio waves falling on Earth every moment in search of a signal that is perhaps artificially produced. After using Suitcase SETI at the giant Arecibo Observatory, in Puerto Rico, Horowitz realized that Harvard University had an unused radio telescope sitting out in the Massachusetts countryside that could be refurbished and dedicated to SETI. With a Planetary Society grant, this was done and Project Sentinel was born.

Sentinel scanned the skies with 131,072 channels in the most advanced radio search of its time. But the inventive Horowitz soon came up with an even more powerful idea—a *mega*channel spectrum analyzer. This system could survey 8.4 million simultaneous frequency channels. It would require a custom supercomputer, with 144 fast parallel processors, 20,000 back-plane connections, and half a million solder joints (all done by hand). It would also cost $95,000. The film producer Steven Spielberg provided a generous grant so that META could be built and put into operation, and it is now the most advanced SETI program in the world.

Ironically, in 1981, at the very time when Horowitz conceived his SETI program, funds for NASA's SETI program were deleted by Congress, despite serious scientific interest and broad public enthusiasm. So the Planetary Society stepped in to fill the void. With its support, some federal funds were eventually restored, permitting NASA scientists to keep working. They have been testing powerful new equipment. But, so far, they have been denied authorization and funds to go operational. In the meantime, the Planetary Society is still leading the world in SETI research.

Project META is not the only SETI program that the society supports. At the Hay River Observatory, in Canada, the society has helped a group of amateur astronomers get another SETI program off the ground. Since META covers the skies only from the Northern Hemisphere, it cannot see most of the stars in our Galaxy, which are visible from the Southern Hemisphere. To expand the search coverage, the society is setting up a SETI program in Argentina, using equipment designed by Horowitz. When this search begins, nearly the entire sky visible from Earth will be covered in an international program, created by the Planetary Society.

The Planetary Society has focused much of its activity on the exploration of Mars. It has subsidized the three "Case for Mars" conferences, held at the University of Colorado, Boulder. There leading scientists and engineers eager to renew the exploration of Mars that came to an abrupt halt with the end of the

Viking mission gather every couple of years. At the latest conference, in July 1987, called "Case for Mars III: Strategies for Exploration," the society held a "Spacebridge" satellite conference between Soviet and American researchers and policymakers. This conference served as the basis for a Public Broadcasting System special, entitled "Together to Mars?" and broadcast in late 1987. It was aired in the USSR and in Japan as well.

The Mars Institute is an informal society program designed to promote the study of the human exploration of Mars at colleges around the world. The institute provides course materials, bibliographies, and encouragement to instructors and students working on the problems involved in sending humans to another planet. Each year it holds a student contest to stimulate young people to research these problems.

To heighten the general public's interest in the planet Mars, the Planetary Society initiated Mars Watch '88, to take advantage of the unusually favorable 1988 perihelic opposition. Educational workshops, observing parties, lectures, and other activities helped the public share the scientists' enthusiasm during the outstanding observing conditions. There were opportunities for amateur astronomers to contribute to scientific advances. Mars is a planet subject to seasonal changes and global storms, two phenomena that are visible through telescopes on Earth. The long-term water and carbon dioxide cycles manifested by changes in the polar ice caps and the dust storms that can enshroud the planet must both be better understood before we send humans to set up bases on Mars. With the help of the Association of Lunar and Planetary Observers (ALPO), the Planetary Society coordinated the viewing activities of professional and amateur observers, in the expectation that the observations in 1988 will advance our understanding of Mars.

The society has also taken on the publication of the *Mars Underground News,* a quarterly newsletter written for and by the community of people actively working toward landing humans on the Red Planet. In the early 1980s a small group of people, calling themselves the Mars Underground, began to agitate for renewed human exploration of Mars. The Case for Mars conferences grew out of their efforts. The purpose of this newsletter is to keep the members of the Mars Underground in contact with one another and to increase interest in their work. Since members of the Planetary Society finance it, they may subscribe for only a small fee.

The society also promotes electronic communications among people interested in Mars. For several years it has provided grants to support the Space Network, an electronic bulletin board and data base for students and researchers working on the future exploration of Mars. The network serves as an electronic mail service, as a forum for debate, and as an information resource.

Several small research projects are now under way to develop the technology to enable humans to live on Mars. At the University of Virginia researchers are developing a prototype of a system that would use indigenous Martian materials to produce rocket propellant on Mars. In cooperation with NASA, the Planetary Society is supporting this work.

An innovative idea of how Mars might be explored has recently captured society attention and support. The society adviser Jacques Blamont, fresh from the success of leading the French-Soviet balloon probes released during the Soviet VEGA mission to Venus, proposed that a similar technique can be used on Mars. The Soviet Union is planning to send a spacecraft to the Red Planet in 1994. Once again a French-Soviet team is studying instrumented balloons to explore the planet's surface and atmosphere. Recognizing the value and urgency of the idea—and aware that neither Soviet nor American space officials seemed overly excited by it—the Planetary Society stepped in and launched its own international study of the balloons. Joined by researchers from JPL, Caltech, Utah State University, the University of British Columbia, Ball Aerospace, and Titan Systems, and working closely with researchers from the French National Space Agency and from the Institute for Space Research in Moscow, the society has played a crucial role in moving the concept toward a practical realization.

Scientific and technological discoveries are not always shared with people in the Third World. Therefore, the Planetary Society has begun an educational outreach to help bring the excitement of planetary exploration to students in these countries. The first program was held in Mexico City in 1987, where leading planetary scientists were brought together with Mexican educators to share their knowledge of the Solar System. The lectures and demonstrations were televised and will be available in other Spanish-speaking countries. This successful program will itself soon be repeated in other regions. The Planetary Society is now investigating ways to aid the nascent Pan-American Space Organization (PASO), a group seeking to bring together nations in North, Central, and South America so that the benefits of space technology and exploration are available in all the member countries. The first project will be the launch and tracking of an Argentinian communications satellite. PASO hopes to find a launch vehicle for the satellite and to put together a tracking network of antenna facilities within member nations.

The Planetary Society has made a special effort to reach young people. It gives a series of scholarships annually to promising students in the planetary sciences. For several years it has supported a National Merit Scholarship for outstanding students just entering college, which is now augmented by the New Millennium Committee scholarships, funded by the society's largest individual donors. We have recently awarded five college fellowships to students already well on their way to careers in planetary science. Teachers also receive our support. We have sponsored a series of workshops around the United States to help educators in their continuing efforts to improve science education.

In Canada we have cosponsored essay contests and donated prizes for the winners, including a satellite dish for their schools. From around the world we have collected and archived children's drawings of the 1986 apparition of Halley's comet, and we will keep them for the children who will see the next apparition, in order to let them know how their parents and grandparents saw the comet. Before the comet arrived in Earth's neighborhood, the society sponsored a PBS television special, "Halley's Comet: Here It Comes Again," which helped prepare a national audience for this once-in-a-lifetime event.

One of our most visible educational efforts is the "Explorer's Guide to Mars," a large map and informational poster that we have distributed to schools and institutions around the world. The detailed captions and charts printed on the map are augmented by a booklet detailing suggested educational activities. This accurate map can be found not only on the walls of schools but also at NASA, JPL, and the Johnson Space Center, and in the offices of European space agencies and in space laboratories in Moscow.

The Planetary Society is especially active in Canada, which boasts the second-largest number of society members. In addition to its support of the Hay River Observatory for SETI studies, the society has funded an astrometric research program at the University of Victoria, in British Columbia. Research scientists have there developed new techniques for precisely measuring asteroids' positions, in order to increase our knowledge of their orbits, which will aid in planning missions to these small, dark bodies. As an extra benefit, in the course of making measurements, the research team discovered a new asteroid.

The Planetary Society's substantial support for worthwhile research projects has been widely reported in the media, as has its promotion of the Mars goal described in chapter 22. The organization has established its credentials as a reliable, informative, and "quotable" source for reporters working on stories about the space program. Sagan, Friedman, and I frequently appear on the network morning news shows, the "MacNeil/Lehrer News Hour," "Nightline," and other broadcasts. The society distributes "fact sheets" covering planetary missions by many space-faring nations, including efforts by ESA, ISAS, and Intercosmos. Science writers find this service extremely helpful.

Congress, too, has turned to the Planetary Society for expert testimony. For example, in September 1984 Friedman delivered to the Senate Committee on Foreign Relations a statement supporting cooperation in space science between the United States and the Soviet Union. In May 1987 Sagan, Friedman, and I testified on the topic "A Space Station Worth the Cost" before the Senate Appropriations Committee. That year the society gave several congressional briefings as part of its effort to promote the Mars Observer mission. Sagan has often appeared before congressional groups supporting the scientific exploration of space. Most recently, in July 1988, he spoke to the Senate Air and Space Caucus.

The Planetary Society also informs the public about the benefits of planetary exploration through its public events. Planetfest '81, our first major effort in this field, was a conspicuous success. It was held in conjunction with the *Voyager 2* encounter with Saturn. We believe that it was the largest event ever held to honor the achievements of planetary exploration. Over 15,000 people witnessed live transmission of images from Saturn, listened to a selection of panel discussions, enjoyed a concert by the composer and conductor John Williams, and toured exhibits of other worlds and the Jet Propulsion Laboratory. Planetfest '89 will celebrate the Voyage encounter with Neptune.

We continued in this celebratory mode in 1982, when the society hosted a Washington, D.C., commemoration of the first successful planetary mission, *Mariner 2* to Venus. In 1985 we brought together the Apollo-Soyuz astronauts on the tenth anniversary of their cooperative mission in Earth orbit and cosponsored, with the American Institute of Aeronautics and Astronautics (AIAA), a "Steps to Mars" conference to explore other possible directions for joint U.S.-USSR exploration. Recognizing that the growing effort directed to the Strategic Defense Initiative would affect planetary exploration, the Planetary Society in 1985 also sponsored a conference entitled "The Potential Effects of Space Weapons on the Civilian Uses of Space," held in the auditorium of the National Academy of Sciences, in Washington, D.C.

The Planetary Society has often served as a bridge between the public and the professional, scientific, and engineering organizations involved in space exploration. It has hosted many public sessions at meetings of the AIAA, the Division for Planetary Sciences of the American Astronomical Society, and the American Association for the Advancement of Science (AAAS). The society is an affiliate organization of the AAAS and a member of the International Astronautical Federation. It has set up an official liaison with the Hungarian Astronautical Society and has helped sponsor a report on international cooperation in space produced by the United Nations Association of the U.S.A.

We in the society especially value these international contacts and have been pleased to be able to facilitate cooperation among scientists even when their nations were not on friendly terms. In 1984, while the cooperation agreement between the United States and the Soviet Union was still in limbo, we held a meeting in Graz, Austria, for nineteen scientists from the American, Soviet, and European space programs. We made it possible for several American scientists to attend a meeting on the search for extraterrestrial intelligence held in Tallinn, in the Soviet republic of Estonia.

The Planetary Society represents no one economic, social, or political group; it puts no national, professional, or attitudinal restrictions on membership. Rather, its constituency springs from the imagination and hope stirring in individuals all over the world, longing to make dreams possible. We believe that nuclear swords can be converted into great rocket engines to explore Mars, that the nations of the world can work together to monitor from space—and someday wisely manage together—our common planet, and that humankind should use its burgeoning communications technology to pursue the most important question of all: Are we alone in the universe?

Acknowledgments

MANY PEOPLE GAVE GENEROUSLY of their time to improve the accuracy and clarity of this book. I am especially indebted to five individuals who read the entire manuscript in semifinal form and made important technical, organizational, and stylistic suggestions. They are Jacques Blamont, counselor, Centre National d'Etudes Spatiales of France, distinguished visiting scientist at JPL, and veteran experimentalist on European, Soviet, and American space missions; Elizabeth A. R. Brown, professor of history, Brooklyn College, City University of New York; John Gardner, former secretary of health, education, and welfare, and founder of Common Cause; John Logsdon, professor and director of the Graduate Program in Science, Technology, and Public Policy, George Washington University; and Carl Sagan, professor of astronomy and space sciences and director of the Laboratory for Planetary Studies, Cornell University. In addition, Edwin Barber, of W. W. Norton and Company, showed special dedication in editorial review—from early drafts through final manuscript. Edward Hutchings, former editor of *Engineering and Science,* Caltech, provided substantial help in matters of style for the complete manuscript, as did Otto Sonntag, of W. W. Norton.

Thirty-eight other persons reviewed one or more preliminary parts of the manuscript for technical and historical accuracy. I am pleased to acknowledge the important contributions of Robert Allnutt, an official at NASA headquarters during the 1960s and the early 1980s; Philip Barnett, longtime launch vehicle expert for JPL; John C. Beckman, manager, Advanced Programs Office, JPL; Harold Brown, former president of Caltech and former secretary of defense; James D. Burke, JPL space pioneer and enthusiast; John Casani, deputy assistant laboratory director for flight projects, JPL; Stewart ("Andy") Collins, TV imaging expert, JPL; N. W. ("Bill") Cunningham, formerly NASA program manager for the Mercury/Venus mission; G. Edward Danielson, member of the professional staff, Caltech, and member of several Mariner imaging teams and of the Voyager imaging team; Merton Davies, Rand Corporation, member of the imaging team for Voyager and for nearly all of the Mariner missions; Walter Downhower, longtime staff analyst, JPL; Donald Fowler, general counsel, Caltech; Louis Friedman, executive director, the Planetary Society, and an important JPL mission designer in the 1970s; Eugene Fubini, former chairman of the Defense Science Board; Walker E. ("Gene") Giberson, project manager for *Mariner 10;* Norman Haynes, current Voyager project manager, JPL; Raymond Heacock, former Voyager project manager, JPL; James Head, professor, Department of Geological Sciences, Brown University, and pioneer in U.S.-USSR cooperation; Norman Horowitz, professor emeritus of biology, Caltech, and a prominent scientist on the Viking Lander enterprise; Torrence Johnson, project scientist for the JPL Galileo mission; Charles Kohlhase, JPL, Voyager mission designer; S. M. Krimigis, former chief scientist, Applied Physics Laboratory, Johns Hopkins University; Robert Leighton, professor emeritus of physics, Caltech, and leader of the *Mariner 4, 6,* and *7* imaging teams; John McLucas, former secretary of the Air Force; Duane Muhleman, professor of planetary science, Caltech; Gerry Neugebauer, professor of physics, Caltech, and director of Mt. Palomar Observatory; Robert Parks, former assistant laboratory director for flight projects, JPL; Donna Pivorotto, JPL, mission analyst on *Mariner 10;* Stanley Rawn, Jr., member of the board of trustees, Caltech; Eberhardt Rechtin, former president, Aerospace Corporation; the late Fred Scarf, TRW, Inc., veteran space physicist on U.S. planetary missions; Harris M. ("Bud") Schurmeier, formerly a key project manager at JPL; David Stevenson, professor of planetary science, Caltech; Edward Stone, chairman, Division of Physics, Mathematics, and Astronomy, Caltech, and project scientist, Voyager mission; Lieutenant General Charles H. Terhune, Jr., formerly deputy director of JPL; James A. Van Allen, professor of physics, University of Iowa, Iowa City, pioneering space physicist; J. P. A. ("Ox") Van Hoften, manager, Space Programs, Bechtel Corporation, and former astronaut once scheduled to fly on the Shuttle/Centaur launch of Galileo; Craig Waff, historian of U.S. planetary activities.

Early drafts of the appendix were prepared mostly by Charlene Anderson, the editor of the *Planetary Report,* and I appreciate her expert assistance.

Victoria Hester was especially diligent in preparation of the numerous drafts of all the chapters and sections, ably assisted toward the end by JoAnn Boyd, of Caltech. Jurrie Van Der Woude, of JPL, who has helped me and others for twenty-five years, was again very helpful in suggesting and locating key photographs. Audrey Steffan, of JPL, provided key insights and suggestions for the layout and composition of the chronologies. Lorna Griffith, my long-suffering secretary at Caltech, provided great help in pulling together diverse pieces of this enterprise.

Once again, I would like to express my great appreciation to all those mentioned above for their specialized help in the creation of this book.

References and Notes

AN EXCELLENT RECENT REFERENCE WORK TO THE PLANETS is David Morrison and Tobian Owen, *The Planetary System* (Reading, Mass.: Addison-Wesley, 1988). Another good source is *The New Solar System*, ed. J. Kelley Beatty, Brian O'Leary, and Andrew Chaikin (Cambridge, Mass.: Sky Publishing, 1981).

Scientific American articles about the planets have been republished as *The Planets* (San Francisco: W. H. Freeman, 1983). Carl Sagan's *Cosmos* (New York: Random House 1980) is an exceptionally successful exposition of the facts and intellectual contexts of Earth's cosmic setting.

An excellent compendium to the first two decades of space activity is Kenneth Gatland et al., *The Illustrated Encyclopedia of Space Technology* (London: Salamander Books, 1980).

The Planetary Society offers a variety of educational materials about the planets and space exploration for both educators and the general public. For information, contact The Planetary Society, 65 North Catalina Avenue, Pasadena, CA 91106; 818-793-5100 or 800-255-2001.

A number of the references here are to NASA SP publications and to U.S. government publications. Information on how to obtain such publications is available from the Superintendent of Documents, Government Printing Office, Washington, D.C. More specific information about the NASA histories mentioned here (and others) can be obtained from the NASA History Office, Code XH, NASA headquarters, Washington, D.C. 20546.

Prologue

The belligerent beginning and the steps to the Apollo Moon landing on July 20, 1969, are well described in a highly regarded recent history by Walter A.

McDougall, ... *The Heavens and the Earth* (New York: Basic Books, 1985).

The U-2 incident and related events are explored in detail in Michael R. Beschloss, *Mayday: Eisenhower, Khrushchev and the U-2 Affair* (New York: Harper & Row, 1986).

An official NASA history of the Apollo is Arnold S. Levine, *Managing NASA in the Apollo Era,* NASA SP-4102 (Washington, D.C.: NASA, 1982).

John M. Logsdon, *The Decision to Go to the Moon: Project Apollo in the National Interest* (Chicago: University of Chicago Press, 1970), treats the Apollo program skillfully from the viewpoint of national policy. The President's Science Advisory Committee (PSAC) produced sound recommendations in February 1967 and again in March 1970: *The Space Program in the Post-Apollo Period* and *The Next Decade in Space,* published as White House reports. Both still make interesting reading and provide a hint of how space policy might have been different if the White House had remained interested in space achievements after Apollo.

For the history of JPL, see Clayton R. Koppes, *JPL and the American Space Program: A History of the Jet Propulsion Laboratory* (New Haven: Yale University Press, 1982), which treats the Pickering years fully and accurately. The earlier days of JPL, in which Frank Malina was the critical figure, can now be studied more fully because Malina's papers are cataloged and available for scholarly analysis in the Caltech Archives. The period after 1976 was not researched thoroughly by Koppes.

A history of the Deep Space Net is under preparation by Craig Waff under the auspices of the NASA History Office.

The early development of space physics is chronicled in James A. Van Allen, *Origins of Magnetospheric Physics* (Washington, D.C.: Smithsonian Institution Press, 1983).

References to the *Challenger* disaster and ensuing events are listed under part 5.

O N E   The Search for Life on Mars

A good review of the transition from Martian myth to Martian reality is provided in Norman H. Horowitz, *To Utopia and Back: The Search for Life in the Solar System* (New York: W. H. Freeman, 1986). A popular account of the Viking phase of the search is Henry S. F. Cooper, Jr., *The Search for Life on Mars* (New York: Holt, Rinehart and Winston, 1980).

An interesting account of attempts during the late sixties to start a piloted mission to Mars is Edward Ezell, "Humans to Mars," *Planetary Report* 8, no. 4 (July–August 1988): 10–14.

There are many books on the nature of Mars and its surface features, in addition to the general ones cited above. A standard reference work on the nature of Mars's surface and history is Michael Carr, *The Surface of Mars* (New Haven: Yale University Press, 1981). Mars is compared with other terrestrial planets in Bruce Murray, Michael Malin, and Ronald Greeley, *Earthlike Planets* (San Francisco: W. H. Freeman, 1981).

Soviet efforts to explore Mars are described well in U.S. Congress, Senate Committee on Commerce, Science, and Transportation, *Soviet Space Programs, 1976–1980 (with Supplementary Data through 1983),* pt. 3, *Unmanned Space Activities,* 99th Cong., 1st sess., 1985.

[1] Bruce Murray and Merton E. Davies, "A Comparison of U.S. and Soviet Efforts to Explore Mars," *Science* 151 (1966): 945–54.

[2] Robert Leighton, Bruce Murray, Robert P. Sharp, J. Denton Allen, and Richard K. Sloan, "Mariner IV Photography of Mars: Initial Results," *Science* 149 (1965): 627–30.

[3] Mert Davies has the unusual distinction of holding a patent on part of a reconnaissance satellite design—U.S. Patent No. 3,143,048, granted August 4, 1964, covering "photographic apparatus." For fascinating new details of the early years, see Merton E. Davies and William R. Harris, *Rand's Role in the Evolution of Balloon and Satellite Observation Systems and Related U.S. Space Technology* (Santa Monica: Rand Corporation, 1988).

[4] *The View from Space* (New York: Columbia University Press, 1971).

[5] Ray Bradbury, Arthur C. Clarke, Bruce Murray, Carl Sagan, and Walter Sullivan, *Mars and the Mind of Man* (New York: Harper & Row, 1973).

[6] Five years and two coauthors later, the book finally appeared as *Earthlike Planets.*

[7] *Space Research: Directions for the Future,* Report by the Space Science Board, Woods Hole, Mass. (Washington, D. C.: National Academy of Sciences–National Research Council, 1966).

[8] *Planetary Exploration: The Condon Lectures* (Eugene: University of Oregon Press, 1970).

[9] Frank T. Kyte et al., in *Science* 241 (1988): 63–65, discuss the possibility that the Pleistocene glaciation was triggered by an asteroid or cometary impact. However, even so, Earth's climatic circumstances must already have been close to glacial conditions in order for the relatively slight trauma of an impact to cause a permanent switch in global climate.

[10] See the appendix for a discussion of the Planetary Society's Search for Extraterrestrial Intelligence efforts.

T W O Probing Warmer Worlds

The circumstances leading to the *Mariner 10* mission to Venus and Mercury and the results obtained are described in Bruce Murray and Eric Burgess, *Flight to Mercury* (New York: Columbia University Press, 1977). The official NASA history is Eric Burgess and James A. Dunne, *The Voyage of Mariner 10: Mission to Venus and Mercury,* NASA SP-424 (Washington, D.C.: NASA, 1978).

The history of Venus exploration is outlined in Jacques Blamont, *Vénus dévoilée: Voyage autour d'une planète* (Paris: Editions Odile Jacob, 1987).

Information about Venus and Mercury can be found in *The Planetary System, The New Solar System, The Planets,* and *Earthlike Planets,* all cited in part 1.

Don Wilhelms, John F. McCauley, and Newell J. Trask, *The Geologic History of the Moon,* USGS Professional Paper 1348 (Washington D.C.: U.S. Geological Survey, 1987), is the definitive work on the subject.

[1] Audouin Dollfus, in *Planets and Satellites,* vol. 3 of *The Solar System,* ed. Gerard P. Kuiper and Barbara M. Middlehurst (Chicago: University of Chicago Press, 1961), 550.

[2] The first detection of midinfrared radiation from an object outside the Solar System was reported in Robert L. Wildey and Bruce Murray, "Ten Micron Photometry of 25 Stars from B8 to M7," *Astrophysical Journal* 139 (1964): 435–41. A broader account of the emergence of the field of midinfrared astronomy is James A. Westphal and Bruce Murray, "Infrared Astronomy," *Scientific American,* August 1965, 20–29.

T H R E E   Voyager and the Grandest Tour Ever

The Voyager mission has been the most successful robotic exploration to date. Surprisingly, there exists no official NASA or JPL history of this long, complex, and meritorious endeavor. The earlier *Pioneer 10* and *11* missions have been well described in Richard O. Fimmel, William Swindell, and Eric Burgess, *Pioneer Odyssey: Encounter with a Giant,* NASA SP-349 (Washington, D.C.: NASA, 1974).

The encounters of *Voyagers 1* and *2* with Jupiter are popularly described and handsomely illustrated in David Morrison and Jane Samz, *Voyage to Jupiter,* NASA SP-439 (Washington, D.C.: NASA, 1983). See also David Morrison, *Voyages to Saturn,* NASA SP-451 (Washington, D.C.: NASA, 1982).

A popular account of the imaging of Saturn is Henry S. F. Cooper, Jr., *Imaging Saturn: The Voyager Flights to Saturn* (New York: Holt, Rinehart and Winston, 1982). Morrison and Owen, *The Planetary System,* summarizes Voyager's findings well. Morrison and Owen are both members of the Voyager scientific team.

Craig Waff, in the course of preparing an official JPL history of Project Galileo, has also explored the early history of the Grand Tour. He summarizes some of his findings in "Searching for an Outer Planet Exploration Strategy: NASA and Its Scientific Advisory Groups, 1965–71," *Issues in the History of*

Space Astronomy, IAU General Assembly, Baltimore, August 6, 1988 (preprint).

[1] Morrison and Samz, *Voyage to Jupiter,* 122.

[2] The reader may ask, "But what about Pluto? That's the most distant planet." Pluto has a rather elliptical orbit. At its greatest distance from the Sun (aphelion), it is farther from the Sun than Neptune is. At present, however, on its way to perihelion, it is traveling closer to the Sun than Neptune is. Thus, for Voyager, Neptune is indeed the last landfall of recognized objects orbiting the Sun, at least until Voyager reaches the zone of pristine comets ("Oorts Cloud") believed to surround the Sun at a distance of hundreds or thousands of astronomical units.

F O U R Lost in Space

A detailed and authoritative chronology of the Galileo mission is being prepared by Craig Waff under contract from JPL and the NASA History Office. Entitled *Jovian Odyssey,* the book will be available through that office.

Prior to the *Challenger* accident the Space Shuttle received uncritical and unduly supportive treatment by nearly all journalists and reporters. (The *New York Times* and the "News and Comment" section of the weekly journal *Science* did report most major events objectively, but piecemeal.) The fantasy of the Shuttle as a safe, cheap, and effective means for Americans to enter the new frontier of space was so appealing to most audiences that there was simply no journalistic market for naysayers. Most serious for informed American opinion, the media were not committed enough to the public interest. The print media and, especially, television did little more than tell the public what it wanted to hear. In contrast, malfeasance and mediocrity in the Pentagon have always been considered newsworthy and, thus, are widely publicized. Yet, until *Challenger,* there was rarely any incisive reporting about NASA's shortcomings. Even the serious print media generally failed to provide balanced coverage of the Shuttle *program* to complement the thrilling entertainment afforded by the prospect of human-piloted Shuttle *flights.*

Congress, likewise, sensed the powerful political appeal of the Shuttle's promise and, thus, remained virtually mute in exercising its overview function. As a consequence, those American leaders (like Ronald Reagan) without either technical training or an insightful staff tended to acquire their views on the Shuttle and on space from the general media. In the absence of more rigorous or critical sources, they believed too long that the Shuttle's only problem was that it ran behind schedule, which NASA glibly blamed on "underfunding" by previous administrations.

For these same reasons, there are few truly objective published sources about the Shuttle program until after January 28, 1986. The *New York Times* then focused its large and excellent staff on both news coverage and journalistic analysis of the Shuttle and, more generally, of NASA. The resulting stream of news articles, feature stories, and editorial opinion constitutes the best single source of information and perspective from then through the present (August 1988).

Joseph J. Trento, *Prescription for Disaster: From the Glory of Apollo to the Betrayal of the Shuttle* (New York: Crown, 1986), is a quickly prepared compilation of interviews and journalistic interpretation. It exposes, somewhat unevenly, some of the problems, misrepresentations, and compromises that led to the programmatic disaster.

The report of the thirteen-member presidential commission assessing the *Challenger* accident, under the direction of former Secretary of State William Rogers, was released on June 9, 1986. It provided a detailed analysis of the proximate causes of that disaster. It also offered important recommendations that have both directed and, to some extent, constrained NASA's subsequent actions. That report, however, did not delve deeply enough into the institutional and political setting that made the Shuttle's programmatic failure inevitable and disaster likely. The late Richard Feynman, a member of the commission, did develop an interpretation of how the bureaucratic mentality within NASA combined with the external political environment to set the stage for catastrophe. Feynman's views were published in Caltech's *Engineering and Science* 51, no. 1 (Fall 1987): 6–10, and in his *What Do You Care What Other People Think? Further Adventures of a Curious Character* (New York: W. W. Norton, 1988), especially pp. 212–19.

Hans Mark's *The Space Station: A Personal Journey* (Durham, N.C.: Duke University Press, 1987) includes many fascinating details that vividly document his unquestioning commitment to the Shuttle and to the Space Station as *the* NASA goals. He recounts with pride some of his personal actions to influence national policy and actions in that regard. He tends to portray himself as an enthusiastic follower of other major figures like von Braun, Teller, and Harold Brown. My impression is that he was much more the éminence grise and less the naive follower in the Washington power scene that his memoir suggests.

Underlying the whole Shuttle story and debate is the issue of the value and purposes of manned space flight. Michael Collins, *Apollo 11* astronaut and former director of the Smithsonian National Air and Space Museum, eloquently argues the case for the usefulness of humans in space in *Liftoff: The Story of America's Adventure in Space* (New York: Grove Press, 1988). *Pioneering the Space Frontier: A Report by the National Commission on Space* (New York: Bantam Books, 1986) expands on such views.

Most scientists, however, do not feel that humans can perform functions in space as well or as cheaply as properly designed automated systems. James A. Van Allen has argued this for many years, most recently in ''Myths and Realities of Space Flight,'' *Science* 232 (1986): 1075. I share most of Van Allen's views on the greater cost-effectiveness of robots over humans in performing utilitarian and scientific tasks in space. (There may be some aspects of the exploration of distant planetary surfaces in which man's capabilities in manipulation and pattern recognition offer some special benefits, but those are not the subject of any current NASA efforts.) However, I share Collins's conviction that exploration is a natural and worthwhile human activity. Thus, I support the concept of humans' exploring new places, especially Mars (as does Collins), even while I reject the cost-effectiveness of many of the proposed uses of humans in an Earth-orbital space station for purportedly utilitarian purposes.

[1] Mark, *The Space Station*, 71–73.

[2] On June 19, 1987, the Justice Department quietly announced that it was dropping its fraud case against Beggs and others formerly at General Dynamics. Beggs at the time publicly asked for an apology from Attorney General Meese. However, Assistant Attorney General William F. Weld declined "the invitation to embrace the language of apology," saying that he would trust public opinion to recognize that "in a criminal case, if no conviction is obtained, no blemish should attach." Finally, a year later (see *Los Angeles Times*, July 7, 1988), a few weeks before leaving office because of his own ethics problems, Meese issued "a profound apology" to Beggs, saying that he had an opportunity "to review the circumstances surrounding the wrong indictment" against him and adding, "There is no way to undo the pain you have suffered."

F I V E Comet Tales

A good, readable, and well-illustrated discussion of comets is Carl Sagan and Ann Druyan, *Comet* (New York: Random House, 1985).

For popular accounts of the Halley encounter of 1986, see Rich Gore, "Much More Than Meets the Eye—Halley's Comet '86," *National Geographic*, December 1986, and *Planetary Report* 6, no. 3 (May–June 1986); 7, no. 2 (March–April 1987); and 7, no. 5 (September–October 1987). For a nicely illustrated article about the ICE mission, see *Planetary Report* 5, no. 3 (May–June 1985). For professional papers from the ICE mission, see *Science* 232, no. 4748 (1986). Technical analyses of Halley are more scattered because of the international character of the mission and the absence of major U.S. scientific participation. An important early collection of papers is in *Nature* 321, no. 6067 (May 15–21, 1986).

Solar sailing is described in an interesting manner in Louis Friedman, *Star Sailing: Solar Sails and Interstellar Travel* (New York: John Wiley, 1988).

Events surrounding the international exploration of Halley's comet are discussed in detail from the European viewpoint in Jacques Blamont, *Vénus dévoilée*.

A detailed, accurate, and very useful narrative of the events surrounding the United States's nonstart in the exploration of Halley's comet is provided by John

M. Logsdon. His work was first prepared for the NASA History Office under the title "Why No U.S. Mission to Comet Halley?" and will be published in a forthcoming issue of *Isis,* the journal of the History of Science Society.

[1] Logsdon, "U.S. Mission."

[2] This old idea has some new competition. See "Comet Source: Close to Neptune," in "Research News," *Science* 239 (1988): 1372–73. But Halley's origin is probably not involved in this new debate.

[3] See Kevin D. McKeegan, "Oxygen Isotopes in Refractory Stratospheric Dust Particles: Proof of Extraterrestrial Origin," *Science* 237 (1987): 1468–70, and Michael E. Zolensky, "Refractory Interplanetary Dust Particles," ibid., 466–47.

[4] Jacques Blamont, in a personal communication of fall 1987, stated that the idea originated with the West Germans.

[5] Letter from Joseph Veverka, of Cornell University, to Briggs, January 10, 1975, cited in Logsdon, "U.S. Mission."

[6] Interview with Robert Frosch, cited in Logsdon, "U.S. Mission."

[7] Memorandum from E. A. Trendlenberg to Intending Proposers to the International Comet Mission, January 31, 1980, cited in Logsdon, "U.S. Mission."

[8] Logsdon, "U.S. Mission."

[9] "The Challenge of Success" (Commencement Address, Caltech, 1979), published in *Engineering and Science,* September–October 1979, 2–6.

[10] Van Allen, *Origins of Magnetospheric Physics.*

[11] Interview with Keyworth, cited in Logsdon, "U.S. Mission."

[12] Logsdon, "U.S. Mission."

[13] Ibid.

[14] Letter from Eugene H. Levy, chairman of Committee on Lunar and Planetary Exploration of the Space Science Board, to Stofan, August 7, 1981.

[15] Letter from Beggs to Keyworth, September 16, 1981, with cover note dated September 17, 1981.

[16] Memorandum for file by S. M. Krimigis dated March 26, 1986, entitled "Lunch with President Ronald Reagan in the Roosevelt Room of the White House, 12 Noon to 1:00 p.m., March 26, 1986." A slightly sanitized version of the Krimigis notes was published in the Johns Hopkins *APL Technical Digest* 7, no. 4 (October–December 1986): 384–93. The late Frederick Scarf, also a participant, made available to me his own, similar notes.

S I X To Mars ... Together?

A good general reference for ideas and speculations concerning future planetary exploration, especially of Mars, is *Pioneering the Space Frontier,* cited in part 4. Additional information is available through the Planetary Society.

[1] *Pioneering the Space Frontier.*

[2] See, for example, Roland Huntford, *Shackleton* (New York: Fawcett Columbine, 1985).

[3] See, for example, Stephen J. Pyne, *The Ice: A Journey to Antarctica* (Iowa City: University of Iowa Press, 1986), especially chap. 8.

[4] S. Nedell, S. W. Squyres, and D. W. Anderson, "Origin and Evolution of the Layered Deposits in Valles Marineris, Mars," *Icarus* 70 (1987): 409–41.

[5] Christopher McKay et al., "Thickness of Ice on Perennially Frozen Lakes," *Nature* 313 (1985): 561–62; G. M. Simmons et al., "Sand/Ice Interactions and Sediment Deposition in Perennially Ice-Covered Antarctic Lakes," *U.S. Antarctic Journal* (1986).

[6] An edited transcript of this symposium was published in *Planetary Report* 5, no. 3 (May–June 1985). All the quotations here are from this version.

[7] Carl Sagan first discussed this idea publicly in "The Case for Mars," *Discover*, September 1984, 26.

[8] Sagan, "U.S.A. and the U.S.S.R.: Let's Go to Mars—Together," *Parade*, February 2, 1986.

[9] Both sides of the issue of whether a long-term Mars initiative between the Soviet Union and the United States is feasible are presented in *International Security* 2, no. 4 (Spring 1987). Dr. Albert R. Wheelon argues the negative in his paper, "A Born Again Space Program," while I argue, in "Born Anew versus Born Again," that the program can be organized in such a way that neither party can be blackmailed by the other. The United Nations Association published an important paper entitled *Developing the Final Frontier: International Cooperation in the Peaceful Uses of Outer Space* (Briefing Book by Ann Florini, the Multilateral Project, 1985), available through the Publications Department, UNA-USA, 300 East 42nd Street, New York, NY 10017. It emphasizes the value of participation of third parties such as Eastern and Western Europe and Japan. The financial and programmatic commitment to a Mars mission by additional nations would constitute a further deterrent to either the United States's or the Soviet Union's withdrawing simply to express political displeasure over some future bilateral tension. Collins, *Liftoff*, supports U.S.-USSR Mars exploration provided the United States maintains a capability to proceed on its own if U.S.-Soviet tensions become unmanageable.

[10] The editorials referred to are "Adrift in Space" (January 7, 1986); "Death of a School Teacher" (January 29, 1986); "How to Regain Faith in Space" (May 29, 1986); "Which Stars Are We Aiming For?" (June 15, 1986); "Mars: A New Goal" (December 7, 1986); "Star Truck" (January 23, 1987); "The Poverty of NASA's Dreams" (February 12, 1987); "To Mars ... via Moscow" (December 24, 1987); "The Next President's Choices in Space" (October 8, 1988). The *Los Angeles Times* and many other major U.S. newspapers have also endorsed "To Mars ... Together."

[11] James Burke et al., "Mars Balloon System Study" (38th International Astronautical Federation Congress, Brighton, U.K., October 1, 1987); Eric Gaidos, "Solar Balloon Physics: Mars Exploration Applications" (Senior thesis, Applied Physics, Caltech, June 3, 1988).

[12] Trouble may be brewing for USSR space politics with the availability of Energia, the new Saturn 5–class booster developed by the USSR. In July 1988 Glavcosmos, the Soviet government agency concerned with the development of new launch vehicles, began a campaign to promote the use of Energia for Mars missions in the 1990s, even though many Soviet scientists involved want to continue to use the reliable Proton. Shades of Saturn 5 to Mars in 1965!

[13] Bruce Murray, "Whither America Space?" *Issues in Science and Technology* 2, no. 3 (1986): 22.

[14] Sally K. Ride, "Leadership and America's Future in Space: A Report to the Administrator" (Internal NASA Report, August 1987).

[15] *Pioneering the Space Frontier.*

[16] See, for example, "A Thirty Billion Dollar Space Station?" *Science* 237, (1987): 1403.

[17] This interview was released to the news media on Sunday, May 22, 1988, and appeared in the *New York Times,* the *Los Angeles Times,* and, of course, the *Washington Post,* as well as other major newspapers. It was followed by an article in *Time* magazine, June 6, 1988, entitled "Pros and Cons of a Flight to Mars: A Modest Gorbachev Proposal Gets an Ambivalent U.S. Reception."

[18] Murray, "Whither America in Space?" For an earlier but fuller exposition of my views, see *Navigating the Future* (New York: Harper & Row, 1975).

[19] Murray, "Whither America in Space?"

[20] See, for example, Warren A. Leary, "The White House: Candidates Are Urged to Pick Science Aides," *New York Times,* June 10, 1988, which discusses the new book by William T. Golden, *Science and Technology Advice to the President, Congress and Judiciary.* I think one synergistic combination might involve the National Oceanic and Atmospheric Administration and some of the more focused activities of the National Science foundation. (Discipline-oriented support for scientific research, however, should remain independent at NSF in order to protect small-scale science from big endeavors like space.) In this new, more powerful organization, civilian space would fare better than in the now overbearing federal bureaucracy. It could more effectively attract new blood for the challenge of going to Mars, and it could weld together the diverse Earth-oriented sciences necessary for the international mission to Planet Earth.

Index

Page numbers in *italics* refer to illustrations.